Flow past Highly Compliant Boundaries and in Collapsible Tubes

FLUID MECHANICS AND ITS APPLICATIONS
Volume 72

Series Editor: R. MOREAU
MADYLAM
Ecole Nationale Supérieure d'Hydraulique de Grenoble
Boîte Postale 95
38402 Saint Martin d'Hères Cedex, France

Aims and Scope of the Series

The purpose of this series is to focus on subjects in which fluid mechanics plays a fundamental role.

As well as the more traditional applications of aeronautics, hydraulics, heat and mass transfer etc., books will be published dealing with topics which are currently in a state of rapid development, such as turbulence, suspensions and multiphase fluids, super and hypersonic flows and numerical modelling techniques.

It is a widely held view that it is the interdisciplinary subjects that will receive intense scientific attention, bringing them to the forefront of technological advancement. Fluids have the ability to transport matter and its properties as well as transmit force, therefore fluid mechanics is a subject that is particulary open to cross fertilisation with other sciences and disciplines of engineering. The subject of fluid mechanics will be highly relevant in domains such as chemical, metallurgical, biological and ecological engineering. This series is particularly open to such new multidisciplinary domains.

The median level of presentation is the first year graduate student. Some texts are monographs defining the current state of a field; others are accessible to final year undergraduates; but essentially the emphasis is on readability and clarity.

For a list of related mechanics titles, see final pages.

Flow past Highly Compliant Boundaries and in Collapsible Tubes

Proceedings of the IUTAM Symposium
held at the University of Warwick, United Kingdom,
26–30 March 2001

Edited by

PETER W. CARPENTER
School of Engineering,
University of Warwick,
Coventry, United Kingdom

and

TIMOTHY J. PEDLEY
Department of Applied Mathematics and Theoretical Physics,
Cambridge University,
Cambridge, United Kingdom

KLUWER ACADEMIC PUBLISHERS
DORDRECHT / BOSTON / LONDON

A C.I.P. Catalogue record for this book is available from the Library of Congress.

ISBN 1-4020-1161-X

Published by Kluwer Academic Publishers,
P.O. Box 17, 3300 AA Dordrecht, The Netherlands.

Sold and distributed in North, Central and South America
by Kluwer Academic Publishers,
101 Philip Drive, Norwell, MA 02061, U.S.A.

In all other countries, sold and distributed
by Kluwer Academic Publishers,
P.O. Box 322, 3300 AH Dordrecht, The Netherlands.

Printed on acid-free paper

Printed in the Netherlands.

CONTENTS

PREFACE

The IUTAM Symposium on Flow in Collapsible Tubes and Past Other Highly Compliant Boundaries was held on 26-30 March, 2001, at the University of Warwick. As this was the first scientific meeting of its kind we considered it important to mark the occasion by producing a book. Accordingly, at the end of the Symposium the Scientific Committee met to discuss the most appropriate format for the book. We wished to avoid the format of the conventional conference book consisting of a large number of short articles of varying quality. It was agreed that instead we should produce a limited number of rigorously refereed and edited articles by selected participants who would aim to sum up the state of the art in their particular research area. The outcome is the present book.

Peter W. Carpenter, Warwick
Timothy J. Pedley, Cambridge
May, 2002.

SCIENTIFIC COMMITTEE

ACKNOWLEDGEMENT

We are very grateful to IUTAM and KLUWER for their financial support of this conference. We would also like to thank the Scientific Committee for their invaluable role in making the conference a success and producing this book

Chapter 1

INTRODUCTION

PETER W. CARPENTER
School of Engineering, University of Warwick
Coventry, CV4 7AL, U.K.

TIMOTHY J. PEDLEY
Department of Applied Mathematics and Theoretical Physics
University of Cambridge, Cambridge CB3 0WA, U.K.

Abstract

The motivation for the book and its individual chapters are briefly discussed.

P.W. Carpenter and T.J. Pedley (eds.),
Flow past Highly Compliant Boundaries and in Collapsible Tubes, 1-11.

1. Motivation for the book and its individual chapters

The IUTAM Symposium on *Flow in Collapsible Tubes and Past Other Highly Compliant Boundaries* was held on 26-30 March, 2001, at the University of Warwick. As this was the first scientific meeting of its kind we considered it important to mark the occasion by producing a book. Accordingly, at the end of the Symposium the Scientific Committee met to discuss the most appropriate format for the book. We wished to avoid the format of the conventional conference book consisting of a large number of short articles of varying quality. It was agreed that instead we should produce a limited number of rigorously refereed and edited articles by selected participants who would aim to sum up the state of the art in their particular research area. The outcome is the present book.

There are numerous physiological examples of compliant tubes which collapse because the internal-external pressure difference falls below a critical value close to zero. In such cases the cross-sectional area varies significantly in response to the dynamics of the internal flow (e.g., veins above the heart, arteries under a cuff, large airways in forced expiration, and the urethra during micturition). Laboratory experiments using rubber tubes of finite length reveal a rich variety of self-excited oscillations, indicating a complex underlying dynamical system. Theoretical models have been developed of zero, one, and two space dimensions, including a complete numerical simulation of a realizable two-dimensional system. Nevertheless, the mechanism of the oscillations (the structure of the dynamical system) is still poorly understood despite recent progress. It has been proposed both at the Symposium and previously that the oscillations are 'merely' the result of a hydroelastic instability. But the analogy is certainly not exact because most flutter theories analyse the instability of a parallel undisturbed flow state whereas in very compliant tubes the steady state from which oscillations develop is highly deformed.

More generally there are many other biomechanical, biological and engineering examples of external and internal flows past highly compliant boundaries (e.g., mechanics of snoring, pressure pulse propagation in the cerebrospinal-fluid system in the spine, hydrodynamics of dolphins and other aquatic animals, drag reduction, process engineering, and structure-borne sound). The underlying physical phenomena and mathematical modelling have much in common with those applicable to collapsible tubes. Nevertheless, until the IUTAM Symposium there had been little exchange of ideas between the researchers working in the various areas. Certainly, as this book will make clear, much progress has been made towards a fuller understanding of the interaction of flows with highly compliant boundaries. Nevertheless, many unanswered questions remain. In particular, significant contradictions remain unresolved between the theoretical, computational-simulation and experimental approaches. In addition, much remains to be learnt about nonlinear effects, particularly from a dynamical systems perspective. Recent progress towards understanding the effects of wall compliance on turbulent flow was reported at the Symposium and this, too, is reflected in the present book. It still remains true, however, that the physics involved is poorly understood.

The main part of the book is divided into three sections: *A. Collapsible Tubes*, *B. Instability of Flow Past Compliant Walls*, and *C. Drag Reduction and Turbulence Modification.*

A. FLOW IN COLLAPSIBLE TUBES

A collapsible tube is an elastic tube that is subjected to an external pressure large enough to cause its cross-section to deform significantly in both shape and area, thereby becoming so compliant that flow-induced internal pressure changes can couple strongly to the wall-deformation. The first of the two chapters in this volume on flow in collapsible tubes, chapter 2, is much longer than most of the chapters in this book because the two authors, M. Heil and O. E. Jensen, (with the editors' enthusiastic agreement) decided to join forces and provide a substantial review of both the biological applications (and motivation) of the field and the current state of theoretical modelling. Among other things, they survey applications to:

- blood flow in the cardiovascular system, notably systemic arteries that collapse under a sphygmomanometer cuff, dynamic flow-induced collapse of arteries downstream of atherosclertic stenoses, veins above the level of the heart (notably the jugular veins of a giraffe), and limb veins subjected to external compression in order to inhibit deep vein thrombosis in the bedridden.
- Urodynamics: the urethra collapses during micturition, and the ureters, like the gut, are squeezed by contraction of the smooth muscle in their walls, leading to peristaltic pumping of their contents.
- Air flow in the lungs, where the larger intrathoracic bronchi collapse during forced expiration, sneeze, or cough, and parts of the upper airways (e.g. soft palate and pharynx) of some subjects collapse during respiration, causing snoring. The vocal cords can also be thought of as a collapsible tube in which self-excited oscillations arise.

The relevant presentations at the IUTAM Symposium, referred to in the chapter, were by B. S. Brook and T. J. Pedley (giraffe jugular vein), T. S. Balint and A. D. Lucey (snoring), J. L. van Leeuwen (phonation in birds) and K. Berkouk, P. W. Carpenter and A. D. Lucey (pressure waves in the cerebrospinal fluid system in the spine).

Most of the chapter, however, concerns the theoretical modelling. The authors review the major theoretical and computational developments of the past twenty-five years, with particular focus on the development of self-excited oscillations: lumped-parameter models, one-dimensional models, and the most recent two- and three-dimensional models and simulations. Not content with surveying what has gone before, the authors also provide a taste of the new modelling that is currently in progress, notably by themselves (separately and together) and by X. Y. Luo and T. J. Pedley and their colleagues. The presentations from the Symposium that feature here are the separate ones by O. E. Jensen and M. Heil, and those by Z. X. Cai and X. Y. Luo, T. J. Pedley (and J. C. Guneratne), and C. Davies and P. W. Carpenter.

Chapter 3, by C. D. Bertram, is a wide-ranging historical review of the literature on observations and experiments in collapsible tubes. It begins with the early records of venous collapse by Valsalva around 1700, and the first observations of self-excited oscillations (in the jugular vein of the horse) by Barry in 1824, but notes that the first, or at least most famous, early use of a short length of non-biological collapsible tube in a biomedical context was that by Starling in 1914, giving rise later to the name "Starling resistor" for the standard collapsible-tube experimental apparatus. The chapter explains the differences between early experiments, not always fully recognised by the experimenters, and outlines all the major features that experimentalists (notably Dr Bertram himself) have since sought to measure and to understand:

- the elastic properties of collapsible tubes, often assumed to be given by a "tube-law" relating transmural pressure to local cross-sectional area, and the problems of measuring either pressure or area;
- flow patterns in collapsible tubes, notably flow separation and its importance or otherwise in the generation of self-excited oscillations;
- choking, when the fluid speed becomes equal to the wave speed, and whether the existence of supercritical flow is necessary for self-excited oscillations;
- flow limitation and pressure-drop limitation;
- collapsible stenoses (cf chapter 2);
- collapsible tubes as complex nonlinear dynamical systems.

The chapter concludes with a brief survey of biomedical applications. The presentations from the Symposium that are referred to here are those by K. Ohba et al, by Y. Matsuzaki et al and by C. D. Bertram and N. S. J. Elliott.

B. INSTABILITY OF FLOW PAST COMPLIANT WALLS

This section begins with Chapter 4 – an account of convective and absolute instabilities in flow over compliant walls by C. Davies. For such systems both the fluid flow and the flexible wall are wave-bearing media. The result is the rich spectrum of eigenmodes that is reviewed in this chapter. This proliferation of eigenmodes greatly increases the opportunities for modal interaction and coalescence, making absolute instability far more likely. Much recent progress has been made in understanding the flow physics involved.

Chapter 5 by V. Kumaran describes the recent progress towards understanding the hydrodynamic stability of flow through compliant pipes. Again there is a bewildering variety of different disturbance types that can become unstable. These are described and explained in this chapter.

The complexity of the eigenmode structure and the rich variety of wave types that are so much a feature of the previous two chapters have motivated various approaches and applications. For example, many of the problems studied in the context of structure-borne sound evidently have a close connection to those studied for other applications of compliant walls. In Chapter 6 A.D. Lucey and N. Peake review the recent work of this type on wave excitation of flexible walls in the presence of a fluid flow. Model problems that have been fairly widely studied involve the response of various flexible surfaces to impulsive and harmonic forcing in the presence of a fluid flow. In fact, the impulsively driven flexible plate was one of the very first systems in fluid dynamics to be analysed (Brazier-Smith & Scott 1984) by using the approach of Briggs (1964) and Bers (1983). The corresponding problem with harmonic forcing was initiated by the seminal paper of Crighton & Oswell (1991).

Much of the theoretical work on waves propagating in flows over compliant and other flexible surfaces is based on the assumption of an infinitely long surface. Plainly in real applications and in experimental studies the flexible surfaces are of finite length. Nevertheless, until recently little was known about the behaviour of vortical waves (such as Tollmien-Schlichting waves) when incident on junctions between rigid and compliant walls. The theoretical analysis of longitudinal waves incident on junctions in flexible tubes is well-known and well-understood; see, for example, Lighthill (1978). But these waves are not vortical and the conservation principles applied at the junctions do not generalize to vortical waves. These were apparently first investigated in the numerical-simulation study of Davies and Carpenter (1997). They noted that when a Tollmien-Schlichting wave was incident on the leading edge of a finite compliant panel, its response appeared to correspond to a complex superposition of various eigenmodes of the infinitely long fluid-flow/compliant-wall eigensystem. Given this, the key question is: how do you relate the amplitudes and phases of the eigenmodes on either side of a junction? This question has been investigated by several authors recently and their work is reviewed by P.W. Carpenter and P.K. Sen in Chapter 7.

The compliant disc rotating in water that is otherwise still until disturbed has been the basis of compact and relatively simple apparatus for many investigations into drag reduction and transition delay by compliant walls. Usually in such investigations only the speed of rotation and the driving torque are measured. And, until recently, little was known about the effects of wall compliance on the nature and behaviour of the disturbances in the rotating-disc boundary layer. Over the past few years these disturbances have been extensively investigated in a series of theoretical, computational and experimental studies, mostly carried out at the University of Warwick. This work and the earlier experimental studies are reviewed in Chapter 8 by P.W. Carpenter, P.J. Thomas and M. Nagata. They also review the recent work on the effects of wall compliance on thin-gap rotating flows, done at Marseille, and on Taylor-Görtler instabilities, particularly the recent work at Kyoto on the stability of Taylor-Couette flows.

C. DRAG MODIFICATION AND TURBULENCE MODIFICATION

There are many different, multi-disiciplinary aspects to an assessment of the potential of compliant walls for drag reduction and turbulence modification. In an extended version of his key-note lecture, these are reviewed in Chapter 9 by M. Gad-el-Hak with an emphasis on the practical engineering issues involved.

As will have been made clear in the previous chapter, the effects of wall compliance on flow stability and laminar-turbulent transition are now well-understood.

Furthermore, there is now good agreement between experimental measurements and theoretical predictions. In contrast, our understanding of the effects of wall compliance on turbulent flows is still rather poor. There is a wealth of experimental evidence, much of it from Novosibirsk (Semenov 1991) and Kiev (Babenko *et al.* 1993), that substantial drag reduction and turbulence suppression is possible by use of compliant walls. But we still lack a firm theoretical basis for understanding how wall compliance can modify turbulent flow. For example, it seems from the work of Semenov and others that flexible walls orders of magnitude stiffer than the original compliant coatings of Kramer (1957, 1960) are capable of substantially reducing the level of turbulence and skin friction. Yet, much more flexible walls, similar to those of Kramer, also appear to possess such a capability. This apparent paradox is still unresolved. There has been some recent progress towards a theoretical understanding of the effects of wall compliance on turbulence and this is reviewed in Chapter 10 by D. Rempfer, L. Parsons, S. Xu and J. Lumley. One aspect of the interaction of a flexible wall with turbulent flow that has become better understood in recent years is the effects of turbulence on the flow-induced surface waves. The recent progress on this topic is reviewed by K.-S. Yeo in Chapter 11, while in Chapter 12, K.-S. Choi reviews the recent experimental research on turbulent flow over compliant walls.

Research on the use of compliant walls for drag reduction and laminar-flow control was originally inspired by studies of dolphin hydrodynamics. First there was the study by Gray (1936) of the energetics of dolphin propulsion. Gray concluded that if conventional hydrodynamics were involved, the power output per unit mass of dolphin muscle would need to exceed the mammalian norm sevenfold in order to propel the animal at the swimming speeds commonly observed. This became known as *Gray's Paradox.* It and the remarkable drag reductions achieved by Kramer (1957, 1960) with his artificial dolphin skin led to the many subsequent investigations of the drag-reducing and laminar-flow capabilities of compliant walls. Accordingly, it seems fitting to end the book with a re-assessment of Gray's paradox and review of dolphin hydrodynamics by V.V. Babenko and P.W. Carpenter.

References

Babenko, V.V., Kanarskii, M.B. and Korobov, B.I. (1993) *Boundary layers over elastic walls*, (in Russian), Kiev Naukova Dumka.

Bers, A. (1983) Space-time evolution of plasma instabilities – absolute and convective, In *Handbook of Plasma Physics*, ed M.N. Rosenbluth, R.Z. Sagdeev, North-Holland, 1: 451-517.

Briggs, R.J. (1964) *Electron-stream Interaction with Plasmas*, Monograph No. 20, MIT Press.

Brazier-Smith, P.R. and Scott, J.F. (1984) Stability of fluid flow in the presence of a compliant surface, *Wave Motion* **6**, 547-560.

Crighton, D. G. and Oswell, J.E. (1991) Fluid loading with mean flow, *Phil. Trans. Royal Soc. London A***335**. 557-592.

Davies, C. and Carpenter, P.W. (1997) Numerical simulation of the evolution of Tollmien-Schlichting waves over finite compliant panels, *J. Fluid Mech.* **335**, 361-392.

Gray, J. (1936) Studies in animal locomotion VI: The propulsive powers of the dolphin, *J. Exp. Biology* **13**, 192-199.

Kramer, M.O (1957) Boundary-layer stabilization by distributed damping, *J. Aero. Sci.* **24**, 459.

Kramer, M.O (1960) Boundary-layer stabilization by distributed damping, *J. Amer. Soc. Naval Engrs.* **72**, 25-33; *J. Aero/Space Sci.* **27**, 69.

Lighthill, J. (1978) *Waves in Fluids*, Cambridge University Press.

Semenov, B. N (1991) On conditions of modelling and choice of viscoelastic coatings for drag reduction, In *Recent Developments in Turbulence Management*, Ed K.-S. Choi, Kluwer Academic Publishers, Dordrecht, pp. 241-262.

Appendix

List of presentations at the IUTAM Symposiom on
Flow in Collapsible Tubes and Past Other Highly Compliant Boundaries
26-30 March, 2001, University of Warwick

The number in square brackets refers to the chapter in which the work presented is described and discussed.

Flow in Collapsible Tubes

M. Heil, University of Manchester, U.K.
Keynote Lecture: Flow in collapsible tubes: The biomechanics perspective. [**2**]

C.D. Bertram and N.S.J. Elliott, University of New South Wales, Australia.
Comparison of flow limitation in one tapered and two uniform collapsed tubes. [**2, 3**]

B.S. Brook[1] and T.J. Pedley[2]
[1]University of Sheffield; [2]University of Cambridge, U.K.
The effect of non-uniform mechanical and elastic properties on steady and time-dependent flow in (giraffe jugular) veins. [**2**]

J.C. Guneratne, G. Picard and T.J. Pedley, Cambridge University, U.K.
High-Reynolds-number asymptotics for flow in collapsible channels and tubes. [**2**]

Z.X. Cai[1], X.Y. Luo[1] and T.J. Pedley[2]
[1]University of Sheffield; [2]Cambridge University, U.K.
A localized asymptotic solution for a fluid-beam problem. [**2**]

Z.X. Cai and X.Y. Luo, University of Sheffield, U.K.

A fluid-beam model for collapsible channel flow. **[2]**

M. Hamadiche[1] and M. Gad-el-Hak[2]
[1]École Central de Lyon, France; [2]University of Notre Dame, U.S.A.
Stability of collapsible ducts. **[2]**

O.E. Jensen, University of Nottingham, U.K.
Self-excited oscillations in an elastic-walled channel: insights from a simple asymptotic model. **[2]**

K. Ohba, T. Kamimura, K. Bando and K. Hanazono, Kansai University, Japan.
Distribution of flow velocity and pressure in a largely deformed collapsible tube. **[3]**

R.A. Scroggs, E.A. Patterson and S.B.M. Beck, University of Sheffield, U.K.
Numerical fluid-structure interaction model of a collapsible tube using LS-DYNA. **[3]**

Biomechanical Applications

T.S. Balint and A.D. Lucey, University of Warwick, U.K.
Instability of a flexible cantilevered plate in plane channel flow. **[2]**

M. Bathe, A. Shirai and R.D. Kamm, M.I.T., Cambridge, Mass., U.S.A.
Numerical simulation of neutrophils and other highly compliant cells. **[2]**

K. Berkouk, P.W. Carpenter and A.D. Lucey, University of Warwick, U.K.
Pressure propagation in the spinal CSF system. **[2]**

P.W. Carpenter, University of Warwick, U.K.
Does the dolphin have a secret? The hydrodynamics of dolphin skin re-visited. **[13]**

M. Heil and J.P. White, University of Manchester, U.K.
Airway closure: surface-tension-driven non-axisymmetric instabilities of liquid-lined elastic tubes. **[2]**

J.L. van Leeuwen and C.P.H. Elemans, Wageningen University, The Netherlands.
A biomechanical model of vocalisation in birds. **[2]**

Y. Matsuzaki, M. Watanabe, T. Aomatsu and T. Ikeda, Nagoya University, Japan.

Experiment on flow in a two-dimensional channel with an obstruction oscillating in high frequency: Preliminary study. **[2, 3]**

Effects of wall compliance on boundary-layer stability and transition

M. Gad-el-Hak, University of Notre Dame, U.S.A.
Keynote Lecture: Compliant coatings – What works and what doesn't? **[9]**

V.V. Babenko, Institute of Hydromechanics UAS, Kiev, Ukraine.
External and internal flows on elastic surfaces. **[13]**

A. Cros[1], R. Ali[2], L. Schouveiler[1], P.J. Thomas[2], P. Le Gal[1], M.-P. Chauve[1], P.W. Carpenter[2] and C. Davies[3] [1]IRPHE, Marseille, France; [2]University of Warwick; [3] Cardiff University, U.K.
Transition of torsional Couette flow between a compliant rotating disk and a stationary rigid wall. **[8]**

S. Manuilovich, Central Aero-Hydrodynamics Institute (TsAGI) Moscow, Russia.
Propagation of Tollmien-Schlichting wave in a boundary layer over a flexible path of a wall. **[7]**

Effects of wall compliance on turbulence

K.S. Choi[1], E. Le-Hou[1], B.R. Clayton[1] and M.P. Escudier[2]
[1]University of Nottingham; [2]University of Liverpool, U.K.
Turbulent boundary-layer structure over a compliant surface. **[12]**

E.R. Fitzgerald, Johns Hopkins University, U.S.A.
Influence of viscoelastic material properties on turbulent flow past thin compliant walls. **[8, 13]**

V.M. Kulik and S.L. Morosova, Institute of Thermophysics RAS, Novosibersk, Russia.
Response of a compliant coating to fluctuating wall pressure. **[12]**

L. Parsons[1], D. Rempfer[2] and J.L. Lumley[1]
[1]Cornell University; [2]Illinois Institute of Technology, U.S.A.
Low-dimensional model of the interaction of near-wall turbulence with a compliant boundary. **[10]**

V.P. Reutov and G.V. Rybushkina,
Institute of Physics RAS, Nizhny Novgorod, Russia.

Nonlinear dynamics of hydroelastic waves in a turbulent boundary-layer flow past a compliant wall. **[11]**

B. Semenov, Novosibersk, Institute of Thermophysics RAS, Novosibersk, Russia.
Interference action of compliant boundaries on flow in collapsible tubes. **[10, 12]**

S. Xu[1], D. Rempfer[2] and J.L. Lumley[1]
[1]Cornell University; [2]Illinois Institute of Technology, U.S.A.
Direct numerical simulation of the interaction of near-wall turbulence with a compliant wall. **[10]**

K.S. Yeo, H.Z. Zhao and B.C. Khoo, National University of Singapore.
Turbulent flow over a compliant surface – wave instabilities. **[11]**

Flow-induced waves over flexible walls

C. Davies[1] and P.W. Carpenter[2]
[1]Cardiff University; [2]University of Warwick, U.K.
Numerical simulation of instabilities in boundary layers over a compliant wall. **[4]**

U. Ehrenstein[1] and O. Wiplier[2]
[1]Université de Nice-Sophia Antipolis; [2]Université de Lille, France.
Local and global stability behaviour in a boundary-layer flow with compliant coatings. **[4, 7]**

E. de Langre, LadHyX, Ecole Polytechnique, France.
The effect of structural characteristics on the onset of absolute instability in compliant structures submitted to inviscid flow. **[6]**

A.D. Lucey, University of Warwick, U.K.
Wave excitation and the destabilisation of a flexible surface by a uniform mean flow. **[6]**

N. Peake, Cambridge University, U.K.
Nonlinear stability of a fluid-loaded elastic plate with mean flow. **[6]**

P.K. Sen[1], P.W. Carpenter[2], J.S.B. Gajjar[3] and S. Hegde[1].
[1]IIT Delhi, India; [2]University of Warwick; [3]University of Manchester, U.K.
The jump conditions for instability waves at a rigid-compliant joint. **[7]**

Internal flows with compliant walls

M. Gad-el-Hak[1] and M. Hammadiche[2]
[1]University of Notre Dame, Indiana, U.S.A; [2]École Centrale de Lyon, France.
Temporal stability of flow through viscoelastic tubes. [**5**]

J.S.B. Gajjar[1], A. Gibson[1] and P.K. Sen[2]
[1]University of Manchester, U.K.; [2]IIT Delhi, India.
Instabilities in compliant pipes and related flows. [**5**]

S. Koga and M. Nagata, Kyoto University, Japan.
Stability of Couette flow witha compliant cylinder. [**8**]

V. Kumaran, Indian Institute of Science, Bangalore, India
Classification of instabilites in flow past flexible surfaces. [**5**]

M. Tamilarasan and K.-S. Choi, University of Nottingham, U.K.
Drag reduction in turbulent pipe flows using compliant coating. [**12**]

SECTION A

FLOW IN COLLAPSIBLE TUBES

Chapter 2

FLOWS IN DEFORMABLE TUBES AND CHANNELS

THEORETICAL MODELS AND BIOLOGICAL APPLICATIONS

MATTHIAS HEIL
Department of Mathematics, University of Manchester, Oxford Road, Manchester M13 9PL, UK

OLIVER E. JENSEN
School of Mathematical Sciences, University of Nottingham, University Park, Nottingham NG7 2RD, UK

Abstract

This chapter gives an overview of the main physiological applications of collapsible tube flows and reviews the major theoretical and computational developments of the past twenty-five years, ranging from lumped-parameter models to three-dimensional Navier–Stokes simulations. We also discuss some of the significant questions that, despite substantial progress, still remain open.

P.W. Carpenter and T.J. Pedley (eds.),
Flow past Highly Compliant Boundaries and in Collapsible Tubes, 15-49.

1. Introduction

Many fluid-conveying vessels in the human body are highly elastic and deform substantially in response to the traction (pressure and viscous stress) that the fluid exerts on them. The study of flows in elastic vessels is therefore of considerable interest in many biomechanical and biomedical applications. It also presents an interesting and extremely challenging fluid-mechanical problem in its own right. Experimental studies of flow in collapsible tubes, described in more detail in Chapter 3, show that the interaction between the finite-Reynolds-number internal flow and large deformations of the tube wall can cause the development of large-amplitude self-excited oscillations. Many distinct modes of oscillation have been observed experimentally. Understanding fully the mechanisms driving these oscillations remains a major research goal.

A completely rational theoretical description of flow in collapsible tubes would have to be based on the unsteady 3D Navier-Stokes equations, coupled to the equations of large-displacement shell theory. The numerical solution of this problem would require enormous computational resources, and would yield limited (but of course very valuable) insights into the underlying physics. Much theoretical work on flow in collapsible tubes has therefore concentrated on the development and analysis of simpler models, obtained, for example, by reducing the spatial dimension of the problem. Such simplified models invariably involve a number of *ad hoc* assumptions whose validity must be critically assessed as better models or more detailed experimental data become available.

The problem of flow in collapsible tubes has many obvious similarities to the problem of flow past compliant surfaces and coatings, which is reviewed in other Chapters in this volume. However, there are a few important differences between these problems. First, collapsible tube experiments are performed with finite-length elastic tubes whose upstream and downstream ends are held open. The inertia and resistance of the fluid in the supporting rigid tubes have an important influence on the system's overall dynamics; conversely, most (but not all) theoretical studies of flow past compliant coatings are based on infinite domains. Second, self-excited oscillations in collapsible tubes tend to have large amplitude and can develop from steady states in which the tube is strongly deformed, and in which the viscous pressure drop along the tube is significant; conversely, most stability analyses of flows past compliant coatings are based on perturbations of an initially uniform, undeformed state. Third, self-excited oscillations in collapsible tubes do not always require the presence of wall inertia, whereas most flutter-type analyses consider systems in which wall inertia is significant.

It is the aim of this Chapter to provide a review of previous theoretical and computational analyses of flows in collapsible tubes (earlier review articles on the subject were written by Shapiro 1977, Kamm and Pedley 1989, Grotberg 1994 and Pedley and Luo 1998), to identify the current state of the art in the field, and to outline some open research problems. The Chapter includes brief summaries of relevant talks presented at the IUTAM Symposium. Since many of these talks contained unpublished work, we have included references to papers which, at the time of writing, are in press, under review or even in preparation.

The structure of this Chapter is as follows: Section 2 provides an overview of the major biological applications of collapsible tube flow. Section 3 introduces the model problem considered in most theoretical analyses and discusses its relation to typical laboratory experiments. Sections 4–6 discuss successively 1D, 2D and 3D models of flow in collapsible tubes. Finally, Section 7 provides a brief summary and some suggestions for future work.

2. Biological Background

The propagation of the pulse wave through the arterial system is a well-known and well-understood example of a physiological flow in which wall elasticity plays an important role. One-dimensional models, based on an inviscid description of the fluid dynamics, coupled to a simple description of the wall mechanics, are able to explain satisfactorily many features of this problem (see e.g. McDonald 1974, Lighthill 1975, Pedley 1980). The analysis of pulse-wave propagation is facilitated by the fact that under normal conditions the arteries are subject to a positive transmural (internal minus external) pressure and therefore remain inflated and relatively stiff throughout the pulse cycle. There are, however, many examples of fluid-conveying vessels that are subject to negative (compressive) transmural pressures, which cause the vessels to buckle and collapse non-axisymmetrically. Buckled vessels are very flexible and even small changes in fluid pressure can induce large changes of their cross-sectional area. This leads to a strong interaction between fluid and solid mechanics, which gives rise to many intriguing phenomena, including flow limitation and the propensity to develop large-amplitude self-excited oscillations.

Physiological examples of collapsible tubes are numerous, as are corresponding models, so the following list of topics is far from complete but hopefully reasonably representative. In the cardiovascular system, first of all, spontaneous collapse can occur in veins above the heart and outside the skull (due to hydrostatic reduction of blood pressure). This is particularly important in subjects with long necks, such as giraffes (see Pedley et al. 1996). Flow-induced collapse of blood vessels is believed to play a role in the auto-regulation of blood supply to many internal organs; see Rodbard and Takacs (1966) and Rodbard (1966). Dynamic flow-induced collapse of blood vessels downstream of atherosclerotic stenoses has been proposed as a mechanism of plaque rupture, which can lead to vessel occlusion distally, with potentially serious consequences (Binns and Ku 1989; Ku 1997). Blood vessels can also collapse as a result of external compression. Coronary blood vessels collapse during systole (see e.g. Gregg and Fisher 1963), and collapsible-tube models have been used to understand how compression mediates their delivery of blood to the myocardium (e.g. Guiot et al. 1990). Active compression of veins in the lower limbs is used successfully as a therapy to prevent deep-vein thrombosis; see e.g. Kamm (1982) and Olson et al. (1982). Both veins and arteries also collapse during sphygmomanometry, when an inflatable cuff is placed around the upper arm to measure blood pressure.

During micturition, the urethra behaves like a passive collapsible tube (e.g. Griffiths 1969, 1971), which can exhibit flow limitation effects. In contrast the ureter, and a number of other deformable muscular vessels in the body (particularly in the gut), transport fluid by active peristaltic pumping, a process which couples fluid, solid and muscle mechanics (e.g. Carew and Pedley 1997) in a manner that is only beginning to be properly understood.

Air flow in the lung is strongly affected by the elasticity of the airways. For instance during forced expiration, contraction of the expiratory muscles increases the pleural pressure that drives air out of the peripheral airways. If the pleural pressure increases beyond a critical level, it can initiate the collapse of the proximal airways. The reduction of their cross-sectional area increases the local fluid velocity. The Bernoulli effect, which then reduces the internal fluid pressure, leads to a further increase in compression. This mechanism causes an increasingly strong collapse of the airways and results in so-called flow limitation and possibly 'negative effort dependence,' whereby an increase in expiratory effort (at a given lung volume) beyond a certain level can lead to a reduction in expiratory flow rate.

Self-excited oscillations of collapsible lung airways are believed to give rise to a number of different respiratory noises. Flutter instabilities have been proposed as the origin of respiratory wheezes during forced expiration (e.g. Gavriely et al. 1984, 1989), and speech production involves controlled flow-induced vibrations of the vocal chords, which can be modelled as a collapsible tube system (e.g. Berke et al. 1991). Similarly bird song involves oscillations of a set of membranes in the avian syrinx. Fee et al. (1998) have shown that modulations of bird song display characteristics (such as mode locking and period doubling) of a nonlinear dynamical system, and they provide experimental and theoretical evidence of the primary mechanism being a dynamic flow-structure interaction. Results from a computational model of this process were presented to the IUTAM Symposium by J. L. van Leeuwen.

Snoring sounds in humans during obstructive sleep apnea have their origins in flow-induced deformation of the soft palate and pharyngeal wall. The Bernoulli effect can induce upper airway collapse and closure (characterised using a simple lumped-parameter model by Gavriely and Jensen 1993, for example), and flow-induced instabilities of the airway wall can then lead to noise production; for a distributed collapsible-tube model of this process see Aittokallio et al. (2001). Flow-induced flutter of the soft palate, which can be modelled as a flexible cantilevered elastic plate that can flap as air flows past it (Huang 1995), is an independent source of noise production. In a study presented to the IUTAM Symposium, Balint & Lucey used Navier–Stokes simulations, coupled to plate equations, to show how instabilities of the soft palate can arise either through a flutter mechanism or through so-called static divergence (where the plate undergoes a non-oscillatory sideways displacement); for full details see Balint (2001).

Spontaneous flow-induced oscillations also arise throughout the cardiovascular system. These include cervical venous hum, arising through oscillations of the external jugular vein, which is collapsed because of a low hydrostatic pressure (Danahy and Ronan 1974). Korotkoff sounds heard during sphygmomanometry have been associated with the development of self-excited oscillations of the partially reopened brachial artery (see e.g. Ur and Gordon 1970 and Bertram et al. 1989), although alternative mechanisms of noise production have also been proposed, for example by Shimizu and Tanida (1983). Oscillations in coronary blood vessels have also been observed during open-heart surgery (Tsuji et al. 1978).

3. Background to theoretical models of collapsible-tube flows

Many of the physiologically observed phenomena described above can be reproduced in laboratory experiments using the 'Starling Resistor,' shown in Fig. 1. Inside a pressure chamber, a thin-walled elastic tube (typically made of latex rubber) is mounted on two rigid tubes. Fluid (typically air or water) is driven through the tube, either by applying a controlled pressure drop $p_{entry} - p_{exit}$ between the ends of the rigid tubes or by controlling the flow rate Q. The external pressure, p_{ext}, can be controlled independently. If p_{ext} exceeds the fluid pressure by a sufficiently large amount, the tube buckles non-axisymmetrically, as sketched in Fig. 1. Flow-structure interactions then lead to a strongly nonlinear relation between pressure-drop and flow-rate, depending on which pressure differences are held fixed as the flow increases (e.g. Conrad 1969). Among these are flow limitation (increasing $p_{up} - p_{down}$ while keeping $p_{up} - p_{ext}$ fixed limits the maximum possible flux Q) and pressure-drop limitation (increasing Q with $p_{down} - p_{ext}$ fixed limits the maximum possible $p_{up} - p_{down}$). At sufficiently large Reynolds numbers, the system also readily produces self-excited oscillations, and exhibits hysteresis in transitions between dynamical states, multiple modes of oscillations (each having a distinct frequency range) and rich and complex nonlinear dynamics (Bertram 1986; Bertram et al. 1990). For full details see Chapter 3.

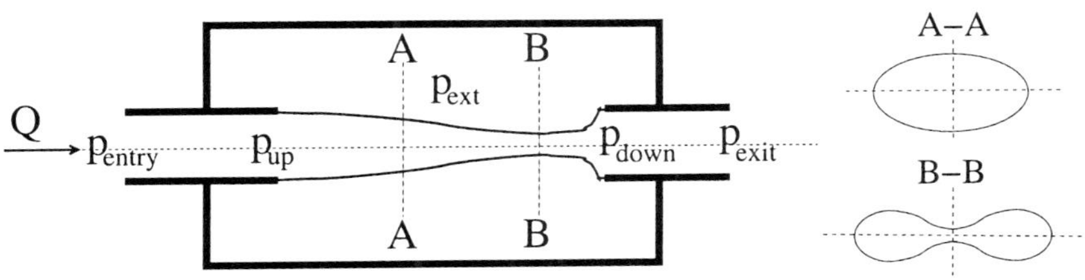

Figure 1. Left: sketch of the typical experimental setup which forms the basis of most theoretical models. Right: tube cross-sections at A–A and B–B.

A closely related physical model system was introduced by Pedley (1992). It consists of a 2D channel, one wall of which has a segment replaced by a membrane under longitudinal tension, as shown in Fig. 2. Viscous flow is driven through the channel by an imposed pressure drop $p_{entry} - p_{exit}$. The external pressure p_{ext} and the internal flow determine the deformation of the membrane. Despite the practical difficulties of producing 2D flows experimentally, this system has attracted considerable theoretical attention since it avoids the complications of fully 3D flows found in the Starling Resistor, while still exhibiting phenomena such as flow limitation and self-excited oscillations.

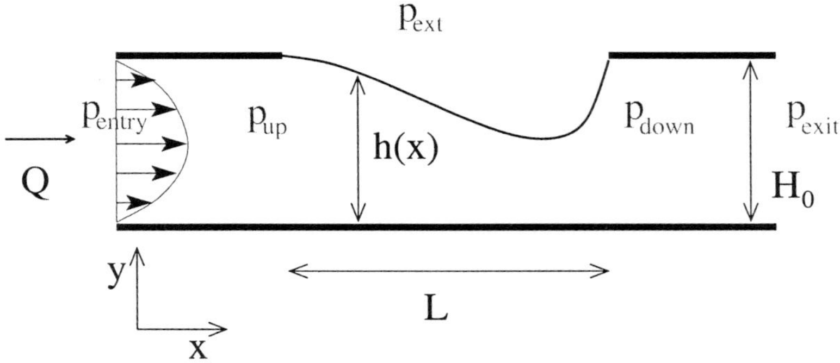

Figure 2. Sketch of the 2D model problem: viscous flow through a channel in which part of one wall has been replaced by an elastic membrane under tension T_0.

The earliest theoretical models of flow in collapsible tubes (e.g. Conrad 1969, Katz et al. 1969, Bertram and Pedley 1982) were lumped-parameter models in which the system's behaviour is characterised by a small number of scalar variables (such as the cross-sectional area, the transmural pressure and the fluid velocity at the point of strongest collapse, etc.). The temporal variations of these quantities are described by nonlinear ODEs whose numerical solutions can exhibit oscillatory behaviour reminiscent of that observed in experiments. While these models successfully capture many important flow features (such as the crucial role played by the fluid in the rigid parts of the system), they fail to describe such key phenomena as wave propagation. A more sophisticated approach is therefore warranted.

4. One-dimensional models

4.1. DERIVATION

Motivated by the successful application of 1D inviscid models for pulse-wave propagation in the arteries, the earliest distributed models of collapsible tube flow were based on a similar mathematical framework. Employing a long-wavelength approximation to the Navier-Stokes equations (including body forces and a 'lumped' representation of small viscous effects), the axial component of the momentum equation is given by

$$\frac{\partial u}{\partial t} + u\,\frac{\partial u}{\partial x} = g - \frac{1}{\rho}\frac{\partial p}{\partial x} - \mathcal{R}u. \tag{1}$$

Here $u = u(x,t)$ is the cross-sectionally averaged axial fluid velocity, $p = p(x,t)$ is the fluid pressure, ρ the fluid density and $\mathcal{R} > 0$ is a friction factor whose value depends on the local cross-sectional area $A(x,t)$ (and also on u in certain circumstances). g represents the gravitational acceleration when the tube is held vertically. The integral continuity equation is

$$\frac{\partial A}{\partial t} + \frac{\partial (uA)}{\partial x} = 0. \tag{2}$$

These equations need to be augmented by a constitutive model for the wall. In the earliest 1D models, this was provided by a functional relationship between the local transmural pressure, $p_{tm}(x,t) = p(x,t) - p_{ext}$, and $A(x,t)$. This relationship is generally referred to as the 'tube law' and has the general form

$$p_{tm}(x,t) = \mathcal{P}(A(x,t),x) \quad \text{or} \quad A(x,t) = \mathcal{A}(p_{tm}(x,t),x), \tag{3}$$

where the vessel's undeformed cross-sectional area, A_0, is characterised by

$$\mathcal{P}(A = A_0) = 0 \quad \text{or} \quad \mathcal{A}(p_{tm} = 0) = A_0. \tag{4}$$

Note that the functional form (3) of the tube law allows for axial variations of the vessel's elastic properties (reflecting, for example, changes in the undeformed cross-sectional area or variations in the vessel's wall thickness and Young's modulus).

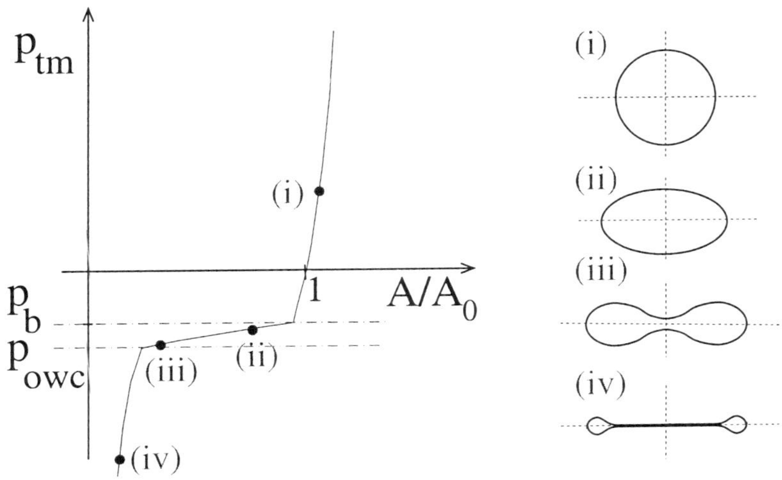

Figure 3. Left: schematic representation of the tube law. Right: typical tube shapes.

The schematic representation of the tube law in Fig. 3 illustrates the drastic change in the wall distensibility (the inverse stiffness)

$$D = \frac{1}{A}\left(\frac{\partial \mathcal{A}}{\partial p}\right) = \frac{1}{A}\left(\frac{\partial \mathcal{P}}{\partial A}\right)^{-1} \tag{5}$$

as the tube's cross-section changes from inflated to collapsed. In the axisymmetrically inflated state, (i) in Fig. 3, any deformation is accompanied by a stretching of the wall. Therefore large changes in transmural pressure are required to induce any change to the vessel's cross-sectional area. The axisymmetric vessel can withstand small compressive loads ($0 > p_{tm} > p_b$). However, when the transmural pressure falls below p_b, the tube buckles non-axisymmetrically, typically to a two-lobed state. Once buckled, only the tube's small bending stiffness resists any further collapse. Hence the vessel can undergo large changes in cross-sectional area when the transmural pressure changes slightly; see (ii)–(iii) in Fig. 3. As the compression increases further, the vessel's opposite walls come into contact (when $p_{tm} < p_{owc}$) and the wall stiffness increases again as the two outer lobes need to be bent strongly in order to further reduce the cross-sectional area, (iv) in Fig. 3. Various curve fits which approximate the $p_{tm} = \mathcal{P}(A)$ relationship sketched in Fig. 3 and which incorporate the correct solid-mechanical behaviour (see Flaherty et al. 1972) have been suggested in the literature; see e.g. Shapiro (1977) or Elad et al. (1987).

4.2. STEADY FLOW: CHOKING, FLOW LIMITATION AND ELASTIC JUMPS

Valuable insight into the behaviour of collapsible tubes can be gained from the steady version of equations (1–3). We will initially ignore the effect of gravity ($g = 0$) and assume that the elastic properties of the vessel (and thus the tube law) are independent of axial position. The three steady equations can then be combined to give

$$\frac{1}{A}\frac{dA}{dx} = -\frac{1}{u}\frac{du}{dx} = \frac{-\mathcal{R}u}{c^2 - u^2}, \qquad (6)$$

where $uA = Q$ is a constant and

$$c = \sqrt{1/(\rho D)} \qquad (7)$$

is the wavespeed of small-amplitude disturbances travelling along the tube (as in pulse-wave propagation). The dependence of the wavespeed c on the tube's stiffness implies that c is much larger in regions where the vessel is inflated than in those where it is collapsed without opposite walls in contact.

Equation (6) predicts an interesting phenomenon, known (in analogy to a related phenomenon in gas dynamics) as 'choking.' Assume that, at the upstream end of the tube, the average fluid velocity u is less than the wavespeed c. Since $\mathcal{R} > 0$, (6) implies that $dA/dx < 0$, i.e. the tube's collapse increases in the direction of the flow. Continuity then requires that $du/dx > 0$, i.e. the flow is accelerated in the streamwise direction. Provided the tube is long enough, we will therefore approach a location at which the 'speed index' $S = u/c \to 1$ and thus $dA/dx \to -\infty$, violating the long-wavelength assumption implicit in the 1D model. Clearly, this situation is physically unrealisable, implying that steady flows are impossible (according to this simple model) if the flow rate becomes so large that u approaches c anywhere along the tube. Alternatively, if $S > 1$ at some point along the tube, then variations in downstream conditions cannot propagate upstream (since small-amplitude waves travel at speeds $u \pm c$); this underlies the wave-speed mechanism of flow limitation that is believed to operate in the large airways of the lung (Dawson and Elliot 1977; Elliot and Dawson 1977). Several authors speculated that the occurrence of supercritical flow ($u > c$) at some point along the tube might coincide with the onset of self-excited oscillations. Experimental evidence (e.g. Brower and Scholten 1975 and Bonis and Ribreau 1978) appeared first to provide partial support for this view. However Gavriely et al. (1989) found that while flow-induced oscillations in a tube with large wall inertia occurred only when the flow was limited, the onset flow speed could be as low as $S \approx 0.3$ (although their definition of wave speed did not take account of wall mass). Later experiments by Bertram and Raymond (1991) and computations by Luo and Pedley (2000) have also cast doubt over a causal link between flow limitation and self-excited oscillations.

Having identified a choking mechanism which is similar to that found in compressible gas flow, one might expect the 1D collapsible tube equations to have shock-like solutions similar to those found in gas flows through Laval nozzles (resembling also hydraulic jumps in shallow-water flows). Experimentally, shock-like structures (known as 'elastic jumps') in collapsible tube flow are easily generated (see e.g. Kececioglu et al. 1981). Within the framework of the 1D model, elastic jumps are predicted to occur in situations where supercritical ($S > 1$) flow is generated somewhere in a collapsible tube. Experimentally this is done most effectively by pinching the tube at a far upstream location; the reduction in cross-sectional area increases the fluid velocity u while the associated reduction in wall stiffness simultaneously reduces the wavespeed c. The ensuing supercritical flow is still governed by (6), which now predicts that $dA/dx > 0$. Hence the cross-sectional area A increases and, provided the tube is long enough, the reduction in fluid velocity u now leads to a situation in which $dA/dx \to +\infty$ at some point. As in the case of choking, this locally violates the model's assumptions. However, a shock-like transition region, in which the flow speed is reduced from super- to subcritical, can instead develop upstream of the point where $dA/dx \to +\infty$ would occur. The application of jump conditions (similar to those used in gas dynamics) across the thickness of the elastic jump (Oates 1975, Shapiro 1977, Cowley 1982) establishes how the flow changes as it passes from the supercritical to the subcritical regime. The location of the elastic jump is determined by the downstream boundary conditions, which can influence only the subcritical region of the flow downstream of the jump. Standing waves can appear either upstream or downstream of elastic jumps through the effects of longitudinal bending or tension, attenuated by viscous effects (McClurken et al. 1981; Cowley 1983).

4.3. EFFECTS OF GRAVITY AND NON-UNIFORM TUBE PROPERTIES

As mentioned in Section 2, the venous system is strongly affected by gravitational pressure variations. Inclusion of gravity for a vertical tube changes the steady equation (6) to

$$\frac{1}{A}\frac{dA}{dx} = -\frac{1}{u}\frac{du}{dx} = \frac{g - \mathcal{R}u}{c^2 - u^2}. \tag{8}$$

In the presence of gravity, choking can be avoided because smooth transitions from sub- to supercritical flows are possible if the flow becomes critical ($u = c$) at a location where $g = \mathcal{R}u$. Shapiro (1977) lists a number of other physiologically relevant scenarios in which smooth transitions through $u = c$ are possible. They include **(i)** axial variations of the vessel's elastic properties; **(ii)** spatial variations in the external pressure (representing, e.g., a localised compression as in sphygmomanometry); and **(iii)** variations in the vessel's undeformed cross-sectional area (corresponding to flow in tapered elastic tubes). Properties **(i)** and **(iii)** in particular have been used (for example by Elad et al. 1987) to describe flow limitation in the lung: during forced expiration, a sub- to supercritical flow transition arises through non-uniform airway properties, and a super- to subcritical transition can occur further downstream via an elastic jump. Changes in the downstream boundary conditions affect the location of the elastic jump but not the overall flow.

A closely related example is provided by the giraffe jugular vein. Experiments suggest it is in a strongly collapsed state when the giraffe is standing upright, and that the flow in it can be supercritical. Pedley et al. (1996) related the position of an elastic jump in the vein to the downstream flow rate Q and the vein's downstream cross-sectional area (which is set at the junction with the distended superior vena cava). Pedley et al. (1996) showed that an increase in Q forces the elastic jump to move further downstream, and concluded that, in steady flow, Q cannot exceed the value $Q_{\max}$ for which the elastic jump occurs at the downstream end of the jugular vein.

4.4. UNSTEADY FLOWS AND WAVE PROPAGATION

The full time-dependent equations (1–3) are of hyperbolic character. Kamm and Shapiro (1979) performed a detailed analysis of the linear and nonlinear waves governed by these equations. Particular attention was paid to the time-dependent collapse and refilling of tubes which are subjected to an instantaneously applied, spatially non-uniform external pressure, a model of compression therapy for deep-vein thrombosis (Kamm 1982, Olson et al. 1982, Dai et al. 1999).

For a particular choice of the tube law $\mathcal{P}(A)$ and the resistance function $\mathcal{R}$, Cowley (1981) was able to show that equations (1–3) have solutions which have the character of roll-waves in open channel flows (analysed by Dressler 1949). Recently, Brook et al. (1999) re-examined the properties of roll-wave solutions in the collapsible-tube equations under less restrictive assumptions. Using a numerical Godunov scheme, they found that roll waves can develop from unstable steady solutions of the 1D collapsible tube equations. Applying time-dependent computations to the problem of blood flow in the giraffe jugular vein described above, Brook and Pedley (2001) showed that for $Q > Q_{\max}$ the flow remains unsteady, but roll waves were not predicted to occur. At the IUTAM Symposium, Brook & Pedley showed that axial variations of the tube properties result in smooth super- to sub-critical transitions that are temporally stable. An increase in flow rate is again found to move the position of the critical point (at which u smoothly falls below c) further downstream. Thus a maximum flow rate is predicted above which the critical point would no longer be located inside the collapsible tube; for flow rates in excess of this value, the flow must therefore be unsteady; see Brook (1997) and Brook and Pedley (2001).

At the IUTAM Symposium, Berkouk, Carpenter & Lucey presented a study of the propagation of pressure waves along the fluid-filled spinal cord (Carpenter et al. 1999, 2001a, 2001b). The model consists of two co-axial tubes: the rigid outer tube was taken to represent the bony part of the spine (the subarachnoid space), while the elastic inner tube represents the spinal cord containing cerebrospinal fluid. It was shown that the propagation of pressure waves in the cerebrospinal fluid is strongly affected by the presence of blockages in the outer tube, a situation that is characteristic of syringomyelia, a disease of the spinal cord.

4.5. FLUTTER AND STATIC-DIVERGENCE INSTABILITIES

The models described so far are relevant for systems in which the inertia of the fluid dominates that of the wall. In experiments mimicking flows in lung airways, however, in which air is blown through thin-walled tubes, wall inertia cannot be neglected. Drawing on studies in aeroelasticity, a number of workers have presented investigations of the stability of potential flows in channels where one or both walls of the channel are some form of heavy spring-backed plate with bending stiffness and subject to longitudinal tension. In these studies it was assumed that the flow domain is unbounded in the streamwise direction, and a linearised (small-amplitude) analysis was used to obtain dispersion relations for waves of a given wavelength, assuming also that the mean velocity profile is uniform across the channel. Grotberg and Davis (1980), for example, showed how such a system, in the absence of any dissipation, exhibits an oscillatory flutter instability, for which a downstream-propagating wave can grow temporally. The introduction of a small amount of wall damping had the dramatic effect of changing the instability to one of static divergence (a non-oscillatory instability in which hydrodynamic pressures overcome the restoring forces in the channel wall), as was also found by Weaver and Païdoussis (1977) and Matsuzaki and Fung (1977). (Choking, described in Section 4.2, can also be regarded as a manifestation of static divergence.) Only by adding dissipation to the fluid (using a Darcy-type term that was later justified by Grotberg and Shee 1985), making pressure and displacement act out of phase with one another so that the fluid could do work on the wall, could an oscillatory travelling-wave flutter (TWF) instability be recovered (Grotberg and Reiss 1984). In this sense TWF is a Class B instability (in the energy classification of Benjamin 1963 & Landahl 1962), in that it is destabilised when energy is transferred from the flow to the wall. Even though these investigations involved 2D potential flows, Walsh (1995) has demonstrated that their long-wave (1D) limit captures the dominant physical behaviour, while showing also that coupling between axial and transverse motion for curved shells (such as the trachea) can be dynamically significant (see also Walsh et al. 1991).

These models capture two fundamental mechanisms of instability in infinitely long fluid-filled channels (static divergence and TWF), and demonstrate the important role of dissipation in both the fluid and the wall, and of the mechanical model chosen for the wall. However to understand the role of such instabilities in the Starling-Resistor, in which the flexible segment has finite length, alternative models and solution techniques are required.

4.6. 1D MODELS OF FLOW IN THE STARLING RESISTOR

4.6.1. *Model development*

To describe flows in the Starling Resistor (Fig. 1), the governing equations (1–3) must be supplemented by boundary conditions describing the rigid parts of the system, which are known both experimentally and from lumped-parameter models to have a major effect on both steady and unsteady flows. Typical pressure-flux relationships that incorporate the effects of viscous resistance (R) and fluid inertia (I) in the upstream and downstream segments are of the form

$$p_{entry} - p_{up} = R_{up} Q_{up} + I_{up} \frac{dQ_{up}}{dt},$$

$$p_{down} - p_{exit} = R_{down} Q_{down} + I_{down} \frac{dQ_{down}}{dt}. \tag{9}$$

The tube area must also be prescribed at either end of the collapsible segment, where it is mounted onto the rigid pipes (e.g. $A_{up} = A_{down} = A_0$, where A_0 is defined in (4)). However an immediate problem with the model given by (1–3) is that it is only possible to prescribe the cross-sectional area at a single (say, upstream) location. To overcome this problem, McClurken et al. (1981) augmented the tube law (which only represents the effect of the tube's transverse bending stiffness) by terms which represent the effects of the axial wall elasticity. Given that in a strongly collapsed state the aspect ratio of the tube's cross-section is large, McClurken et al. (1981) suggested that the flow in such a tube would be similar to the flow between two parallel membranes. Provided the tube is subject to a large axial tension T_0 (as are many biological vessels), and that wall inertia can be neglected, the constitutive equation for the wall can be written as

$$p_{tm} = \mathcal{P}(A) - T_0 \frac{\partial^2 A}{\partial x^2}. \tag{10}$$

Mathematically, the inclusion of the axial-tension term increases the order of the governing equation and thus allows the application of additional boundary conditions. Furthermore, it makes the system dispersive, allowing sufficiently short waves to propagate upstream against any oncoming flow. Representation of tension in this way was one reason the simpler 2D channel flow illustrated in Fig. 2 attracted subsequent interest, and indeed if $\mathcal{P}(A)$ is removed from (10), and A in (2) and (10) is replaced by the width $h(x,t)$, the resulting equations then apply directly to this simpler system, provided membrane slopes are sufficiently small.

A further significant improvement to the 1D model was proposed and analysed by Cancelli and Pedley (1985). In earlier lumped-parameter models (e.g. Bertram and Pedley 1982), it had been found that the energy loss in the separated-flow region downstream of the point of strongest collapse played an important role in the development of self-excited oscillations. In particular, if this energy loss was completely neglected (corresponding to attached, inviscid flow with complete pressure recovery, as in Fig. 4a), the tube was found always to choke if the flux was increased sufficiently (i.e. $A \to 0$ at some point along the length of the

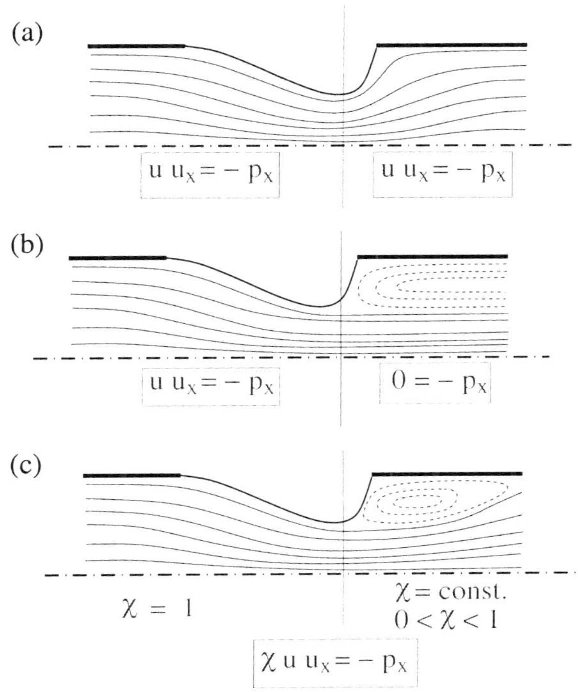

Figure 4. Models of flow separation in a collapsible tube: (a) Attached flow; no pressure loss. (b) Ideal separation; parallel-sided jet downstream of the separation point; no pressure recovery. (c) An intermediate case: $0 < \chi < 1$ downstream of the separation point corresponds to partial pressure recovery.

collapsible tube in finite time via a static divergence instability). Cancelli and Pedley (1985) showed that such behaviour persisted in a 1D inviscid model even when the dispersive effects of longitudinal tension were accounted for. In contrast, if the energy loss associated with flow separation was made as large as possible (corresponding to the formation of a parallel-sided jet beyond the narrowest point in the tube, along which there is no pressure recovery, as in Fig. 4b), Bertram and Pedley (1982) found that steady flow was always possible and no oscillations were predicted. They found that self-excited oscillations arose only if some pressure recovery was allowed. To represent this effect in a 1D framework, Cancelli and Pedley (1985) replaced the inviscid momentum equation (1) by

$$\frac{\partial u}{\partial t} + \chi\, u\, \frac{\partial u}{\partial x} = -\frac{1}{\rho}\frac{\partial p}{\partial x}, \tag{11}$$

where

$$\left.\begin{array}{ll} \chi = 1 & \text{upstream} \\ 0 < \chi < 1 & \text{downstream} \end{array}\right\} \text{ of the separation point.} \tag{12}$$

(The small viscous losses represented by $\mathcal{R}$ in (1) were found to be of minor importance and were therefore neglected.) Fig. 4 illustrates the idea behind this approach and shows that (11) and (12) are intermediate between the extreme cases of no separation ($\chi \equiv 1$) and perfect separation ($\chi = 0$ downstream of the separation point). Estimates for χ were based on the steady Borda–Carnot condition. Cancelli and Pedley's (1985) theory also allowed for the motion of the separation and attachment points by linking their position to the value of the pressure gradient (although the only way for the dissipative term $(1 - \chi)uu_x$ to vary smoothly across the separation point is to choose this point to coincide with that at which $u_x = 0$). Significantly, they found that fixing the location of the separation point was sufficient to suppress oscillations.

4.6.2. *Computational results*

Cancelli and Pedley's (1985) numerical solution of the new time-dependent equations showed that, as in the lumped parameter model, self-excited oscillations occurred only if $0 < \chi < 1$ downstream of the assumed point of flow separation. Following Cancelli and Pedley's (1985) initial computations, Reyn (1987) and Jensen and Pedley (1989) used phase-plane techniques to investigate systematically the existence and non-existence of steady solutions of this model, and demonstrated the existence of multiple steady states. An important effect of dissipation due to flow separation was to reduce the tendency of the tube to choke, and to increase the range of parameters over which a steady solution could exist. Typical steady flow states are illustrated in Fig. 5. Jensen (1990) then analysed the stability of the steady solutions and identified distinct modes of oscillation, each with its own range of frequencies, capturing a key qualitative feature of experiments (Bertram et al. 1990, 1991). A weakly nonlinear analysis was employed to explore interactions between these different modes.

In addition, many authors (e.g. Matsuzaki and Matsumoto 1989, Jensen 1992, Matsuzaki et al. 1994, Hayashi et al. 1998, Ikeda and Matsuzaki 1999) performed detailed numerical simulations to explore the time-dependent behaviour of this and closely related models. Matsuzaki and co-workers refined the representation of energy losses in the region of separated flow; Hayashi and co-workers replaced the dissipative term with a distributed frictional term similar to that in (1). All these numerical studies exhibited self-excited oscillations of considerable complexity. While care is required in solving this problem numerically (to avoid generating spurious numerical oscillations), there is convincing evidence that these studies captured genuine oscillatory behaviour, and there is even evidence of chaotic dynamics (Jensen 1992). Nonlinear periodic oscillations obtained from computations are illustrated in Fig. 6.

Despite the use of *ad hoc* assumptions, 1D models can reproduce steady behaviour with reasonable quantitative accuracy. Fig. 7, for example, shows how such a model predicts pressure-drop limitation, and the corresponding experimental results. Major qualitative features of unsteady flows, such as multiple modes of oscillation, can also be simulated with qualitative success. The key ingredients for the prediction of low-frequency oscillations in both lumped and 1D models

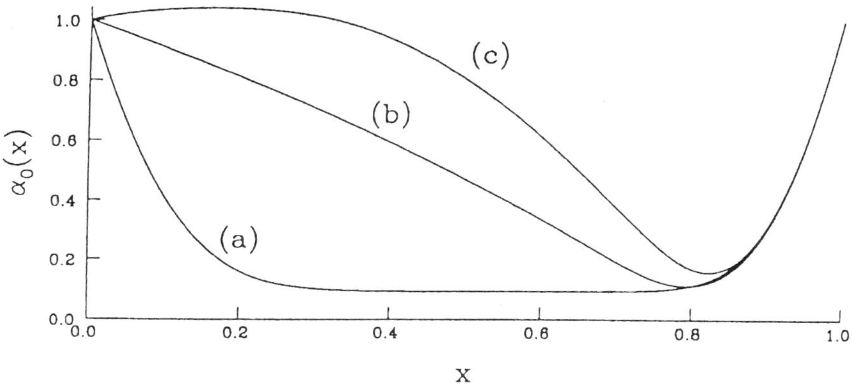

Figure 5. Wall shapes predicted by Cancelli & Pedley's 1D model: an increase in flow rate ($a \to b \to c$) at constant downstream transmural pressure reopens the tube. The flow is assumed to separate beyond the constriction. From Jensen (1992).

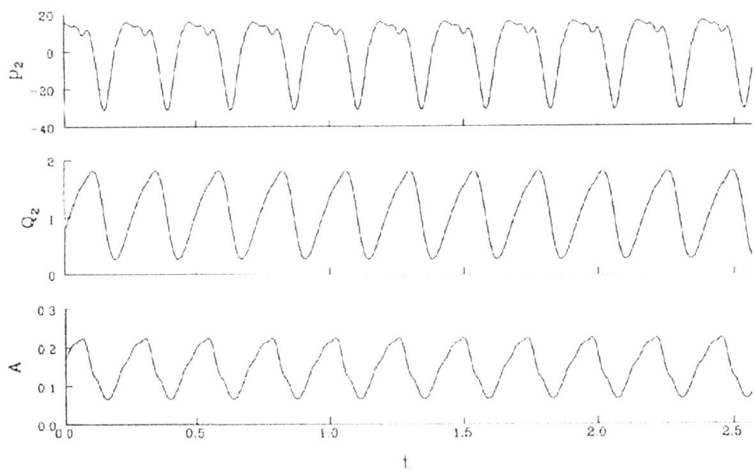

Figure 6. Time-traces of the downstream pressure, flow rate and the minimum cross-sectional area during a self-excited oscillation. From Jensen (1992).

therefore appear to be **(i)** representation of fluid inertia (introduction of wall inertia can introduce additional higher frequency modes); **(ii)** representation of dissipation in the system, either distributed along the collapsible tube or confined to the separated flow region (either of which can introduce a phase difference between pressure fluctuations and wall displacement, overcoming the tendency of

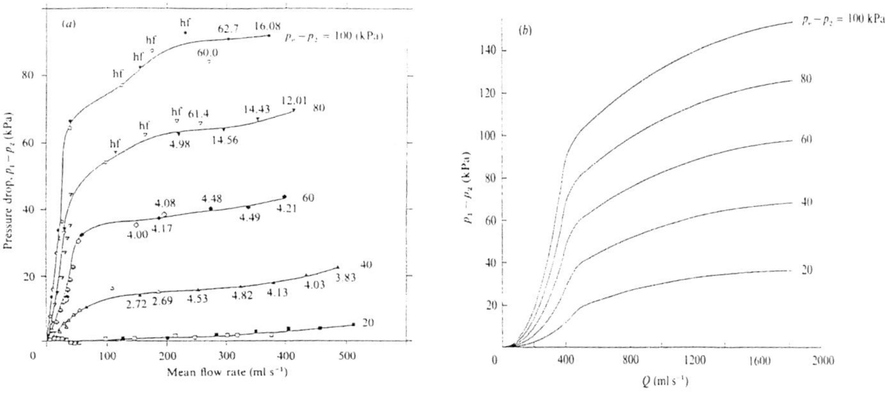

Figure 7. Pressure drop through the tube vs. flow rate for fixed downstream transmural pressure. (a) Experimental results of Bertram (1986) (numbers adjacent to symbols denote frequency of oscillations; for these cases time-averaged pressures are recorded); (b) numerical predictions from the 1D model of Jensen and Pedley (1989).

the tube to collapse directly via static divergence); and **(iii)** strong coupling of the flow in the collapsible tube to the rigid upstream and downstream segments via (9).

4.6.3. *Links with flutter instabilities*

The connection between oscillations in the Starling resistor and travelling-wave-flutter identified in potential-flow calculations is not directly obvious, because of the neglect of wall inertia in many of the computational studies and the profound influence of the rigid parts of the system. However some insight into their relationship can be obtained by examining the situation in which the longitudinal tension T_0 in (10) is large, specifically

$$\mathcal{T} \equiv \frac{T_0 A_0}{\rho U_0^2 L_0^2} \gg 1, \quad \text{where} \quad U_0 \equiv \frac{p_{entry} - p_{exit}}{\rho \mathcal{R} L_0 + (R_{up} + R_{down}) A_0} \tag{13}$$

is the speed of the mean flow through the system assuming $A = A_0$ along the length L_0 of the collapsible segment (the total pressure drop balancing the viscous resistances defined in (1) and (9)). The large tension ensures that the tube remains almost uniform ($A \approx A_0$ in $0 \le x \le L_0$), that its shape is insensitive to the magnitude of the external pressure (unless of course $|p_{ext}|$ is very large) and that transverse elastic stresses (represented by $\mathcal{P}(A)$ in (10)) are not significant. Rather like a stretched string, the Starling Resistor then admits a family of discrete, high-frequency normal modes in which the tube wall oscillates in and out (with one or more half-wavelengths along its length) and the fluid inside the tube oscillates backwards and forwards. To illustrate, we write $A = A_0 + \mathrm{Re}(\hat{A}(X)E)$,

$u = \mathrm{Re}(\hat{u}(X)E)$, where $x = L_0 X$, $E \equiv \exp(i\beta(A_0 T_0/\rho L_0^4)^{1/2} t)$ and β is a dimensionless frequency, and then linearise (1, 2, 9, 10) about the uniform state; the dominant terms at high frequencies are unsteady fluid inertia in both the compliant and rigid parts of the system and membrane tension (neglecting gravity in (1) and $\mathcal{P}(A)$ in (10)). The perturbation velocity along the collapsible segment then turns out to satisfy

$$\hat{u}'''' - \beta^2 \hat{u} = 0 (0 \leq X \leq 1), \tag{14}$$

$$\hat{u}'(0) = 0, \hat{u}'''(0) - \beta^2 (A_0 I_{up}/\rho L_0)\hat{u}(0) = 0, \tag{15}$$

$$\hat{u}'(1) = 0, \hat{u}'''(1) + \beta^2 (A_0 I_{down}/\rho L_0)\hat{u}(1) = 0, \tag{16}$$

with area fluctuations given by $\hat{A}(X) = -A_0 \hat{u}'(X)/i\beta L_0$. This fourth-order problem can be solved to obtain a family of eigenmodes with discrete frequencies $0 < \beta_1 < \beta_2 < \ldots$. Taking the first mode plus a mean flow as the leading-order term in an expansion in powers of $\mathcal{T}^{-1/2}$, one obtains at the following order the linear system (14)-(16) with forcing terms arising from advective inertia and viscous dissipation. Applying a solvability condition leads to a condition on the growth rate of the normal modes: the growth rate is positive when

$$\hat{u}^2(0) - \hat{u}^2(1) > \frac{R_{up} A_0}{\rho U_0} \hat{u}^2(0) + \frac{R_{down} A_0}{\rho U_0} \hat{u}^2(1) + \frac{\mathcal{R} L_0}{U_0} \int_0^1 \hat{u}^2 \, \mathrm{d}X. \tag{17}$$

Provided $\hat{u}^2(0) > \hat{u}^2(1)$ (which computations indicate is the case for $I_{down} > I_{up}$, since large downstream inertia suppresses velocity fluctuations at $X = 1$), as the pressure drop across the system increases, U_0 increases through the threshold defined by (17) and normal mode oscillations can become unstable, extracting more energy from the mean flow than is lost to viscous dissipation throughout the system. These oscillations can therefore be regarded as a form of flutter, although in this spatially bounded system they are distinct from convectively unstable travelling waves that arise also in either potential-flow models (e.g. Grotberg and Reiss 1984) or in studies of thin falling liquid films, which are governed by a closely related set of PDEs (Chang and Demekhin 1996).

The analysis leading to (17) was presented to the IUTAM Symposium by Jensen (2001) (for a very similar set of governing equations), who showed also that (17) agrees well with both a stability analysis for arbitrary T_0 and numerical solutions of the full equations. At lower tensions, Jensen's (2001) 1D model also exhibits multiple nonlinear static solutions (arising via static divergence instability), and remarkably intricate oscillations exhibiting, for example, a well-defined period-doubling cascade towards chaos. However, while the stability threshold (17) is effective in understanding 1D model systems in the high-tension limit, it is physically inconsistent: at high frequencies, dissipation is likely to be confined to thin viscous boundary layers that are very poorly described by quasi-steady dissipative terms as simple as that in (1). This reflects the deficiencies present in any 1D model of a high-Reynolds-number flow, and motivates the analysis of more rational 2D systems.

5. Two-dimensional models

5.1. THE MODEL PROBLEM

Given the large resources required for the numerical solution of the 3D problem, much recent activity has focussed on the 2D deformable channel problem illustrated in Fig. 2. The flow is governed by the 2D Navier–Stokes equations

$$\rho\left(\frac{\partial u_i}{\partial t}+u_j\frac{\partial u_i}{\partial x_j}\right)=-\frac{\partial p}{\partial x_i}+\mu\frac{\partial^2 u_i}{\partial x_j^2}\qquad\text{and}\qquad\frac{\partial u_j}{\partial x_j}=0,\tag{18}$$

where μ is the fluid's viscosity and $\mathbf{u}(\mathbf{x},t)$ the 2D velocity field. No-slip and kinematic conditions are imposed at the walls and a prescribed parabolic velocity profile may be imposed at the inlet. The wall is typically modelled as a membrane under axial tension T_0, whose deformation (characterised by the variable channel width $h(x)$) is assumed to be governed by

$$p-p_{ext}=T_0\;\kappa,\tag{19}$$

where κ is the wall curvature, which can be approximated by $\kappa\approx -d^2h/dx^2$ if the membrane slope is small, as in (10). The flow can then be parameterised by a Reynolds number $Re=\rho U_0 H_0/\mu$ based on the inlet channel width H_0 and an inlet flow speed U_0, and a dimensionless tension parameter $T=T_0/H_0\rho U_0^2$.

Pedley (1992) studied this problem in the lubrication-theory limit (assuming small membrane slope and low Re), and showed how steady solutions may break down if the fluid shear stress reduces the wall tension to zero at some point along the wall. Matsuzaki and Fujimura (1995) showed later that inclusion of bending stiffness in (19) ensures the existence of steady solutions for all values of the tension. Jensen (1998) also used lubrication theory to obtain explicit pressure-flow relations for a highly collapsed channel that exhibits flow and pressure-drop limitation through purely viscous mechanisms.

The remaining studies of this problem have employed either direct numerical methods or high-Reynolds-number asymptotics. In numerical studies (details of which are given below), a moving mesh was employed to discretise the 'Arbitrary Lagrangian Eulerian' form of the Navier-Stokes equations by finite elements. The fluid-mesh update in response to the movement of the wall was typically performed by variants of Kistler and Scriven's (1983) 'Method of Spines'. Two different methodologies were employed to obtain coupled solutions of the fluid and solid equations. In the segregated approaches of Lowe and Pedley (1995), Luo and Pedley (1995) and Liang et al. (1997), the fluid and solid equations were discretised independently and a coupled solution was obtained by a fixed-point iteration between the two domains. All subsequent work was based on a fully-coupled approach in which a global Newton–Raphson method was employed to solve the large system of nonlinear algebraic equations which arises from a coupled discretisation of both sets of equations. The fully-coupled approach is more difficult to implement but tends to provide a much more robust procedure. In fact, most authors found the use of a fully-coupled approach to be necessary to obtain converged solutions for the simulation of time-dependent problems with significant fluid-structure interaction.

5.2. STEADY FLOW

5.2.1. *Numerical studies*

Following Lowe and Pedley's (1995) initial study of zero-Reynolds-number flow in collapsible channels, Rast (1994) and Luo and Pedley (1995) performed detailed studies of the system's behaviour at finite Reynolds number. Fig. 8 illustrates some of Rast's (1994) results and shows the streamlines for the flow through a strongly collapsed channel. Downstream of the point of strongest collapse the flow field is similar to that found in the flow over a backward facing step (Armaly et al. 1983). As *Re* increases, a sequence of recirculating eddies develops on alternate sides of the channel. Note that the flow field immediately downstream of the separation point is not dissimilar to that assumed in Cancelli and Pedley's (1985) 1D representation of flow separation.

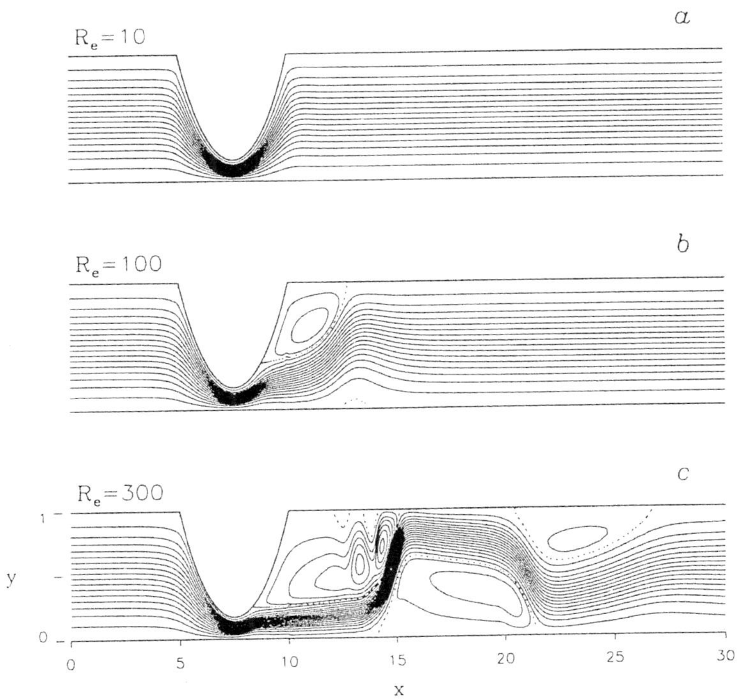

Figure 8. Streamlines for steady finite Reynolds number flow through a collapsible channel; reproduced from Rast (1994) with permission.

When studying the system's response to changes in the wall tension T, Luo and Pedley (1995) noticed that it became increasingly difficult to obtain converged solutions as the tension was reduced. No solutions could be obtained if the tension was decreased below some critical value T_c, which was dependent on Re. It was observed that T_c was close to the tension T_b at which the wall began to bulge out near the upstream end of the elastic segment (resulting in a wall shape similar to that shown in Fig. 9a below). Bulging of the wall presents computational difficulties, since the fluid pressure and the vorticity develop singularities at the convex upstream corner. In the context of the collapsible-channel problem, these singularities were first analysed by Lowe and Pedley (1995). The development of sharp corners (and the associated singularities in the flow field) can be suppressed by including bending stiffness into the wall model. This was first done by Liang et al. (1997), whose linear Eulerian wall model represented the elastic section as a pre-stressed elastic beam. At the IUTAM Symposium, Cai and Luo (2001) presented a model which incorporated bending and extensional stiffness into a geometrically nonlinear Lagrangian representation of the wall; they also analysed how bending stiffness affects the structure of the flow near the ends of the flexible segment.

5.2.2. *Asymptotic models*

Only a few studies are available in which high-Reynolds-number asymptotic analysis has been applied to steady flows in deformable tubes and channels. Tutty (1984) solved the boundary-layer equations coupled to a tube law to integrate into a singular choke point over a long $O(Re)$ lengthscale (recovering the basic choking phenomenon described in Section 4.2), while Tutty (1989) used interactive triple-deck theory to describe so-called 'free interaction' solutions over shorter lengthscales for nearly uniform symmetric channels. This work builds on the framework of Smith, Smith (1976a, 1976b), who showed how to couple viscous boundary layers on either wall of a long channel to an inviscid rotational core flow between them. More recent work in this direction by Guneratne (1999) and Guneratne and Pedley (2001) was reported to the IUTAM Symposium by Pedley. At high T or small transmural pressure, the membrane deflection is weak enough that the wall displacements are of comparable magnitude to the boundary-layer thickness. Displacement of the wall, induced either by the external pressure p_{ext} or by flow-induced pressure variations, causes a sideways displacement of the inviscid core flow. Guneratne solved the interactive boundary-layer equations to determine steady flow configurations, formally assuming that the channel length $L_0 = H_0\lambda$ is large ($Re^{1/7} \ll \lambda \ll Re$) and the boundary layers are narrow (having thickness of order $H_0(\lambda/Re)^{1/3}$). The special case $\lambda = O(Re^{1/7})$ captures the effect of a cross-stream pressure gradient, although this turns out not to be particularly significant to the overall solution structure. With $p_{ext} = 0$, the uniform state loses stability to eigensolutions in which the membrane is either weakly indented or dilated, or both. Stability is lost through a sequence of transcritical bifurcations, with membrane tension a bifurcation parameter. Imposition of nonzero p_{ext} acts as a symmetry-breaking effect. When the upstream transmural pressure is held constant the system exhibits abrupt jumps between solutions and there are even ranges of T for which no steady solution could be found. Guneratne demonstrated that the system exhibits a rich variety of nonuniform steady states, reminiscent of those identified in the inviscid 1D models of Jensen and Pedley (1989) and Reyn (1987), and which presumably are once again a nonlinear manifestation of static divergence instability. Although the stability of these solutions to unsteady disturbances was not determined, their richness points to even greater complexity in the full dynamic problem.

5.3. UNSTEADY FLOW AND SELF-EXCITED OSCILLATION

5.3.1. *Fundamental modes of instability*

Before discussing computational studies of unsteady 2D flows, it is helpful to digress briefly to mention studies of the linear stability of 2D flows in unbounded channels having deformable walls. For more detailed reviews of much of this work see Davies and Carpenter (1997a) and Chapter 4 of this volume. These studies, extending the simpler potential-flow models described in Section 4.5, identify three primary modes of instability for channels with spring-backed walls having bending stiffness, longitudinal tension, damping and inertia. As we have seen already, travelling-wave flutter (TWF) is a convective, oscillatory Class B instability (in the Benjamin-Landahl energy classification), destabilised when energy is transferred from the flow to the wall. These waves are stabilised by wall damping. In their long-wave analysis of the Orr-Sommerfeld problem for disturbances to Poiseuille flow in a compliant channel, Davies and Carpenter (1997a) show how TWF is destabilised by effects confined to critical layers (which put pressure and wall displacement out of phase, allowing the flow to do work on the wall, an effect described also by Huang 1998), but is stabilised by viscous effects in wall (Stokes) layers. Tollmien–Schlichting (TS) waves are intrinsic hydrodynamic modes of instability that are present even when the channel walls are rigid; these are Class A waves (stabilised when energy is transferred from the flow to the wall, and destabilised by wall damping).

Finally, static divergence is a direct, non-oscillatory (or very low frequency) instability that is Class A or C (C being one that is insensitive to energy transfer between the flow and the wall). In addition to these three primary modes, further modes can arise through mode coalescence, so that TWF and TS modes can combine to give rise to a powerful Class C flutter instability. Larose and Grotberg (1997) also identified an apparently distinct long-wave instability of developing flow in a compliant channel, and Davies and Carpenter (1997b) showed how energy can be transferred from TS waves to flow-induced surface instabilities at junctions in flows over compliant panels of finite length. Huang (2001) has also examined the linear stability of the configuration shown in Fig. 2 when the membrane is nearly flat, and shown how both flutter and divergence modes can arise when the membrane has significant inertia. It should be borne in mind that **(i)** the strongly nonlinear phenomena to be described below presumably have their origins in these fundamental instabilities and **(ii)** neglecting effects such as bending stiffness or wall damping may be dangerous because these can be singular limits of the full stability problem.

5.3.2. *Computational studies*

Unsteady finite-Reynolds-number flows in collapsible channels have been computed by Luo and Pedley (1996, 1998, 2000), Liang et al. (1997) and Pedrizzetti (1998). For time-dependent problems, the application of the no-slip condition on the channel wall poses a subtle problem because the Eulerian wall model (19) determines the wall's overall shape but not the position of individual material particles. Therefore certain *ad hoc* assumptions regarding the wall displacement field have to be made when matching the fluid and wall velocities. Luo and Pedley (1996) considered two reasonable assumptions (wall displacements either in the vertical direction or in the direction normal to the wall's instantaneous shape) and showed that the results obtained with the two boundary conditions differed only slightly. Cai and Luo's (2001) new Lagrangian wall model will avoid this ambiguity in future time-dependent simulations. Given the important role that the wall tension T appeared to play for steady flows, Luo and Pedley (1996) investigated the system's time-dependent behaviour for different values of T. They found that, for sufficiently large Reynolds number, self-excited oscillations developed when the tension was reduced below a critical value, T_u. They reported that T_u tended to be close to but not identical to the value T_b at which the upstream end of the collapsible segment began to bulge out, although without varying many of the remaining parameters in the problem the significance of bulging is hard to estimate. For values of T just below T_u, the wall performed sustained harmonic oscillations. As the tension was reduced further, the system appeared to undergo several period-doubling bifurcations and carried out increasingly complex oscillations, many features of which were reminiscent of those observed in the experiments of Bertram et al. (1990).

5.3.3. *Vorticity waves*

Fig. 9 shows the wall shapes and instantaneous streamlines in a channel which is undergoing such large-amplitude self-excited oscillations. One of the key features in the flow field is the generation of travelling waves downstream of the oscillating wall. The waves are very similar to the 'vorticity waves' which are generated in 2D channels in which an elastic segment of one wall is oscillated in a prescribed manner (Stephanoff et al. 1983; Pedley and Stephanoff 1985; Ralph and Pedley 1988, 1989, 1990). Comparison in these studies between experiment and a weakly nonlinear theoretical analysis supports the view that small-amplitude vorticity waves are sustained through an inviscid mechanism, involving perturbations of the vorticity in the oncoming Poiseuille flow, and that they have a wavelength $O(St^{-1/3})$ that is approximately independent of Re, where $St = (H_0^2\omega/\mu) \ll 1$ is the Strouhal number of the oscillation (ω being the oscillatory frequency). In a similar way, flow forced by unsteady pressure gradients through channels with fixed asymmetric indentations also generates vorticity waves (Sobey 1985; Tutty 1992; Tutty and Pedley 1993; Rosenfeld 1995), provided the forcing frequency is in an appropriate range.

The origin of vorticity waves is of fundamental interest to the present problem. The analysis of Bogdanova and Rhyzov (1983) of TS waves in plane Poiseuille flow,

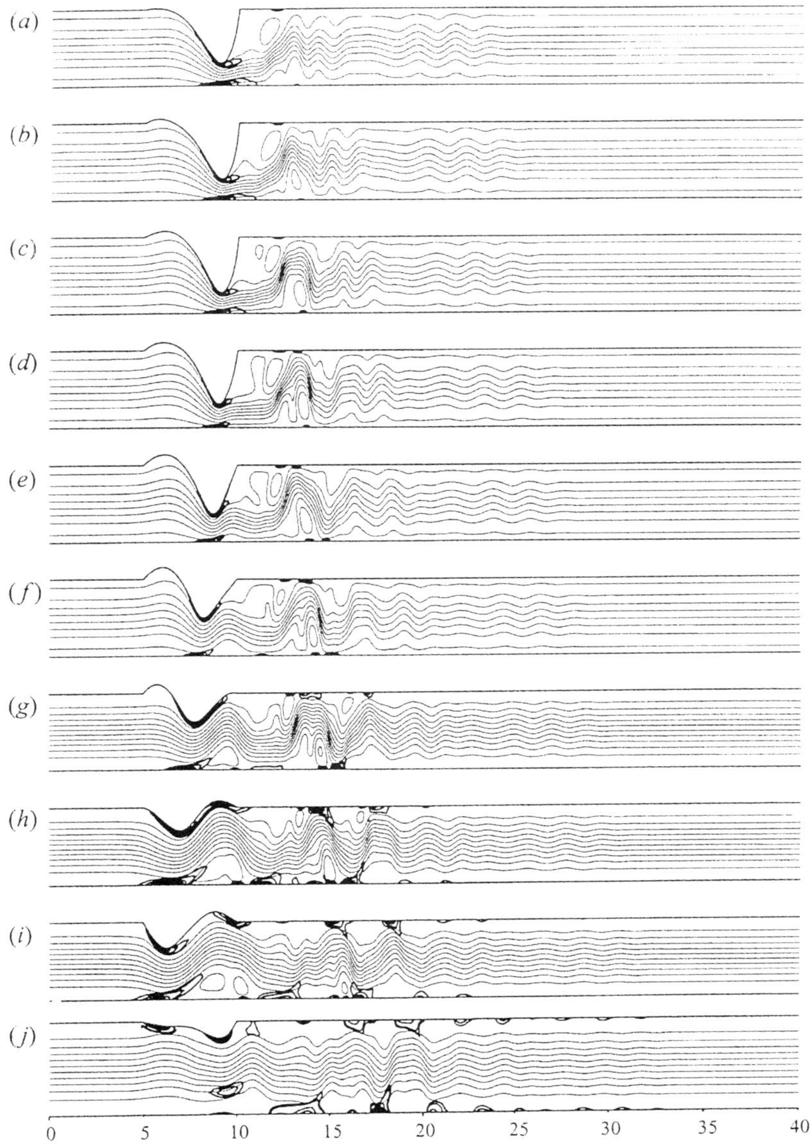

Figure 9. Wall shapes and instantaneous streamlines during self-excited oscillations of a collapsible channel; from Luo and Pedley (2000).

using the interactive boundary-layer framework developed by Smith, Smith (1976a, 1976b), shows that downstream-propagating TS waves with wavelength $aRe^{1/7}$

can grow spatially if $StRe^{3/7}$ exceeds a critical value; F. T. Smith and Burggraf (1985) went on to suggest a link between large amplitude lower-branch TS waves and vorticity waves. These observations are consistent with the computations of Rosenfeld (1995) who shows how, as St increases with Re fixed for oscillatory flow through a channel with one indented wall, waves (presumably vorticity waves) are generated when St is of magnitude comparable to $Re^{-3/7}$. The waves change character as St increases, suggesting alternative mechanisms are operating, and are suppressed as St rises above unity.

These observations raise some immediate questions. Are the self-excited oscillations computed by Luo and Pedley (1996) generated by a primary flutter mechanism, which then forces secondary vorticity (TS) waves, or do vorticity waves generate the primary instability? Might a combined TS/TWF mode (such as reported by Davies and Carpenter 1997a) be driving the oscillations? And if any one of these mechanisms operates, how universal is it as membrane properties and upstream and downstream flow conditions are varied?

While we are far from a satisfactory answer to these questions, some useful evidence is provided by results of a recent combined asymptotic and computational study (Jensen and Heil 2002) suggesting that the high-tension flutter instability described in a 1D model by Jensen (2001) (see Section 4.6.3) has a direct analogue in two dimensions. A formal asymptotic approximation of the 2D problem illustrated in Fig. 2, assuming $Re \gg 1$, $T \gg 1$ and membrane length $L_0 \gg H_0$, predicts that under suitable conditions the normal modes of the system described by (14)-(16) can become unstable and grow into self-excited oscillations when the Re exceeds a threshold Re_c of the form

$$Re_c \propto T^{1/2}\beta \left[\frac{\int_0^1 \hat{A}_X^2 \, \mathrm{d}X}{\hat{A}_X^2(1) - \hat{A}_X^2(0)} \right]^2 . \tag{20}$$

Here β and $\hat{A}(X)$ are exactly the eigenfrequency and area perturbation computed using the 1D model via (14)-(16) (with I_{up} and I_{down} defined appropriately). Again, the analysis indicates that a necessary condition for this instability to occur is that $I_{down} > I_{up}$. Since dissipation is confined to unsteady Stokes layers at high frequencies, Re_c scales with $T^{1/2}$, an important feature that was not captured in the analogous 1D result (17). It is also striking that dissipation in the rigid parts of the system does not contribute to the stability boundary in this limit. Predictions such as (20) show good agreement with Navier–Stokes computations (for details see Jensen and Heil 2002), supporting the view that flutter alone can provide the primary mechanism of instability, at least when membrane tension is large.

5.3.4. *Dissipation and flow limitation*

Luo and Pedley (1996) computed the rate of dissipation in the fluid in order to check Cancelli and Pedley's (1985) assumption that most of the energy loss, which had been found to be crucial for the development of self-excited oscillations, occurred in the separated-flow region downstream of the point of strongest collapse. Interestingly, their computations did not confirm this assumption but showed that most of the energy is dissipated in the boundary layers which form on the walls *upstream* of the point of strongest collapse. Pedley and Luo (1998) attempted to employ a von Karman–Pohlhausen technique (similar to the one initially suggested by Ikeda and Matsuzaki 1999) to incorporate these findings into an improved 1D model. Unfortunately, the predictions from this model agreed only poorly with those from the 2D computations.

In a later study, Luo and Pedley (1998) showed that the inclusion of wall inertia into the system allows the wall to perform flutter-type oscillations of relatively high frequency. During the early stages of the self-excited oscillations, these higher-frequency oscillations were found to be merely superimposed on to the lower-frequency oscillations which develop in the absence of wall inertia. However, the higher-frequency oscillations were found to grow in amplitude so that they ultimately dominated the system's behaviour. Attempts were made to relate these observations to the predictions from studies of flutter-type instabilities in flows past compliant boundaries (Grotberg and Reiss 1984; Davies and Carpenter 1997a, 1997b; Larose and Grotberg 1997). Some common trends were observed but overall the agreement tended to be no more than qualitative, presumably because of the significant differences between the systems considered in the different studies (large amplitude oscillations of a finite-length elastic wall in Luo and Pedley (1998) vs. small amplitude oscillations of (usually) infinite, initially undisturbed panels in the flutter-type studies). It was established, however, that in all cases the unstable growth of the high-frequency oscillations was caused by the energy transfer from the fluid into the wall, which is possible because of the inertially-induced phase differences between the fluid loading and the wall displacement field. This behaviour is typical of a class-B flutter instability.

In all the computational studies reviewed so far, the transmural pressure at the far downstream end of the rigid tube ($p_{tm}^{(exit)} = p_{exit} - p_{ext}$ in Fig. 2) was held constant while other parameters (such as the flow rate or the wall tension) were varied. Under these conditions, an increase in the driving pressure drop, $\Delta p = p_{entry} - p_{exit}$ can be caused only by an increase in the upstream transmural pressure $p_{tm}^{(entry)} = p_{entry} - p_{ext}$ which causes the wall to bulge out. Hence the channel's flow resistance decreases as the flow rate is increased, leading to pressure-drop limitation. Physiologically more relevant is the case in which the upstream transmural pressure $p_{tm}^{(entry)}$ is held constant (corresponding to the conditions during forced expiration from the lung, for example). An increase in Δp (by lowering the downstream transmural pressure) now increases the collapse of the wall and thus increases the overall flow resistance. Luo and Pedley (2000) confirmed the expected occurrence of flow limitation in the 2D model but failed to confirm any correlation between flow limitation and the onset of self-excited oscillations. Both stable and unstable solutions were found in the flow-limited regime. The stability properties of the steady solutions were also found to be strongly dependent on the boundary conditions and in particular on the position at which pressures and fluxes were prescribed. This is consistent with the findings in Pedrizzetti's (1998) study of forced oscillations in axisymmetric collapsible tubes.

6. Three-dimensional models

The 2D physical model discussed in the previous section is a rational and (at least in principle) realisable system which can be interpreted as an approximation to the flow in a strongly collapsed 3D tube. However, it is clear that the model ignores many potentially important 3D effects such as **(i)** the strong three-dimensionality of the flow field during phases when the tube collapse is moderate; **(ii)** the significant differences between flow separation in 2D and 3D flows (see e.g. Tobak and Peake 1982); and **(iii)** the drastic changes in wall stiffness as the tube changes from an axisymmetric to a non-axisymmetrically buckled state.

6.1. THE MODEL PROBLEM

The first step towards the development of a rational computational model of flow in 3D collapsible tubes was undertaken in Heil's (1995) PhD thesis. The collapsible tube was modelled as a thin-walled elastic shell (of length L_0, radius R_0 and wall thickness h) whose deformation was described by large-displacement, geometrically nonlinear shell theory. It is known from the experiments of, e.g., Elad et al. (1992) that the non-axisymmetric collapse of thin-walled elastic tubes only induces small extensional deformations of the tube wall. Therefore, a linear stress-strain relationship with two elastic constants (Young's modulus E and the Poisson ratio ν), was used as the constitutive equation.

The tube's undeformed geometry was described by the position vector $\mathbf{r}_w(\zeta^1, \zeta^2)$ to the tube's midplane which was parametrised by two Lagrangian coordinates ζ^1 and ζ^2. The vector field $\mathbf{v}(\zeta^1, \zeta^2)$ which displaces material particles to their new positions $\mathbf{R}_w = \mathbf{r}_w + \mathbf{v}$ was then determined from the principle of virtual displacements

$$\int_0^{2\pi} \int_0^{L_0/R_0} \left[E^{\alpha\beta\gamma\delta} \left(\gamma_{\alpha\beta}\ \delta\gamma_{\gamma\delta} + \frac{1}{12} \left(\frac{h}{R_0} \right)^2 \kappa_{\alpha\beta}\ \delta\kappa_{\gamma\delta} \right) - \left(\frac{R_0}{h} \right) (\mathbf{f} \cdot \delta\mathbf{R}_w) \right] d\zeta^1 d\zeta^2 = 0, \qquad (21)$$

where $\gamma_{\alpha\beta}$ and $\kappa_{\alpha\beta}$ are the midplane strain and bending tensors, respectively, and $E^{\alpha\beta\gamma\delta}$ represents the fourth order tensor of elastic constants; see Heil and Pedley (1996) for details. The load vector $\mathbf{f}$ is given by the combination of the external pressure and the traction that the fluid exerts onto the tube wall. A finite-element method was used to discretise the variational equation (21).

6.2. STEADY FLOWS IN COLLAPSIBLE TUBES

In a series of papers, Heil and Pedley, Heil and Pedley (1995, 1996) and Heil (1996) coupled the wall model (21) to a lubrication-theory-based description of the fluid mechanics and investigated **(i)** the flow through axisymmetric tubes,

(ii) the tubes' linear stability to non-axisymmetric perturbations and **(iii)** their large-displacement postbuckling behaviour, respectively.

For strongly buckled tubes, the wall slope is no longer small and the applicability of lubrication theory must be questioned. Therefore, Heil (1997) replaced the lubrication-theory model by a solution of the 3D steady Stokes equations and performed extensive parameter studies to investigate the system's behaviour under conditions which correspond to typical experimental procedures. The computational study was accompanied by an experimental investigation and good agreement between the numerical predictions and the experimental results was reported. Interestingly, the predictions from the lubrication-theory model were found to agree very well with those from the full Stokes equations, even in cases where the tube was strongly collapsed.

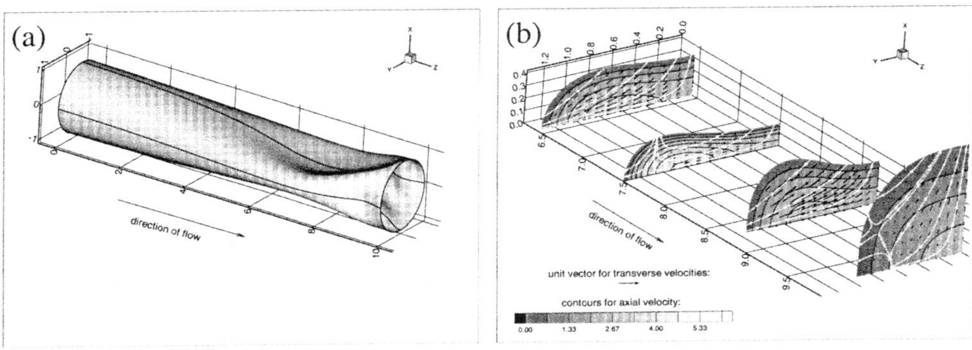

Figure 10. Strongly collapsed tube conveying viscous flow at zero Reynolds number. (a) Wall shape. (b) Flow field in four cross-sections in the most strongly collapsed part of the tube. The white lines indicate the streamlines of the transverse velocity field. After Heil & Pedley (1996).

Figure 10 illustrates the wall deformation and the flow field inside a strongly buckled elastic tube which conveys viscous fluid at zero Reynolds number. The noticeable asymmetry of the tube's deformation reflects the strong interaction between the fluid and solid mechanics. The tube's upstream end is distended by a large positive fluid pressure. In the absence of fluid inertia, the fluid pressure decreases continuously in the streamwise direction. This leads to an increasingly strong compression of the tube wall which causes the tube's downstream end to buckle strongly. The increased flow resistance in the most strongly collapsed part of the tube creates a large local pressure drop. This compresses the tube's downstream end so much that two small regions on the tube's sidewall buckle inwards, creating two 'dimples' near the horizontal plane of symmetry.

Figure 10 highlights a number of important differences between flows in collapsible channels and tubes: **(i)** The flow in the most strongly collapsed part of the tube has little resemblance to the flow between two parallel membranes. The contours of the axial velocity show that, long before the first occurrence of opposite wall contact, the axial flow begins to split up into two separate branches as it passes around the most strongly collapsed region near the tube's centreline.

Recent computational studies of finite-Reynolds-number flows in collapsible tubes (Hazel and Heil 2002) confirmed the development of similar flow structures in the presence of fluid inertia. Two distinct 'jets' develop in the outer lobes when the fluid passes through the most strongly collapsed cross-section. The jets are deflected by the two 'dimples' in the sidewall and persist for a significant axial length until secondary flows and transverse diffusion of momentum smooth out the peaks in the axial velocity. The structure of the separated-flow region (consisting of a relatively passive separation bubble in the centre of the tube) was found to be fundamentally different from that observed in collapsible channels. These findings are consistent with the experimental observations of Bertram and Godbole (1997) and Moore et al. (1995). **(ii)** Even though the tube's upstream end is strongly pressurised, its large extensional stiffness prevents significant axisymmetric inflation; consequently very little bulging is observed. This is in line with experimental observations but is in stark contrast to the predictions from the 2D model. **(iii)** Given the wall deformation shown in Fig. 10, is not clear in what way axial tension should be represented in a tube-law-based model of the wall mechanics. The tension term in the modified tube law (10) is based on the assumption that the tensioned wall collapses radially inwards as sketched in Fig. 2. However, Fig. 10(a) shows that this assumption is only justified in the tube's vertical plane of symmetry where little flow takes place. A significant part of the tube wall (in the vicinity of the horizontal plane of symmetry) bulges radially outwards. The convex wall curvature in this region implies that axial tension would manifest itself in a relation similar to (10) but with a plus rather than a minus sign. These observations raise serious doubts as to whether it is reasonable to assume that flow in a collapsible tube is similar to flow between two parallel tensioned membranes. Furthermore, while axial tension represents an essential ingredient for the formulation of the 2D model, Heil and Pedley (1996) showed that it is only of minor importance for the deformation of 3D tubes. This observation is also consistent with experimental observations. Most investigators subject their tubes to a certain amount of axial tension, primarily to keep the tube from sagging under its own weight. Despite the fact that it is difficult to prescribe precisely the axial pre-stretch in an experiment, the tube's behaviour does not seem to be significantly affected by this lack of control.

Figure 11 illustrates how the nondimensional volume flux, $q = \frac{8\mu Q L_0}{\pi R_0^4 E}$, varies with the driving pressure drop, $\Delta p = (p_{up} - p_{down})/E$ when the upstream transmural pressure $p_{tm}^{(up)} = p_{up} - p_{ext}$ is held at a constant value (again with $Re = 0$). Since the flow is driven by a reduction in the downstream transmural pressure, an increase in Δp increases the compressive load on the initially axisymmetric tube. In the axisymmetric state (represented by the dashed line), the tube's deformation remains very small and the relation between flow rate and pressure drop is essentially linear. However, when the compression of the tube's downstream end exceeds a critical value (at point A), the axisymmetric state loses its stability and the tube buckles non-axisymmetrically. Fig. 11 shows that buckling occurs via a subcritical bifurcation. If the pressure drop Δp is held constant during buckling, the tube jumps dynamically to a new strongly collapsed equilibrium state (point B) on the post-buckled branch (represented by the solid line). In the collapsed

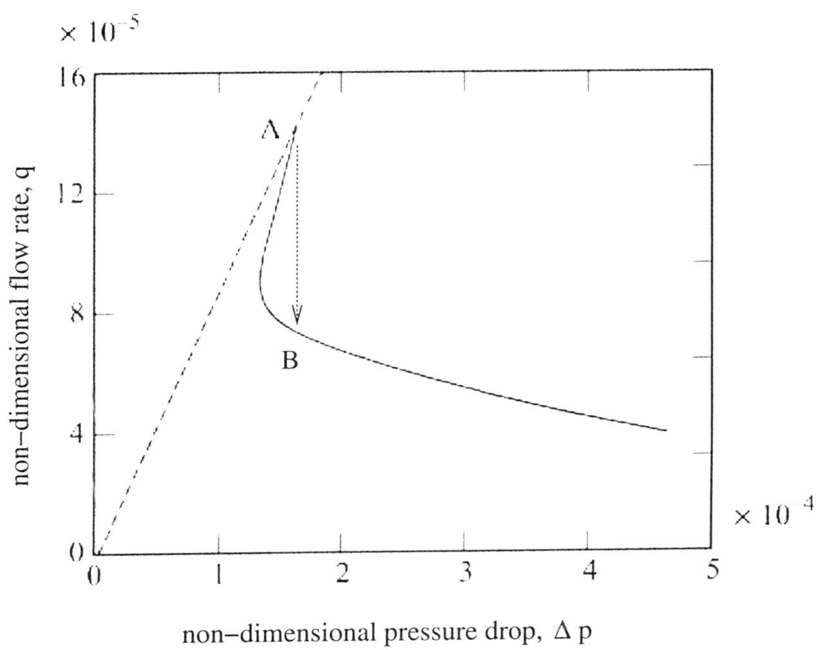

Figure 11. Volume flux as a function of the driving pressure drop through the tube for constant upstream transmural pressure $p_{tm}^{(up)} = 0$. After Heil & Pedley (1996).

state, the tube's flow resistance is strongly increased, and therefore the flow rate drops significantly during buckling. A further increase in the driving pressure drop leads to a further reduction in flow, illustrating the occurrence of negative effort dependence through a purely viscous mechanism.

Fig. 11 demonstrates that the transition between axisymmetric and non-axisymmetric configurations (which cannot easily be incorporated into lower dimensional models) plays a crucial role in the system's behaviour. In experimental studies, the existence of small imperfections will make the transition between axisymmetric and non-axisymmetric states less abrupt. However, Heil and Pedley's (1996) computations with a circumferentially varying external pressure showed that the system's overall behaviour is not strongly affected by the presence of such imperfections.

7. Summary and suggestions for future work

The theoretical investigations reviewed in this Chapter have provided valuable insight into the behaviour of collapsible tubes, both in laboratory experiments and in a variety of physiological applications. The detailed analyses of relatively simple 1D models have allowed the identification of a number of mechanisms which provide qualitative (and occasionally quantitative) explanations for the experimentally observed behaviour of collapsible tubes. However, almost all 1D models involve a certain number of *ad hoc* assumptions whose validity needs to be assessed critically. The analysis of self-excited oscillations in 2D collapsible channels suggests that it might be beneficial to develop improved 1D models which incorporate features such as dissipation in boundary layers and possibly the interaction between wall oscillation and vorticity waves.

The computational and asymptotic analyses of flow in 2D collapsible channels have provided much insight into the details of the complex fluid-structure interaction in this system. Our understanding could be further improved by systematic analyses of the stability of steady flows. Asymptotic approaches (similar to those employed by Guneratne 1999 and Jensen and Heil 2002) will be most appropriate for the analysis of flows in slightly deformed channels, whereas the stability of flows in strongly collapsed channels will probably have to be determined numerically. Further interaction with researchers interested in the stability of flows past compliant walls (see Chapters 4 and 5) can be expected to be particularly beneficial for the study of this problem.

Detailed numerical investigations of steady finite-Reynolds-number flows in 3D collapsible tubes are currently in progress and have already revealed many features which cannot be predicted by any of the lower-dimensional models. The extension of such studies to unsteady flows will demand significant computational resources but would, for example, reveal whether vorticity waves, which appear to play a significant role in the large-amplitude self-excited oscillations in collapsible channels, have an equivalent (or even any) relevance in 3D flows. The development of large-Reynolds-number asymptotic analyses of flows in slightly buckled elastic tubes can be expected to provide useful insight into the structure of flow separation and the development of flow instabilities in this system.

The challenge of understanding self-excited oscillations in collapsible tubes, and in related problems involving flows over compliant surfaces, remains substantial. As the studies reviewed here have demonstrated, however, the combined use of theoretical, computational and experimental techniques can provide significant advances in areas that at first seemed intractable. We have good reason to be optimistic about significant future progress in this field.

References

Aittokallio, T., M. Gyllenberg, and O. Polo: 2001, 'A model of a snorer's upper airway'. *Math. Biosci.* **170**, 79–90.

Armaly, B. F., F. J. Durst, C. F. Pereira, and B. Schonung: 1983, 'Experimental and theoretical investigation of backward-facing step flow'. *Journal of Fluid Mechanics* **127**, 473–496.

Balint, T. S.: 2001, 'Dynamics of the upper airway'. Ph.D. thesis, University of Warwick, Warwick.

Benjamin, T.: 1963, 'The threefold classification for unstable disturbances in flexible surfaces bounding inviscid flows'. *Journal of Fluid Mechanics* **16**, 436–450.

Berke, G. S., D. C. Green, M. E. Smith, D. P. Arnstein, V. Honrubia, M. Natividad, and W. A. Conrad: 1991, 'Experimental-Evidence in the Invivo Canine for the Collapsible Tube Model of Phonation'. *Journal of the Acoustical Society of America* **89**, 1358–1363.

Bertram, C.: 1986, 'Unstable equilibrium behavior in collapsible tubes'. *Journal of Biomechanics* **19**, 61–69.

Bertram, C. D. and S. A. Godbole: 1997, 'LDA measurements of velocities in a simulated collapsed tube'. *ASME Journal of Biomechanical Engineering* **119**, 357-363.

Bertram, C. D. and T. J. Pedley: 1982, 'A mathematical model of unsteady collapsible tube behaviour'. *Journal of Biomechanics* **15**, 39–50.

Bertram, C. D. and C. J. Raymond: 1991, 'Measurements of wave speed and compliance in a collapsible tube during self-excited oscillations: a test of the choking hypothesis'. *Medical and Biological Engineering and Computing* **29**, 493–500.

Bertram, C. D., C. J. Raymond, and K. S. A. Butcher: 1989, 'Oscillations in a collapsed-tube analog of the brachial artery under a sphygmomanometer cuff'. *ASME Journal of Biomechanical Engineering* **111**, 185–191.

Bertram, C. D., C. J. Raymond, and T. J. Pedley: 1990, 'Mapping of instabilities for flow through collapsible tubes of differering length'. *Journal of Fluids and Structures* **4**, 125–153.

Bertram, C. D., C. J. Raymond, and T. J. Pedley: 1991, 'Application of nonlinear dynamics concepts to the analysis of self-excited oscillations of a collapsible tube conveying a fluid'. *Journal of Fluids and Structures* **5**, 391–426.

Binns, R. L. and D. N. Ku: 1989, 'Effect of stenosis on wall motion - a possible mechanism of stroke and transient ischemic attack'. *Arteriosclerosis* **9**, 842–847.

Bogdanova, E. V. and O. S. Rhyzov: 1983, 'Free and induced oscillations in Poiseuille flow'. *Quarterly Journal of Mechanics and Applied Mathematics* **36**, 271–287.

Bonis, M. and C. Ribreau: 1978, 'Etude de quelques proprietes de l'ecoulement dans une conduite collabable'. *La Houille Blanche* **3/4**, 165–173.

Brook, B. S.: 1997, 'The effect of gravity on the haemodynamics of the giraffe jugular vein'. Ph.D. thesis, University of Leeds, Leeds.

Brook, B. S., S. A. E. G. Falle, and T. J. Pedley: 1999, 'Numerical solutions for unsteady gravity-driven flows in collapsible tubes: evolution and roll-wave instability of a steady state'. *Journal of Fluid Mechanics* **396**, 223–256.

Brook, B. S. and T. J. Pedley: 2001, 'A model for time-dependent flow in (giraffe jugular) veins: Uniform tube properties'. *Journal of Biomechanics.* (in print).

Brower, R. W. and C. Scholten: 1975, 'Experimental evidence on the mechanism for the instability of flow in collapsible vessels'. *Medical and Biological Engineering* **13**, 839–845.

Cai, Z. and X. Y. Luo: 2001, 'A fluid-beam model for flow in collapsible channels'. (Preprint).

Cancelli, C. and T. J. Pedley: 1985, 'A separated-flow model for collapsible-tube oscillations'. *Journal of Fluid Mechanics* **157**, 375–404.

Carew, E. O. and T. J. Pedley: 1997, 'An active membrane model for peristaltic pumping .1. Periodic activation waves in an infinite tube'. *ASME Journal of Biomechanical Engineering* **119**, 66–76.

Carpenter, P. W., K. Berkouk, and A. D. Lucey: 1999, 'A theoretical model of pressure propagation in the human spinal CSF system'. *Engineering Mechanics* **6**, 213–228.

Carpenter, P. W., K. Berkouk, and A. D. Lucey: 2001a, 'Pressure wave propagation in fluid-filled co-axial elastic tubes. Part 1: Basic theory'. (Submitted).

Carpenter, P. W., K. Berkouk, and A. D. Lucey: 2001b, 'Pressure wave propagation in fluid-filled co-axial elastic tubes. Part 2: Mechanisms for the pathogenesis of syringomyelia'. (Submitted).

Chang, H.-C. and E. A. Demekhin: 1996, 'Solitary wave formation and dynamics on falling films'. *Adv. Appl. Mech.* **32**, 1–58.

Conrad, W. A.: 1969, 'Pressure-Flow relationship in collapsible tubes'. *IEEE Transactions on bio-medical engineering* **BME-16**, 284–295.

Cowley, S. J.: 1981, 'High Reynolds number flows through channels and tubes'. Ph.D. thesis, University of Cambridge, Cambridge.

Cowley, S. J.: 1982, 'Elastic jumps on fluid-filled elastic tubes'. *Journal of Fluid Mechanics* **116**, 459–473.

Cowley, S. J.: 1983, 'On the wavetrains associated with elastic jumps on fluid-filled elastic tubes'. *Q. J. Mech. Appl. Math.* **36**, 289–312.

Dai, G., J. P. Gertler, and R. D. Kamm: 1999, 'The effects of external compression on venous blood flow and tissue deformation in the lower leg'. *ASME Journal of Biomechanical Engineering* **121**, 557–564.

Danahy, D. T. and J. A. Ronan: 1974, 'Cervical venous hum in patients on chronic hemodialysis'. *New England Journal of Medicine* **291**, 237–239.

Davies, C. and P. W. Carpenter: 1997a, 'Instabilities in a plane channel flow between compliant walls'. *Journal of Fluid Mechanics* **352**, 205–243.

Davies, C. and P. W. Carpenter: 1997b, 'Numerical simulation of the evolution of Tollmien-Schlichting waves over finite compliant panels'. *Journal of Fluid Mechanics* **335**, 361–392.

Dawson, S. V. and E. A. Elliot: 1977, 'Wave-speed limitation on expiratory flow – a unifying concept'. *Journal of Applied Physiology* **43**, 498–515.

Dressler, R. F.: 1949, 'Mathematical solution of the problem of roll-waves in inclined open channels'. *Communications on Pure and Applied Mathematics* **2**, 149–194.

Elad, D., R. D. Kamm, and A. H. Shapiro: 1987, 'Choking Phenomena in a lung-like model'. *ASME Journal of Biomechanical Engineering* **109**, 1–9.

Elad, D., M. Sahar, J. M. Avidor, and S. Einav: 1992, 'Steady flow through collapsible tubes: measurements of flow and geometry'. *ASME Journal of Biomechanical Engineering* **114**, 84–91.

Elliot, E. A. and S. V. Dawson: 1977, 'Test of wave-speed theory of flow limitation in elastic tubes'. *Journal of Applied Physiology* **43**, 516–522.

F. T. Smith, F. T. and O. R. Burggraf: 1985, 'On the development of large-sized short-scaled disturbances in boundary layers'. *Proceedings of the Royal Society of London* **A 399**, 25–55.

Fee, M. S., B. Shraiman, B. Pesaran, and P. Mitra: 1998, 'The role of nonlinear dynamics of the syrinx in the vocalizations of a songbird'. *Nature* **395**, 67–71.

Flaherty, J. E., J. B. Keller, and S. I. Rubinow: 1972, 'Post buckling behavior of elastic tubes and rings with opposite sides in contact'. *SIAM Journal of Applied Mathematics* **23**, 446–455.

Gavriely, N. and O. E. Jensen: 1993, 'Theory and measurements of snores'. *Journal of Applied Physiology* **74**, 2828–2837.

Gavriely, N., Y. Palti, G. Alroy, and J. B. Grotberg: 1984, 'Measurement and theory of wheezing breath sounds'. *Journal of Applied Physiology* **57**, 481–492.

Gavriely, N., T. R. Shee, D. W. Cugell, and J. B. Grotberg: 1989, 'Flutter in flow-limited collapsible tubes: a mechanism for the generation of wheezes'. *Journal of Applied Physiology* **66**, 2251–2261.

Gregg, D. E. and L. C. Fisher: 1963, 'Blood supply to the heart'. In: W. F. Hamilton and P. Dow (eds.): *Handbook of Physiology.* Section 2: Circulation, Volume II. Washington D. C.: American Physiological Society.

Griffiths, D. J.: 1969, 'Urethral elasticity and micturition hydrodynamics in females'. *Medical and biological engineering and computing* **7**, 201–215.

Griffiths, D. J.: 1971, 'Hydrodynamics of male micturition – I: Theory of steady flow through elastic-walled tubes'. *Medical and biological engineering and computing* **9**, 581–588.

Grotberg, J. B.: 1994, 'Pulmonary flow and transport phenomena'. *Annual Review of Fluid Mechanics* **26**, 529–571.

Grotberg, J. B. and S. H. Davis: 1980, 'Fluid-dynamical flapping of a collapsible channel: sound generation and flow limitation'. *Journal of Biomechanics* **13**, 219–230.

Grotberg, J. B. and E. L. Reiss: 1984, 'Subsonic flapping flutter'. *Journal of Sound and Vibration* **92**, 349–361.

Grotberg, J. B. and T. R. Shee: 1985, 'Compressible-flow channel flutter'. *Journal of Fluid Mechanics* **159**, 175–193.

Guiot, C., P. G. Pianta, C. Cancelli, and T. J. Pedley: 1990, 'Prediction of coronary blood-flow with a numerical-model based on collapsible tube'. *American Journal of Physiology* **258**, H1606–H1614.

Guneratne, J. C.: 1999, 'High-Reynolds number flow in a collapsible channel'. Ph.D. thesis, Cambridge University, Cambridge.

Guneratne, J. C. and T. J. Pedley: 2001, 'High Reynolds number flow in a collapsible channel'. (in preparation).

Hayashi, S., T. Hazase, and H. Kawamura: 1998, 'Numerical analysis for stability and self-excited oscillation in collapsible tube flow'. *ASME Journal of Biomechanical Engineering* **120**, 468–475.

Hazel, A. L. and M. Heil: 2002, 'Steady finite Reynolds number flow in three-dimensional collapsible tubes'. (Submitted).

Heil, M.: 1995, 'Large deformations of cylindrical shells conveying viscous flow'. Ph.D. thesis, University of Leeds, Leeds.
Heil, M.: 1996, 'The Stability of Cylindrical Shells Conveying Viscous Flow'. *Journal of Fluids and Structures* **10**, 173–196.
Heil, M.: 1997, 'Stokes flow in collapsible tubes: computation and experiment'. *Journal of Fluid Mechanics* **353**, 285–312.
Heil, M. and T. J. Pedley: 1995, 'Large Axisymmetric Deformations of Cylindrical Shells Conveying Viscous Flow'. *Journal of Fluids and Structures* **9**, 237–256.
Heil, M. and T. J. Pedley: 1996, 'Large Post-Buckling Deformations of Cylindrical Shells Conveying Viscous Flow'. *Journal of Fluids and Structures* **10**, 565–599.
Huang, L.: 1995, 'Flutter of cantilevered plates in axial flow'. *Journal of Fluids and Structures* **9**, 127–147.
Huang, L.: 1998, 'Reversal of the Bernoulli effect and channel flutter'. *Journal of Fluids and Structures* **12**, 131–151.
Huang, L.: 2001, 'Viscous flutter of a finite elastic membrane in Poiseuille flow'. *Journal of Fluids and Structures* **15**, 1060–1088.
Ikeda, T. and Y. Matsuzaki: 1999, 'A one-dimensional unsteady separable and reattachable flow model for collapsible tube-flow analysis'. *ASME Journal of Biomechanical Engineering* **121**, 153–159.
Jensen, O. E.: 1990, 'Instabilities of flow in a collapsed tube'. *Journal of Fluid Mechanics* **220**, 623–659.
Jensen, O. E.: 1992, 'Chaotic oscillations in a simple collapsible-tube model'. *ASME Journal of Biomechanical Engineering* **114**, 55–59.
Jensen, O. E.: 1998, 'An asymptotic model of viscous flow limitation in a highly collapsed channel'. *ASME Journal of Biomechanical Engineering* **120**, 544–547.
Jensen, O. E.: 2001, 'Self-excited oscillations in a collapsible channel: insights from a simple asymptotic model'. (Preprint).
Jensen, O. E. and M. Heil: 2002, 'High-frequency self-excited oscillations in a collapsible-channel flow'. *Journal of Fluid Mechanics.* (Submitted).
Jensen, O. E. and T. J. Pedley: 1989, 'The existence of steady flow in a collapsed tube'. *Journal of Fluid Mechanics* **206**, 339–374.
Kamm, R. D.: 1982, 'Bioengineering studies of periodic external compression as prophylaxis against deep vein thrombosis – Part I: numerical studies'. *ASME Journal of Biomechanical Engineering* **104**, 87–95.
Kamm, R. D. and T. J. Pedley: 1989, 'Flow in collapsible tubes: a brief review'. *ASME Journal of Biomechanical Engineering* **111**, 177–179.
Kamm, R. D. and A. H. Shapiro: 1979, 'Unsteady flow in a collapsible tube subjected to external pressure or body forces'. *Journal of Fluid Mechanics* **95**, 1–78.
Katz, A. I., Y. Chen, and A. H. Moreno: 1969, 'Flow through a collapsible tube – Experimental analysis and mathematical model'. *Biophysical Journal* **9**, 1261–1279.
Kececioglu, I., M. E. McClurken, R. D. Kamm, and A. H. Shapiro: 1981, 'Steady, supercritical flow in collapsible tubes. Part 1. Experimental observations'. *Journal of Fluid Mechanics* **109**, 367–389.
Kistler, S. F. and L. E. Scriven: 1983, 'Coating Flows'. In: J. Pearson and S. Richardson (eds.): *Computational Analysis of Polymer Processing.* London: Applied Science Publishers.
Ku, D. N.: 1997, 'Blood flow in arteries'. *Annual Review of Fluid Mechanics* **29**, 399–434.
Landahl, M.: 1962, 'On the stability of a laminar incompressible bounadry layer over a flexible surface'. *Journal of Fluid Mechanics* **13**, 607–632.
Larose, P. G. and J. B. Grotberg: 1997, 'Flutter and long-wave instabilities in compliant channels conveying developing flows'. *Journal of Fluid Mechanics* **331**, 37–58.
Liang, S.-J., G. P. Neitzel, and C. K. Aidun: 1997, 'Finite element computations for unsteady fluid and elastic membrane interaction problems'. *International Journal for Numerical Methods in Fluids* **24**, 1091–1110.
Lighthill, J.: 1975, *Mathematical biofluiddynamics.* Philadelphia: SIAM.
Lowe, T. W. and T. J. Pedley: 1995, 'Computation of Stokes flow in a channel with a collapsible segment'. *Journal of Fluids and Structures* **9**, 885–905.

Luo, X. Y. and T. J. Pedley: 1995, 'Numerical Simulation of Steady Flow in a 2-D Collapsible Channel'. *Journal of Fluids and Structures* **9**, 149–197.

Luo, X. Y. and T. J. Pedley: 1996, 'A numerical simulation of unsteady flow in a two-dimensional collapsible channel'. *Journal of Fluids Mechanics* **314**, 191–225.

Luo, X. Y. and T. J. Pedley: 1998, 'The effects of wall inertia on flow in a two-dimensional collapsible channel'. *Journal of Fluid Mechanics* **363**, 253–280.

Luo, X. Y. and T. J. Pedley: 2000, 'Multiple solutions and flow limitation in collapsible channel flows'. *Journal of Fluid Mechanics* **420**, 301–324.

Matsuzaki, Y. and K. Fujimura: 1995, 'Reexamination of steady solutions of a collapsible channel conveying fluid'. *ASME Journal of Biomechanical Engineering* **117**, 492–494.

Matsuzaki, Y. and Y. Fung: 1977, 'Stability analysis of straight and buckled two-dimensional channels conveying an incompressible flow'. *Trans. ASME E: J. Appl. Mech.* **44**, 548–552.

Matsuzaki, Y., T. Ikeda, T. Kitagawa, and S. Sakata: 1994, 'Analysis of flow in a two-dimensional collapsible channel using universal "tube" law'. *ASME Journal of Biomechanical Engineering* **116**, 469–476.

Matsuzaki, Y. and T. Matsumoto: 1989, 'Flow in two-dimensional collapsible channel with rigid inlet and outlet'. *ASME Journal of Biomechanical Engineering* **111**, 180–184.

McClurken, M. E., I. Kececioglu, R. D. Kamm, and A. H. Shapiro: 1981, 'Steady, supercritical flow in collapsible tubes. Part 2. Theoretical studies'. *Journal of Fluid Mechanics* **109**, 391–415.

McDonald, D. A.: 1974, *Blood flow in arteries.* London: Edward Arnold, second edition.

Moore, J. E., N. Stergiopulos, X. Golay, D. N. Ku, and J.-J. Meister: 1995, 'Flow measurements in collapsed stenotic arterial models'. In: R. M. Hochmuth, N. A. Langrana, and M. S. Hefzy (eds.): *Proceedings of the 1995 Bioengineering Conference*, Vol. ASME-BED 29. pp. 229–230.

Oates, G. C.: 1975, 'Fluid flow in soft-walled tubes: I. Steady flow'. *Medical and Biological Engineering* **13**, 773–778.

Olson, D. A., R. D. Kamm, and A. H. Shapiro: 1982, 'Bioengineering studies of periodic external compression as prophylaxis against deep vein thrombosis – Part II: experimental studies on a simulated leg'. *ASME Journal of Biomechanical Engineering* **104**, 96–104.

Pedley, T. J.: 1980, *The fluid mechanics of large blood vessels.* Cambridge: Cambridge University Press.

Pedley, T. J.: 1992, 'Longitudinal tension variation in collapsible channels: A new mechanism for the breakdown of steady flow'. *ASME Journal of Biomechanical Engineering* **114**, 60–67.

Pedley, T. J., B. S. Brook, and R. S. Seymour: 1996, 'Blood pressure and flow rate in the giraffe jugular vein'. *Philosophical Transactions of the Royal Society of London* **B 351**, 855–866.

Pedley, T. J. and X. Y. Luo: 1998, 'Modelling Flow and Oscillations in Collapsible Tubes'. *Theoretical and Computational Fluid Dynamics* **10**, 277–294.

Pedley, T. J. and K. D. Stephanoff: 1985, 'Flow along a channel with a time-dependent indentation in one wall: the generation of vorticity waves'. *Journal of Fluid Mechanics* **160**, 337–367.

Pedrizzetti, G.: 1998, 'Fluid flow in a tube with an elastic membrane insertion'. *Journal of Fluid Mechanics* **375**, 39–64.

Ralph, M. E. and T. J. Pedley: 1988, 'Flow in a channel with a moving indentation'. *Journal of Fluid Mechanics* **190**, 87–112.

Ralph, M. E. and T. J. Pedley: 1989, 'Viscous and inviscid flow in a channel with a moving indentation'. *Journal of Fluid Mechanics* **209**, 543–566.

Ralph, M. E. and T. J. Pedley: 1990, 'Flow in a channel with a moving indentation in one wall'. *ASME Journal of Fluids Engineering* **112**, 468–475.

Rast, M. P.: 1994, 'Simultaneous Solution of the Navier-Stokes and Elastic Membrane Equations by a Finite-element Method'. *International Journal for Numerical Methods in Fluids* **19**, 1115 – 1135.

Reyn, J. W.: 1987, 'Multiple solutions and flow limitation for steady flow through a collapsible tube held open at the ends'. *Journal of Fluid Mechanics* **174**, 467–493.

Rodbard, S.: 1966, 'A hydrodynamics mechanism for autoregulation of flow'. *Cardiologia* **48**, 532–535.

Rodbard, S. and L. Takacs: 1966, 'Hydrodynamics of autoregulation'. *Cardiologia* **48**, 433–440.

Rosenfeld, M.: 1995, 'A numerical study of pulsating flow behind a constriction'. *Journal of Fluid Mechanics* **301**, 203–223.

Shapiro, A. H.: 1977, 'Steady flow in collapsible tubes'. *ASME Journal of Biomechanical Engineering* **99**, 126–147.
Shimizu, M. and Y. Tanida: 1983, 'On the mechanism of Korotkoff sound generation at diastole'. *Journal of Fluid Mechanics* **23**, 299–312.
Smith, F. T.: 1976a, 'Flow through constricted or dilated pipes and channels. Part 1'. *Q. J. Mech. Appl. Math.* **21**, 343–364.
Smith, F. T.: 1976b, 'Flow through constricted or dilated pipes and channels. Part 2'. *Q. J. Mech. Appl. Math.* **21**, 365–379.
Sobey, I.: 1985, 'Observations of waves during oscillatory channel flow'. *Journal of Fluid Mechanics* **151**, 395–426.
Stephanoff, K. D., T. J. Pedley, C. J. Lawrence, and T. W. Secomb: 1983, 'Fluid flow along a channel with an asymmetric oscillating constriction'. *Nature* **305**, 692–695.
Tobak, M. and D. J. Peake: 1982, 'Topology of three-dimensional separated flows'. *Annual Review of Fluid Mechanics* **14**, 61–85.
Tsuji, T., K. Nakajim, Y. Takeuchi, K. Inoue, K. Shioma, Y. Koyama, K. Tokuchi, T. Yoshikawa, and K. Suma: 1978, 'Study on haemodynamics during cardiopulmonary bypass (in Japanese)'. *Artificial Organs* **7**, 435–438.
Tutty, O.: 1984, 'High-Reynolds-number viscous flow in collapsible tubes'. *Journal of Fluid Mechanics* **146**, 451–469.
Tutty, O.: 1989, 'The viscous flow through symmetric collapsible channels'. *Mathematika* **36**, 153–181.
Tutty, O. R.: 1992, 'Pulsatile flow in a constricted channel'. *ASME Journal of Biomechanical Engineering* **114**, 50–54.
Tutty, O. R. and T. J. Pedley: 1993, 'Oscillatory flow in a stepped channel'. *Journal of Fluid Mechanics* **247**, 179–204.
Ur, A. and M. Gordon: 1970, 'Origin of Korotkoff sounds'. *American Journal of Physiology* **218**, 524–529.
Walsh, C.: 1995, 'Flutter in one-dimensional collapsible tubes'. *Journal of Fluids and Structures* **9**, 393–408.
Walsh, C., P. Sullivan, and J. Hanson: 1991, 'Subcritical flutter in collapsible tube flow: a model of expiratory flow in the trachea'. *J. Biomech. Engng.* **113**, 21–26.
Weaver, D. and M. Païdoussis: 1977, 'On the collapse and flutter phenomena in thin tubes conveying fluid'. *Journal of Sound and Vibration* **50**, 117–132.

Chapter 3

EXPERIMENTAL STUDIES OF COLLAPSIBLE TUBES

C. D. BERTRAM
Graduate School of Biomedical Engineering
University of New South Wales
Sydney, Australia 2052

Abstract
The experimental literature on collapsible tubes is reviewed with reference to historical landmarks, ways of measuring important variables, the main observed phenomena, and the applications in medical therapy and technology. Mechanistic phenomena covered include transitions to and from supercritical flow, prediction of the onset of oscillation, flow-rate limitation, and the collapsible tube as a nonlinear dynamical system.

P.W. Carpenter and T.J. Pedley (eds.),
Flow past Highly Compliant Boundaries and in Collapsible Tubes, 51-65.

Symbols: p = pressure, Q = flow-rate, A = cross-sectional area, suffix 1 = at the upstream end, 2 = at the downstream end, e = external to the tube, lim = when flow-limited.

1 Historical introduction

Experimental studies of collapsible tubes are sometimes considered to begin with the 1969 paper by Conrad [1], so it is appropriate to remember that the subject has a rather longer history. With such an essentially simple device (in one sense, although it turns out to have rather complex dynamics) as a flexible tube acted upon by negative transmural pressure, no true beginning is easy to trace. Brecher [2] reviews the early physiological observations of collapse of exposed veins, by among others Valsalva (1666–1723). These include what is probably the first documented observation of venous self-excited oscillation, by D. Barry in 1824, who

> "inserted into the jugular vein of a horse a glass bulb in which the jet of blood could be watched, and observed an acceleration of blood flow with each inspiration when the horse was lying down and breathing quietly. When the horse stood up the jets of blood falling through the glass bulb were not synchronous with respiration. Instead they were of a greater frequency than the arterial pulsations. Upon compression of the carotid artery they became less conspicuous."

But a famous early landmark [3] is the use of a collapsible tube by Starling and his colleagues to provide an afterload for an isolated heart. The aim was to mimic the normal systemic circulatory characteristic of maintaining some pressure up to the time of the next aortic valve opening. Applying a typical diastolic pressure to the outside of a tube in a chamber achieved this without excessive impedance to left ventricular ejection. Varying the pressure provided an easy way of varying 'peripheral resistance', insofar as a higher peripheral resistance means a higher end-diastolic pressure, all other things being equal. The tube in a chamber thus acquired a still-popular name: Starling resistor.

Starling seems to have relied on an intuitive knowledge of how the device would work. What appears to be the first systematic investigation of its properties, by Holt [4], arose out of interest in the way in which collapsible veins communicate capillary pressure to the right atrium. He clearly identified the flow limitation property: "the rate of flow did not increase when p_2 was decreased, once the tube was partially collapsed". He found pulsation as soon as the tube collapsed, but that flow limitation continued when p_2 was lowered further and pulsation ceased. The collapsibility of veins, and its paramount importance in coupling thoracic and venous pressures, was also championed by Duomarco *et al.* [5]. Whereas Holt measured only pressures *in vivo*, Brecher [6] measured both pressure drop and flow-rate, showing clearly how jugular venous collapse enforced a maximum flow-rate as downstream pressure was lowered. The flow limitation property was firmly enshrined as the basis of systemic 'venous return' by Guyton and his colleagues [7], and the implications of pulmonary airway collapse for expiratory flow limitation were recognised in the same era [8]. Flow limitation was also identified as a possible mechanism underlying vascular autoregulation by Rodbard and co-workers [9, 10], who paid particular attention to a device called the 'capillaron', in which pressure in the closed chamber was derived from upstream of the tube. In a collapsible-tube analog of the postulated capillarons in the internal organs, flow limitation was again demonstrated.

Thus Conrad's [1] influential study of pressure-flow relationships was built on a foundation of considerable physiological interest in collapsing vessels over the preceding quarter century. The paper has enjoyed enduring relevance for three reasons: (i) systematic data on pressure drop vs. flow-rate with variation of either external pressure or downstream resistance, (ii) identification of a criterion for the existence of self-excited oscillations, and (iii) an accompanying analytical model which was shown to reduce to van der Pol's oscillator. He showed that the relation between pressure drop and flow-rate is sigmoidal, with a middle section of negative slope representing negative incremental resistance. Oscillations were predicted to occur when this negative resistance was greater than the positive resistance of all other elements in the flow path, and in his experiments, did occur only when the operating point lay on the negative slope. Conrad also provided photographs of the tube at various points along the sigmoidal characteristic; these are reproduced better than in the original journal in a later book [11].

It was left to Brower [12] to promulgate the third major way of presenting collapsible-tube pressure-drop/flow-rate relations. In each case curves are drawn on axes of flow-rate Q and pressure drop $p_1 - p_2$. Flow-rate limitation is demonstrated if the experiment is done so that $p_1 - p_e$ is kept constant; the sigmoidal characteristic is found if p_e is constant. If $p_2 - p_e$ is kept constant, the pressure drop reaches an approximately constant level that depends on the setting of $p_2 - p_e$ but is independent of flow-rate; this is called pressure-drop limitation. In a most useful contribution, Brower & Noordergraaf [13] showed that the various ways in which earlier researchers had presented their results were all reducible to one of these three basic forms. Brower's own data on $p_1 - p_2$ vs. Q are best seen in an otherwise theoretical paper [14], where the regions of oscillatory operation are clearly marked.

Beyond this point, the number of contributions multiplies, and a chronological organisation, or one based on the contributions of individuals, is less apt than one based on topic. In the following sections, an attempt will be made to list important experiments according to category. The choice of categories and inclusions is of course arbitrary and individual; I accept blame, and ask forgiveness, for all omissions. Alternative recent reviews of the subject of collapsible-tube flow in general are given by Thiriet *et al.* [15] and Païdoussis [16].

2 Experiments and measurement techniques

2.1 PRESSURE-AREA RELATIONS

The 'tube law' is accepted jargon for the relation between transmural pressure and tube cross-sectional area based purely on local bending stiffness in the wall. Other, non-local influences come into play when the tube is of finite length and under finite axial tension. The simplest technique for measuring the shape-independent cross-sectional area (change) is liquid volume displacement; end effects may be dealt with by measurement of two similar tubes of different lengths. This method is only accurate for long uniform tubes. It has been used on thick-walled [17] and thin-walled [18-21] silicone-rubber tubes and on latex[1] tubes [24-28] and microtubes [29]. Brower & Scholten [30] also showed a tube law for Penrose[2] (latex rubber) tube, but without specifying how area was measured.

An alternative technique, which measures local shape-independent cross-sectional area, is conductimetry, introduced to this field by McClurken, and used by him to measure the tube law of rather large-diameter thin-walled latex tubes [31]. Since the measurement is local, it can also be used to measure the complete area profile of steady tubes [31, 32], the changes in the apparent tube law when end effects are significant [22], and the variation in the tube law along tubes of axially varying stiffness [33, 34]. The technique can also be used to measure unsteady area profiles of periodic oscillations, by making successive measurements of area as a function of time at many sites along the tube and aligning them in phase with respect to some invariant waveform such as pressure at the tube exit [35, 36]. Shimizu [37] adapted the technique for non-invasive simultaneous semi-quantitative measurements of collapsing arterial volume at four sites under a compressive arm cuff.

Circumstances arise where the local area cannot be relied upon to indicate the shape of the local cross-section. For instance very thin-walled tubes (and very short ones) will adopt a collapse mode involving more than the usual two lobes close to where they are joined to a rigid circular pipe [38-41]. Furthermore, simultaneous measurement of area all the way along a tube is necessary to obtain unsteady area profiles when the tube motion is not strictly periodic. Both these needs are answered by oblique video imaging of the outside of a tube on which a grid has been either scribed or projected [42-44]. However, only static tube-law measurements and shape profiles for latex tubes during steady flow have appeared in the literature to date from such stereographic techniques.

Besides these main three, other methods have occasionally been deployed. Shape and area information can be obtained by making casts of the tube at a given transmural pressure [22, 41, 45, 46], taking care that the casting material does not interfere with the natural tube shape. After embedding a fine wire as a single-turn primary coil in the tube wall, transformer techniques will yield cross-sectional area [47], but the method has many limitations. Ribreau and co-workers favour ultrasonic sector-scan imaging of the tube cross-section [48]; this technique requires the tube to be immersed in liquid. Similarly, Kamimura *et al.*, working with silicone-rubber tubes of up to 0.6mm wall thickness surrounded by sucrose solution, have photographed tube cross-sections illuminated by a laser light sheet [49]. Such imaging methods do not interfere mechanically with the tube in any way, and can be used during unsteady flow. However, while it may advantageously support the tube, the external liquid impedes

---

[1] The distinction is important, because silicone rubber exhibits significant hysteresis, i.e. a permanent (non-fading) memory of previous strain state [22]. In situations where the strain is continuously cyclic, hysteresis is sometimes confused with viscoelasticity (fading memory). Silicone rubber is also more viscoelastic than latex [23], i.e. exhibits a greater (frequency-dependent) phase lag between stress and strain. Hysteresis must also be distinguished from plasticity, which property could not be the basis of a stable hysteresis loop and is not exhibited by silicone rubber.

[2] The term 'Penrose tube', regrettably often the only description given, is ambiguous in respect of material. A Penrose tube is essentially a surgical drain tube, typically of 6 mm or 12 mm diameter and wall thickness less than 0.3 mm. They are usually made of natural latex rubber, but in Japan at least it seems that Penrose tubes can be of silicone rubber [19].

tube motions, and negates the hydrostatic effects of tube inclination[3]. Furthermore, derivation from images of quantitative variables such as cross-sectional area (and synchronisation with other measurements) is not simple. Nevertheless, the proliferation and increasing ability of imaging techniques in medicine suggest that wider use will be made of such tools in this field, even if some are anything but portable.

2.2 FLOW VISUALISATION AND SEPARATION

Since modelling pointed to the importance of unsteady flow separation in the dynamics of an oscillating tube (see the chapter in this volume by Heil & Jensen), experimenters have investigated this aspect, using three main approaches. The development of separation past a fixed asymmetrical indentation in a rigid channel was studied using impulsively started flow [50]. Later, the indentation itself was made to move so that the degree of indentation varied, and vortex waves downstream were discovered [51], representing a significantly new fluid-dynamic phenomenon. Although these waves seem unlikely in the basically symmetrically indented situation of a collapsible tube, they certainly play a significant role in the energy budget of a channel with a compliant (collapsible) section in one wall. These and later experiments [52] were conducted with only modest rates of indentation motion, but Matsuzaki *et al.* have also extended their studies of oscillating obstructions to frequencies corresponding to high Strouhal numbers [53, 54].

The understandable urge on the part of theoreticians to reduce the highly three-dimensional problem of flow through a collapsible tube to two dimensions has also led to at least one attempt to realize a collapsible channel experimentally [55]. The technical difficulties of leakage and attaining two-dimensional membrane motion are formidable, and it was acknowledged that they had not been satisfactorily overcome. Rather more success has been had with the 'mechanical trachea', consisting of a stretched membrane forming the fourth side of an otherwise rigid rectangular channel [56, 57][4]. This has many advantages, particularly in terms of relating the cross-sectional area available for flow to the constitutive properties of the membrane material, and allowing pressure tappings in the rigid walls, but is not a two-dimensional geometry.

Flow visualisation or measurement in a collapsible tube executing flow-induced oscillations is another difficult technical challenge. Direct flow visualisation of an oscillating tube has been attempted [59], but the results were too indistinct to lend strong support to the authors' hypothesis regarding self-excited oscillation. Bertram *et al.* took the alternative approach of making detailed measurements of velocity first in a rigid model of a collapsed tube [60, 61], then in the first diameter of rigid pipe downstream of a tube during flow-induced oscillation [62]. They argued that events in the oscillating tube could be inferred from those observed just downstream, with the aid of the steady rigid-tube velocity fields. Flow separation caused in the tube certainly extended into the downstream region, as did secondary-flow swirl. However, the impetus for further such investigations is weakened by the theoretical finding [63] that

---

[3] Inclination and wall thickness taper are two roughly equivalent ways to achieve a gradient along the tube of transmural pressure relative to that needed for collapse. Inclined tubes have been extensively investigated by Ribreau et al. (see the review by Thiriet *et al.* [15]).

[4] A version which was called an artificial trachea, consisting of a rubber membrane stretched over a channel of semi-circular cross-section, was used earlier by Elliott & Dawson [58].

collapsible-tube oscillation may not be as dependent on flow separation as had for many years been thought.

A further experimental alternative is to focus on extremely thin transparent collapsible tubes submerged in a specially chosen liquid such as sucrose solution which is also the working fluid within. By means of laser Doppler anemometry, velocity profiles have been measured inside collapsible tubes for this situation [64, 65]. Laser Doppler measurements of axial velocity at several sites were also made by Shimizu [66], again with an immersed tube. Similar restrictions on the nature of the dynamical system will be necessary to make use of the technique of particle image velocimetry.

2.3 DISTRIBUTED MEASUREMENTS OF OTHER VARIABLES

Pressure within a collapsed tube has been measured by wall puncture [67] and by catheter, either a hollow one leading to a fluid column [27] or remote transducer [31, 58, 68, 69] or an active one bearing a catheter-tip transducer [36]. There is no way of measuring intra-luminal pressure which does not potentially entail some small degree of mechanical interference with the tube flow and configuration. The error is normally small, but could become appreciable for extremely thin-walled tubes. For this reason, Ohba *et al.* [70, 71] choose to infer pressure from optical measurements of flow velocity[5], tube area and cross-sectional shape, used as inputs to a steady-flow model. Yamane *et al.* have shown that insertion of a rod into a collapsible tube has no effect on self-excited oscillation as long as there is no contact between the rod and the tube wall [69].

Measurement of the varying distance of the tube surface from a fixed external probe is a semi-quantitative approach, yielding useful data about wave speed if two or more sites are compared [30, 72], or about the steady-flow collapse profile if many measurements are made along the tube [20]. A picture of the time-varying collapse profile all along the tube may also be built up, but only for repetitive oscillations [20]. Alternatively, even less quantitative collapse profiles (because they no longer respect the collapsed-tube concavity) can be obtained for steady flow by photography and for tubes in motion by video recording or synchronised flash photography. In an experiment at the remarkably low Reynolds number[6] of 0.0014, Heil [74] obtained very satisfactory qualitative correspondence between obliquely photographed collapsed-tube shape and a computer-rendered predicted tube surface.

3 Observed phenomena and mechanistic analysis

3.1 CHOKING AND SUPERCRITICAL FLOW

Griffiths [75] demonstrated a stationary shock-like transition, upstream of which the tube was partially collapsed and supercritical flow was assumed to exist. Brower & Scholten [30] showed that oscillatory instability was coincident in their experiments with fluid speed reaching the wave speed, i.e. attaining a speed index of one. Elliott & Dawson [27] demonstrated supercritical flows upstream of an elastic jump, with speed

---

[5] The volume flow-rate measured beyond the tube suffices except where there is the possibility of separation and reversed flow; in that case the model requires adjustment, and laser Doppler velocimetry is used to find out where reverse flow occurs.

[6] For those interested in this extreme, at least three other low-Reynolds-number experiments have been conducted [29, 38, 73]. The low-Re experiments are also unique in having been conducted with an imposed flow-rate rather than an imposed pressure head.

index as high as ten. Kececioglu *et al.* [31] showed that stationary precursor waves could be radiated upstream of the jump, while dispersive waves ahead of a travelling shock-wave front were recorded by Flaud *et al.* [72, 76] and Jan *et al.* [77]. Testing predictions of alternative, non-shock-like ways of returning to subcritical flow downstream in non-uniform tubes, Kamm *et al.* [78] found only qualitative agreement with their theory. They concluded that while their observations were consistent with the predicted smooth transition, they might also be explained on the basis of a rather gradual conventional elastic jump.

All considerations of speed index are dependent on a measurement of pressure wave speed. There are two main ways: calculation from the slope of the measured tube law [20, 31], and direct measurement of the propagation of either single wave fronts or imposed wave trains [30, 79]. Comparison of the two methods [18, 80] shows good agreement under favourable circumstances, but there are problems with either in tubes of modest length where reflections and axial curvature effects cannot be readily excluded. The wave speed varies complexly with both time and axial position in a tube executing self-excited oscillation.

3.2 INITIATION OF OSCILLATION

Ohba and colleagues rejected the Conrad criterion for self-excited oscillation onset, because they observed oscillations both when the entire relation between pressure drop and (aqueous) flow-rate was of positive slope[7], and on the positive-slope portion of relations which included a negative-slope segment [28, 81]. By hydrodynamic stability analysis of a 1D viscous flow model, they were able to predict graphically the flow-rate at which oscillation would occur, obtaining reasonably good agreement with experiment over several silicone and natural rubber tubes [82]. However the method underestimated the threshold flow-rate when applied to air flow through a silicone-rubber tube [19]. Yamane & Orita revisited the question, finding only negative-slope oscillations in a silicone-rubber tube at high downstream resistance, but oscillations in the absence of a negative slope at low downstream resistance [83]. They also compared wave speed and flow speed (calculated as Q/A), concluding that oscillations could occur at low downstream resistance in the absence of supercritical flow. However, they cautioned that Q/A could underestimate the true flow speed and thus the speed index.

The lowest Reynolds number at which self-excited oscillations have been reported [73] is about 20, under an imposed flow-rate. When flow is the result of an imposed pressure head, the minimum reported Reynolds number jumps to over 200 [64, 84].

3.3 FLOW LIMITATION

Besides those mentioned in section 1, flow limitation has been studied explicitly by Bonis [85], Vischer [40], Fry *et al.* [86], Sullivan et al. [56, 57], Gavriely *et al.* [87], Kamm *et al.* [88], Low *et al.* [89, 90] and Bertram *et al.* [34, 91, 92]. There are of course many investigations of pulmonary airway flow limitation as well. Current questions in this area relate to the extent to which flow-limited flow-rate (Q_{lim}) is under

---

[7] Unlike the other two ways of presenting relationships between pressure drop and flow-rate, the sigmoidal shape that is usually found when p_e is maintained constant does depend on factors such as flow-resistance up- or downstream of the tube. Hence such external matters can determine whether a negative-resistance region is found; there is no one unique family of characteristic curves corresponding to a family of settings of p_e.

the control of transmural pressure at the upstream end of the tube. The traditional view, which can be backed up by simple modelling, has been that Q_{lim} increases monotonically[8] with $p_1 - p_e$. However, for air flow through a tube of tapering wall thickness, Kamm *et al.* found up to three widely separated Q_{lim} values [93]. Their studies of aqueous flow through such tapered-stiffness tubes [78] had not revealed any equivalent behaviour. Re-examining the latter situation, Bertram & Chen [34] found instances of multiple Q_{lim} values, but the extent of the change in Q_{lim} was small. Similar multiple Q_{lim} values, again with only small variations in Q_{lim}, have been observed for aqueous flow in a uniform tube [92].

Almost all of the work by Bertram *et al.* relates to rather thick-walled silicone-rubber tubes. Working instead with 6mm latex-rubber Penrose tubes, they have recently reported [84] observations on water flow which may correspond to the puzzling findings on air flow of Kamm *et al.* Via a modified protocol, they found many Q_{lim} values at the one $p_1 - p_e$. Both sets of observations constitute an unsolved conundrum for modellers at present.

3.4 COLLAPSIBLE STENOSES AND RELATED PARADOXES

Arterial collapse has been postulated to contribute to both coronary [94] and carotid embolic events. *In vitro* experiments on stenosed flexible tubes show that vessels stenosed by atherosclerotic plaque are at increased risk of transient collapse [95-100], because of the extent to which the high blood velocity through the stenosis lowers the local pressure, particularly in systole. If transient collapse does occur, it will be repeated with each heart beat and is then a possible mechanism of plaque fracture and subsequent embolization, leading to infarction or stroke. The experiments showed that this scenario could well happen under haemodynamic conditions which would be unusual but certainly not impossible in the unlimited spectrum of patient pathologies.

Collapse during systole, normally thought of as a time of increased vascular distension, is certainly contrary to naïve thinking, and can be viewed as paradoxical if one is so inclined. A further notable part of the mechanics of such a stenosed flexible vessel is the fact that increased flow-rate can result from increased downstream resistance and *vice versa*. This was known and understood by Rodbard as long ago as 1955 [101], but the apparent paradox has been repeatedly rediscovered [102-104].

3.5 THE TUBE AS A DYNAMICAL SYSTEM

The initial stimulus to analysis of an oscillating collapsible tube as a nonlinear dynamical system was the observation of an unexpectedly large number of oscillation modes [105]. Perhaps as a result of exploring high Reynolds numbers, Bertram *et al.* have consistently seen a wider range of modes in both thick-walled [106] and thin-walled [107] tubes than other groups, but more than one mode has been a common finding. A further stimulus to application of dynamical systems thinking was the observation of large-amplitude aperiodic [105, 108] and compound [105, 109] oscillations.

Once oscillations start, the behaviour is no longer uniquely determined by position on a set of curves of flow-rate limitation or pressure-drop limitation. Factors external to the

---

[8] Q_{lim} usually also varies inversely with $p_1 - p_2$ ('negative effort dependence'), so the dependence on $p_1 - p_e$ must be examined at a single $p_1 - p_2$.

tube are involved in the dynamics [110], so other forms of data mapping may be preferable. While the strict dynamical systems approach would be to use the parameters under the experimenter's direct control as axes, alternative coordinate systems have been found better suited to need here. In rough functional equivalence, both the pressure driving the flow vs. $p_e - \overline{p}_2$ [105], and flow-rate Q itself vs. p_e [111], have been used to map oscillation modes as regions in two-dimensional parameter space. The advantage of $p_e - \overline{p}_2$ as opposed to p_e is that regions of divergent instability can also be mapped. Time-averaged flow-rate is a better parameter than pressure head in that the tube directly experiences it, but it is difficult to measure with sufficiently high precision for dynamical systems purposes, and is not under direct control in most experiments.

Nonlinear systems characteristically display hysteresis, so that the behaviour at a given operating point depends on the route of arrival at those parameter settings, as well as the settings themselves, and this has been a prominent feature of several experiments [111-113]. The established methods of dynamical analysis have also both focussed attention on, and allowed a degree of explanation of, many other aspects of the flow-induced oscillations, such as their sudden onset [105, 112].

However, despite the deployment of a range of tools from the dynamical systems arsenal, Bertram *et al.* did not find unequivocal evidence for aperiodic oscillations being the outcome of low-dimensional chaos [105, 108]. The situation is different when a collapsible tube otherwise executing periodic flow-induced oscillations is subjected to oscillatory forcing from an external source such as a reciprocating-piston pump upstream. Mode-locking [114], in which the tube oscillation is coerced into a mode that is co-periodic with the external forcing over one or more cycles of that forcing, is the habitual response. Aperiodic responses are also observed, and these can be shown to correspond to established indicators of low-dimensional chaos [115].

4 Applications in medicine and technology

The motivation behind almost all collapsible-tube research is the desire to understand phenomena which occur in the human body. This section concentrates on other motivations and applications; a majority of these still involve the human body, but now in the context of medical therapies rather than natural physiology. With this restriction, probably the most widespread application of experimental collapsible-tube research so far has been to understanding Korotkov sounds and related aspects of the Riva-Rocci method of measuring blood pressure (sphygmomanometry) [32, 66, 67, 116, 117]. Causing vascular collapse by limb compression also has found extensive application to prevention of deep vein thrombosis during prolonged bed-rest, by means of inflatable boots [77]. Given uniform external pressure, a collapsing vessel will always collapse first and most at the downstream end, and the boots are therefore better designed to apply a spatial gradient of compression to the limb by means of multiple segments, encouraging collapse from the upstream (distal) end first. The same principle can be applied to external (i.e. non-invasive) cardiac assist, but this has not been as widely adopted. Blood pumps for cardiac assist based on pneumatic compression of collapsing tubes in parallel have also been proposed [17, 118]. External chest massage for emergency resuscitation depends on the collapsibility of both the great vessels and the heart chambers themselves, as has been convincingly shown by *in vitro* experiments [119]. Considerations of collapse also arise in some of the devices currently under development for cardiac assist by means of skeletal-muscle-powered pumping [120].

Surprisingly few direct applications in engineered devices have been made so far of the collapsed-tube flow-limitation principle, perhaps because of doubts about tube service life. However, a Swiss engineer has proposed a high-Reynolds-number application of the principle, in maintenance of a constant discharge from a reservoir [40, 121]. Another potential application is to shunt valves for treatment of hydrocephalus [122]. At a yet lower Reynolds number, a device based on a collapsible tube has been proposed to control leak flow of ocular aqueous humour to relieve glaucoma (Ethier, private communication). These examples, spanning a Reynolds number range of six orders of magnitude (0.2 to 2×10^5), emphasise the universality of the principle.

The self-excited oscillation was proposed as the basis for a prosthetic larynx working on principles akin to those of the natural one [123], although the current generation of passively oscillatory voice prostheses owes little to collapsible-tube ideas. Another proposed application of the flow-induced oscillations has been to use a collapsible tube as a simple way to convert an existing steady flow to a pulsatile one where this is advantageous. Such opportunities exist in devices where there are diffusive transport processes, because the pulsation can increase the efficiency of the transport. To this end, a collapsible tube has been tried successfully as a periodic interrupter of the feedstock flow to a cross-flow microfilter [124-128]. Plans are afoot to apply the idea at the smaller pore scale of ultrafiltration, where again the benefits of cross-flow pulsation are already established.

Acknowledgements—I am indebted to Dr. N. Fujisawa for a translation of two papers in Japanese by Yamane *et al.* Experiments by Bertram *et al.* were mostly funded by the Australian Research Council.

References

1. Conrad, W. A. (1969) Pressure-flow relationships in collapsible tubes, *IEEE Transactions on Bio-Medical Engineering* **16**(4), 284–295.
2. Brecher, G. A. (1956) *Venous Return* (Grune & Stratton, New York).
3. Patterson, S. W. & Starling, E. H. (1914) On the mechanical factors which determine the output of the ventricles, *Journal of Physiology* **48**, 357–379.
4. Holt, J. P. (1941) The collapse factor in the measurement of venous pressure – the flow of fluid through collapsible tubes, *American Journal of Physiology* **134**, 292–299.
5. Duomarco, J. L., Recarte, P. & Rimini, R. (1944) Influencia de las presiones abdominal y torácica sobre el retorno venoso en la cava inferior, *Revista Argentina de Cardiologia* **11**, 286–295 (cited by Brecher, 1956).
6. Brecher, G. A. (1952) Mechanism of venous flow under different degrees of aspiration, *American Journal of Physiology* **169**, 423–433.
7. Guyton, A. C. & Adkins, L. H. (1954) Quantitative aspects of the collapse factor in relation to venous return, *American Journal of Physiology* **177**, 523–527.
8. Fry, D. L., Ebert, R. V., Stead, W. W. & Brown, C. C. (1954) The mechanics of pulmonary ventilation in normal subjects and in patients with emphysema, *American Journal of Medicine* **16**, 80–97.
9. Rodbard, S. & Saiki, H. (1953) Flow through collapsible tubes, *American Heart Journal* **46**, 715–725.
10. Rodbard, S. (1966) A hydrodynamic mechanism for autoregulation of flow, *Cardiologia* **48**, 532–535.
11. Fung, Y. C. (1984) *Biodynamics - Circulation* (Springer-Verlag, New York).
12. Brower, R. W. (1970) *Pressure-flow characteristics of collapsible tubes.* Unpublished Ph.D., University of Pennsylvania.

13. Brower, R. W. & Noordergraaf, A. (1973) Pressure-flow characteristics of collapsible tubes: a reconciliation of seemingly contradictory results, *Annals of Biomedical Engineering* **1**, 333–355.
14. Brower, R. W. & Noordergraaf, A. (1978) Theory of steady flow in collapsible tubes and veins, in *Cardiovascular System Dynamics*, eds. Baan, J., Noordergraaf, A. & Raines, J. (MIT Press), pp. 256–265.
15. Thiriet, M., Naili, S., Langlet, A. & Ribreau, C. (2001) Flow in thin-walled collapsible tubes, in *Biofluid Methods in Vascular and Pulmonary Systems*, ed. Leondes, C. T., Vol. 4 of *Biomechanical Systems: Techniques and Applications* (CRC Press).
16. Païdoussis, M. P. (in press) Thin shells containing or immersed in flow, in *Fluid-Structure Interactions: Slender Structures and Axial Flow*, Vol. 2 (Academic Press, San Diego).
17. Weissman, M. & Mockros, L. (1967) The mechanics of a collapsing tube heart pump, *International Journal of Mechanical Science* **9**, 113–121.
18. Flaud, P., Oddou, C. & Geiger, D. (1985) High-amplitude wave propagation in collapsible tube. I. – Relation between rheological properties and wave propagation, *Journal de Physique* **46**(5), 691–698.
19. Sakurai, A. & Ohba, K. (1986) Self-excited oscillation of flow in collapsible tube. III (A resonant oscillation of air flow), *Technology Reports of Kansai University* **27**, 41–48.
20. Yamane, T. & Orita, T. (1994) Relationship of pressure wave velocity to self-excited oscillation of collapsible tube flow, *JSME International Journal Series A* **37**(1), 71–78.
21. Bassez, S., Flaud, P. & Chauveau, M. (2001) Modeling of the deformation of flexible tubes using a single tube law: application to veins of the lower limb in man, *ASME Journal of Biomechanical Engineering* **123**, 58–65.
22. Bertram, C. D. (1987) The effects of wall thickness, axial strain and end proximity on the pressure-area relation of collapsible tubes, *Journal of Biomechanics* **20**, 863–876.
23. Bertram, C. D., Gow, B. S. & Greenwald, S. E. (1997) Comparison of different methods for the determination of the true wave propagation coefficient, in rubber tubes and the canine thoracic aorta, *Medical Engineering and Physics* **19**(3), 212–222.
24. Bonis, M. & Ribreau, C. (1977) Pressure-flow relationships in collapsible tubes, in *Euromech 92: Cardiovascular and Pulmonary Dynamics*, pp. 459–466.
25. Bonis, M. & Ribreau, C. (1978) Etude de quelques propriétés de l'écoulement dans une conduite collabable, *La Houille Blanche* **3**(4), 165–173.
26. Ribreau, C. & Bonis, M. (1978) Propagation et écoulement dans les tubes collabables. Contribution a l'étude des vaisseaux sanguins, *Journal Français de Biophysique et Médecine Nucléaire* **2**(3), 153–158.
27. Elliott, E. A. & Dawson, S. V. (1979) Fluid velocity greater than wavespeed and the transition from supercritical to subcritical flow in elastic tubes, *Medical and Biological Engineering and Computing* **17**, 192–198.
28. Ohba, K., Yoneyama, N., Shimanaka, Y. & Maeda, H. (1984) Self-excited oscillation of flow in collapsible tube, *Technology Reports of Kansai University* **25**, 1–13.
29. Sipkema, P. & Westerhof, N. (1989) Mechanics of a thin walled collapsible microtube, *Annals of Biomedical Engineering* **17**, 203–217.
30. Brower, R. W. & Scholten, C. (1975) Experimental evidence on the mechanism for the instability of flow in collapsible vessels, *Medical and Biological Engineering* **13**, 839–845.
31. Kekecioglu, I., McClurken, M. E., Kamm, R. D. & Shapiro, A. H. (1981) Steady, supercritical flow in collapsible tubes. Part 1. Experimental observations, *Journal of Fluid Mechanics* **109**, 367–389.
32. Bertram, C. D. & Butcher, K. S. A. (1992) Possible sources of discrepancy between sphygmomanometer cuff pressure and blood pressure quantified in a collapsible-tube analog, *ASME Journal of Biomechanical Engineering* **114**, 68–77 (printer's errata corrected in **114**, 201).
33. Jaekle, D. E., jr. (1987) *Critical transitions associated with steady flow in collapsible tubes with varying wall stiffness*. Unpublished S.M., Massachusetts Institute of Technology.
34. Bertram, C. D. & Chen, W. (2000) Aqueous flow limitation in a tapered-stiffness collapsible tube, *Journal of Fluids and Structures* **14**(8), 1195–1214.
35. Bertram, C. D., Sheppeard, M. D. & Jensen, O. E. (1994) Prediction and measurement of the area-distance profile of collapsed tubes during self-excited oscillation, *Journal of Fluids and Structures* **8**, 637–660.
36. Bertram, C. D. & Godbole, S. A. (1995) Area and pressure profiles for collapsible tube oscillations of three types, *Journal of Fluids and Structures* **9**, 257–277.
37. Shimizu, M. (1992) Blood flow in a brachial artery compressed externally by a pneumatic cuff, *ASME Journal of Biomechanical Engineering* **114**, 78–83.
38. Fung, Y. C. & Sobin, S. S. (1972) Pulmonary alveolar blood flow, *Circulation Research* **30**, 470–490.

39. Dion, B., Naili, S., Renaudeaux, J. P. & Ribreau, C. (1995) Buckling of elastic tubes: study of highly compliant device, *Medical and Biological Engineering and Computing* **33**, 196–201.
40. Vischer, D. (1979) Die selbsttätige Schlauchdrossel zur Gewährleistung konstanter Beckenausflüsse, *Wasserwirtschaft* **69**(12), 371–375.
41. Palermo, T. & Flaud, P. (1987) Study of two or three lobe breakdown in elastic tubes, *Journal de Biophysique et Biomécanique* **11**, 105–111.
42. Thiriet, M., Delpuech, C., Piroird, J. M., Magnin, I., Bonnet, J. C. & Walton, C. (1987) Banc de mesure optique de la déformation de conduites souples, *Innovation et Technologie en Biologie et Médecine* **8**(1), 99–107.
43. Elad, D., Sahar, M., Einav, S., Avidor, J. M., Zeltser, R. & Rosenberg, N. (1989) A novel non-contact technique for measuring complex surface shapes under dynamic conditions, *Journal of Physics E: Scientific Instruments* **22**, 279–282.
44. Elad, D., Sahar, M., Avidor, J. M. & Einav, S. (1992) Steady flow through collapsible tubes: measurements of flow and geometry, *ASME Journal of Biomechanical Engineering* **114**(1), 84–91.
45. Nahmias, J. (1978) Collapsible ducts cross section and linear head drop. In *Digest of the 1st International Conference on Mechanics in Medicine and Biology*, Aachen, Germany, 1978), ed. Reul, H., pp. 260–263. Baden-Baden: Verlag Gerhard Witzstrock.
46. Nahmias, J. (1980) Unpublished Thèse de Troisième Cycle, Université Paris VI.
47. Bertram, C. D. & Ribreau, C. (1989) Cross-sectional area measurement in collapsed tubes using the transformer principle, *Medical and Biological Engineering and Computing* **27**, 357–364.
48. Ribreau, C., Merle, D. & Bonis, M. (1986) Determination expérimentale du module d'Young transversal d'une conduite élastique en dépression lors de son aplatissement, conditions d'application aux veines, *Journal de Biophysique et Biomécanique* **10**, 57–62.
49. Kamimura, T., Ohba, K. & Bando, K. (2000) Two-dimensional numerical simulation and experiment on large deformation of collapsible tube, *JSME International Journal Series C* **43**(4), 889–894.
50. Bertram, C. D. & Pedley, T. J. (1983) Steady and unsteady separation in an approximately two-dimensional indented channel, *Journal of Fluid Mechanics* **130**, 315–345.
51. Stephanoff, K. D., Pedley, T. J., Lawrence, C. J. & Secomb, T. W. (1983) Fluid flow along a channel with an asymmetric oscillating constriction, *Nature* **305**, 692–695.
52. Matsuzaki, Y., Ikeda, T., Matsumoto, T. & Kitagawa, T. (1998) Experiments on steady and oscillatory flows at moderate Reynolds numbers in a quasi-two-dimensional channel with a throat, *ASME Journal of Biomechanical Engineering* **120**, 594–601.
53. Ikeda, T., Matsuzaki, Y. & Sasaki, T. (1994) Separated flow in a channel with an oscillating constriction, *Transactions of the Japan Society of Mechanical Engineers B* **60**(571), 750–757.
54. Matsuzaki, Y., Watanabe, M., Aomatsu, T. & Ikeda, T. (2001) Experiment on flow in a two-dimensional channel with an obstruction oscillating in high frequency: preliminary study, in *Abstracts of the IUTAM Symposium on Flow in Collapsible Tubes and Past Other Highly Compliant Boundaries*, 26–30 March, Warwick, England.
55. Ikeda, T., Heil, M., Beaugendre, H. & Pedley, T. J. (1998) Experiments on flow in a two-dimensional collapsible channel, in *Abstracts of the Third World Congress of Biomechanics*, (eds. Matsuzaki, Y., Nakamura, T. & Tanaka, E.), Sapporo, Japan, pp. 39.
56. Webster, P. M., Sawatzky, R. P., Hoffstein, V., LeBlanc, R., Hinchey, M. J. & Sullivan, P. A. (1985) Wall motion in expiratory flow limitation: choke and flutter, *Journal of Applied Physiology* **59**, 1304–1312.
57. Walsh, C., Sullivan, P. A., Hansen, J. S. & Chen, L.-W. (1995) Measurement of wall deformation and flow limitation in a mechanical trachea, *ASME Journal of Biomechanical Engineering* **117**, 146–152.
58. Elliott, E. A. & Dawson, S. V. (1977) Test of wave-speed theory of flow limitation in elastic tubes, *Journal of Applied Physiology* **43**(3), 516–522.
59. Kounanis, K. & Mathioulakis, D. S. (1999) Experimental flow study within a self oscillating collapsible tube, *Journal of Fluids and Structures* **13**, 61–73.
60. Bertram, C. D. & Godbole, S. A. (1997) LDA measurements of velocities in a simulated collapsed tube, *ASME Journal of Biomechanical Engineering* **119**, 357–363.
61. Bertram, C. D., Muller, M., Ramus, F. & Nugent, A. H. (2001) Measurements of steady turbulent flow through a rigid simulated collapsed tube, *Medical and Biological Engineering and Computing* **39**(4), 422–427.
62. Bertram, C. D., Diaz de Tuesta, G. & Nugent, A. H. (2001) Laser Doppler measurements of velocities just downstream of a collapsible tube during flow-induced oscillations, *ASME Journal of Biomechanical Engineering* **123**(5), 493-499.

63. Hayashi, S., Hayase, T. & Kawamura, H. (1998) Numerical analysis for stability and self-excited oscillation in collapsible tube flow, *ASME Journal of Biomechanical Engineering* **120**, 468–475.
64. Ohba, K., Sakurai, A. & Oka, J. (1989) Self-excited oscillation of flow in collapsible tube. IV (Laser Doppler measurement of local flow field), *Technology Reports of Kansai University* **31**, 1–11.
65. Ohba, K., Sakurai, A. & Oka, J. (1997) Laser Doppler measurement of local flow field in collapsible tube during self-excited oscillation, *JSME International Journal Series C* **40**(4), 665–670.
66. Shimizu, M. (1985) Characteristics of pressure-wave propagation in a compliant tube with a fully collapsed segment, *Journal of Fluid Mechanics* **158**, 113–135.
67. Shimizu, M. & Tanida, Y. (1983) On the mechanism of Korotkoff sound generation at diastole, *Journal of Fluid Mechanics* **127**, 315–339.
68. Kamm, R. D. & Shapiro, A. H. (1979) Unsteady flow in a collapsible tube subjected to external pressure or body forces, *Journal of Fluid Mechanics* **95**, 1–78.
69. Yamane, T., Ikeda, T. & Orita, T. (1995) Suppression of self-excited oscillatory flow in collapsible tubes (effect of rod insertion) (in Japanese), *Transactions of the Japan Society of Mechanical Engineers B* **61**(582), 510–516.
70. Ohba, K., Kamimura, T., Bando, K. & Hanazono, K. (2001) Distribution of flow velocity and pressure in a largely deformed collapsible tube, in *Abstracts of the IUTAM Symposium on Flow in Collapsible Tubes and Past Other Highly Compliant Boundaries*, 26–30 March, University of Warwick, England.
71. Kamimura, T., Ohba, K., Bando, K. & Hanazono, K. (2001) Prediction of pressure distribution in highly deformed collapsible tube, *JSME International Journal* **44**(4), (in press).
72. Flaud, P., Geiger, D. & Oddou, C. (1982) Mechanical properties of collapsible tubes and propagation of large amplitude waves, in *Biomechanics: Principles and Applications*, eds. Huiskes, R., van Campen, D. H. & de Wijn, J. R. (Martinus Nijhoff, Den Haag), pp. 373-379.
73. Lyon, C. K., Scott, J. B. & Wang, C. Y. (1980) Flow through collapsible tubes at low Reynolds numbers. Applicability of the waterfall model, *Circulation Research* **47**, 68–73.
74. Heil, M. (1997) Stokes flow in collapsible tubes: computation and experiment, *Journal of Fluid Mechanics* **353**, 285–312.
75. Griffiths, D. J. (1971) Hydrodynamics of male micturition, I. Theory and steady flow through elastic-walled tubes, *Medical and Biological Engineering* **9**, 581–588.
76. Flaud, P., Geiger, D. & Oddou, C. (1986) High amplitude wave propagation in collapsible tubes. II. Forerunners and high amplitude waves, *Journal de Physique* **47**, 773–780.
77. Jan, D. L., Kamm, R. D. & Shapiro, A. H. (1983) Filling of partially collapsed compliant tubes, *ASME Journal of Biomechanical Engineering* **105**, 12–19.
78. Kamm, R. D., Elad, D., Jaekle, D. E., jr. & Shapiro, A. H. (1991) Theory and experiments on smooth transitions through the critical state (S = 1) in collapsible tube flow, in *1991 Advances in Bioengineering (ASME Winter Annual Meeting)*, (ed. Vanderby, R., jr.), Atlanta, Georgia, pp. 329–332.
79. Bonis, M. & Ribreau, C. (1981) Wave speed in noncircular collapsible ducts, *ASME Journal of Biomechanical Engineering* **103**, 27–31.
80. Bertram, C. D. & Raymond, C. J. (1991) Measurements of wave speed and compliance in a collapsible tube during self-excited oscillations: a test of the choking hypothesis, *Medical and Biological Engineering and Computing* **29**, 493–500.
81. Ohba, K., Aono, M., Yoneyama, N. & Sakurai, A. (1989) Flow oscillation and wave propagation in collapsible tube, in *Progress and New Directions of Biomechanics*, ed. Y.-C. Fung et al. (MITA Press), pp. 213–227.
82. Ohba, K., Yoneyama, N., Aono, M. & Sakurai, A. (1985) Self-excited oscillation of flow in collapsible tube. II (Analysis and experiment on initiation of oscillation), *Technology Reports of Kansai University* **26**, 1–6.
83. Yamane, T. & Orita, T. (1992) Self-excited oscillations with and without supercritical flow in collapsible tubes, in *7th International Conference on Biomedical Engineering*, 2–4 December, Singapore, pp. 502-504.
84. Bertram, C. D. & Elliott, N. S. J. (2001) Aqueous flow limitation in uniform collapsible tubes: multiple flow-limited flow-rates at the same pressure drop and upstream transmural pressure, in *Proceedings of the ASME 2001 Bioengineering Conference*, (eds. Kamm, R. D., Schmid-Schonbein, G. W., Ateshian, G. A. & Hefzy, M. S.), Snowbird, Utah, pp. 383–384.
85. Bonis, M. (1979) *Ecoulement visqueux permanent dans un tube collabable elliptique*. Unpublished Thèse de Doctorat d'Etat, Université de Technologie de Compiègne.
86. Fry, D. L., Thomas, L. J. & Greenfield, J. C. (1980) Flow in collapsible tubes, in *Basic Hemodynamics and its Role in Disease Processes*, eds. Patel, D. J. & Vaishnav, R. N. (University Park Press, Baltimore), pp. 407–424.

87. Gavriely, N., Shee, T. R., Cugell, D. W. & Grotberg, J. B. (1989) Flutter in flow-limited collapsible tubes: a mechanism for generation of wheezes, *Journal of Applied Physiology* **66**, 2251–2261.
88. Kamm, R. D., Patel, N. R. & Elad, D. (1993) On the effect of flow-induced flutter on flow rate during a forced vital capacity maneuver, *FASEB Journal* **7**(3), A11.
89. Low, H. T. & Chew, Y. T. (1991) Pressure/flow relationships in collapsible tubes: effects of upstream pressure fluctuations, *Medical and Biological Engineering and Computing* **29**, 217–221.
90. Low, H. T., Chew, Y. T., Winoto, S. H. & Chin, R. (1995) Pressure/flow behaviour in collapsible tube subjected to forced downstream pressure fluctuations, *Medical and Biological Engineering and Computing* **33**, 545–550.
91. Bertram, C. D. & Castles, R. J. (1999) Flow limitation in uniform thick-walled collapsible tubes, *Journal of Fluids and Structures* **13**, 399–418.
92. Bertram, C. D. & Elliott, N. S. J. (2001) Comparison of flow limitation in one tapered and two uniform collapsible tubes, in *Abstracts of the IUTAM Symposium on Flow in Collapsible Tubes and Past Other Highly Compliant Boundaries*, 26–30 March, Warwick, England.
93. Patel, N. R. (1993) *A study of flow limitation and flow-induced oscillations during airflow through collapsible tubing*. Unpublished B.S., Massachusetts Institute of Technology.
94. Gould, K. L. (1980) Dynamic coronary stenosis, *American Journal of Cardiology* **45**, 286–292.
95. Ku, D. N., Ziegler, M., Binns, R. L. & Stewart, M. T. (1990) A study of predicted and experimental wall collapse in models of highly stenotic arteries. In *Biofluid Mechanics – Blood Flow in Large Vessels* (2nd International Symposium on Biofluid Mechanics and Rheology, Munich, 1989), ed. Liepsch, D., pp. 409–415. Berlin: Springer-Verlag.
96. Binns, R. L. & Ku, D. N. (1989) Effect of stenosis on wall motion – a possible mechanism of stroke and transient ischemic attack, *Arteriosclerosis* **9**, 842–847.
97. Powell, B. E. & Ku, D. N. (1991) Contribution of flow choking to dynamic stenosis resistance, in *1991 Advances in Bioengineering (ASME Winter Annual Meeting)*, (ed. Vanderby, R., jr.), December 1–6, Atlanta, Georgia, pp. 337–340.
98. Stergiopulos, N., Moore, J. E., jr., Strässle, A., Ku, D. N. & Meister, J.-J. (1993) Steady flow tests and demonstration of collapse on models of compliant axisymmetric stenoses, in *ASME Winter Annual Meeting*, New Orleans, Louisiana.
99. Moore, J. E., jr., Stergiopulos, N., Golay, X., Ku, D. N. & Meister, J.-J. (1995) Flow measurements in collapsed stenotic arterial models, in *Proceedings of the 1995 ASME Bioengineering Conference*, (eds. Hochmuth, R. M., Langrana, N. A. & Hefzy, M. S.), June 28–July 2, Beaver Creek, Colorado, pp. 229–230.
100. Siebes, M., Campbell, C. S. & D'Argenio, D. Z. (1996) Fluid dynamics of a partially collapsible stenosis in a flow model of the coronary circulation, *ASME Journal of Biomechanical Engineering* **118**, 489–497.
101. Rodbard, S. (1955) Flow through collapsible tubes: augmented flow produced by resistance at the outlet, *Circulation* **6**, 280–287.
102. Schwartz, J. S., Carlyle, P. F. & Cohn, J. N. (1979) Effect of dilation of the distal coronary bed on flow and resistance in severely stenotic coronary arteries in the dog, *American Journal of Cardiology* **43**, 219–224.
103. Walker, J. (1987) The Amateur Scientist – Why a fluid flows faster when a tube is pinched, *Scientific American* **257**(1), 86–89.
104. Judd, R. M. & Mates, R. E. (1990) Pressure-flow relationships in partially-occluded flexible tubes. In *Biofluid Mechanics – Blood Flow in Large Vessels* (2nd International Symposium on Biofluid Mechanics and Biorheology, Munich, Germany, 1989), ed. Liepsch, D., pp. 417–423. Berlin: Springer-Verlag.
105. Bertram, C. D., Raymond, C. J. & Pedley, T. J. (1991) Application of dynamical system concepts to the analysis of self-excited oscillations of a collapsible tube conveying a flow, *Journal of Fluids and Structures* **5**, 391–426.
106. Bertram, C. D. & Butcher, K. S. A. (1992) A collapsible-tube oscillator is not readily enslaved to an external resonator, *Journal of Fluids and Structures* **6**, 163–180.
107. Elliott, N. S. J. (2000) *Aqueous flow-limitation in a uniform thin-walled collapsible tube.* Unpublished B.E., University of New South Wales.
108. Bertram, C. D. (1995) Dynamical system analyses of aperiodic flow-induced oscillations of a collapsible tube, *Journal de Physique III* **5**, 2101–2116.
109. Yahia, H. (1980) *Stabilité des écoulements dans les systèmes comportant un tube collabable.* Unpublished Thèse de Troisième Cycle, Université de Technologie de Compiègne.

110. Conrad, W. A., Cohen, M. L. & McQueen, D. M. (1978) Note on the oscillations of collapsible tubes, *Medical and Biological Engineering and Computing* **76**, 211–214.
111. Yamane, T. & Orita, T. (1994) Hysteresis and multiplicity of collapsible tube flow (in Japanese), *Transactions of the Japan Society of Mechanical Engineers B* **60**(571), 807–812.
112. Yamane, T. & Orita, T. (1990) Stability of the equilibrium for collapsible tube flows, in *JSME Bioengineering Division Conference*, Nagaoka.
113. Bertram, C. D., Raymond, C. J. & Pedley, T. J. (1990) Mapping of instabilities for flow through collapsed tubes of differing length, *Journal of Fluids and Structures* **4**, 125–153.
114. Bertram, C. D. & Sheppeard, M. D. (2000) Interactions of pulsatile upstream forcing with flow-induced oscillations of a collapsed tube: mode-locking, *Medical Engineering and Physics* **22**, 29–37.
115. Bertram, C. D. & She, J. (2000) Chaotic and mode-locked interactions between flow-induced collapsible-tube oscillation and pulsatile upstream forcing. In *Stochastic and Chaotic Dynamics in the Lakes* (Stochaos, Ambleside, England, 1999), eds. Broomhead, D. S., Luchinskaya, E. A., McClintock, P. V. E. & Mullin, T., pp. 510–516. Melville, NY: American Institute of Physics.
116. Kenner, T. (1976) Pulse-wave reflection at the collapsed segment of an artery in Riva-Rocci's method, *Pflügers Archiv* **364**, 285–289.
117. Conrad, W. A., McQueen, D. M. & Yellin, E. L. (1980) Steady pressure flow relations in compressed arteries: possible origin of Korotkoff sounds, *Medical and Biological Engineering and Computing* **18**, 419–426.
118. Mahrenholtz, O. (1974) Zur Pumpwirkung kollabierfähiger ventilloser Schläuche, *Ingenieur-Archiv* **43**, 173–182.
119. Lichtenstein, O. & Dinnar, U. (1990) Experimental analysis of pulsatile flow through elastic collapsible tubes: application to cardiac assist device, *ASME Journal of Biomechanical Engineering* **112**, 75–79.
120. Klapproth, P., Guldner, N. W. & Sievers, H. H. (1997) Stroke volume validation and energy evaluation for the dynamic training of skeletal muscle ventricles, *International Journal of Artificial Organs* **20**, 580–588.
121. Volkart, P. U. & de Vries, F. (1985) Automatic throttle hose—new flow regulator, *ASCE Journal of Irrigation and Drainage Engineering* **111**(3), 247–264.
122. Sood, S., Canady, A. I. & Ham, S. D. (1999) Adjustable antisiphon shunt, *Child's Nervous System* **15**, 246–249.
123. Schoendorfer, D. W. & Shapiro, A. H. (1977) The collapsible tube as a prosthetic vocal source, in *Proceedings of the San Diego Biomedical Symposium*, pp. 349–356.
124. Bertram, C. D., Hoogland, M. R., Li, H., Odell, R. A. & Fane, A. G. (1993) Flux enhancement in crossflow microfiltration using a collapsible-tube pulsation generator, *Journal of Membrane Science* **84**, 279–292.
125. Hadzismajlovic, D. E. & Bertram, C. D. (1996) A collapsible-tube pulsation generator for crossflow microfiltration: fatigue testing of silicone rubber tubes, *Journal of Applied Polymer Science* **61**, 703–713.
126. Hadzismajlovic, D. E. & Bertram, C. D. (1998) Flux enhancement in laminar crossflow microfiltration using a collapsible-tube pulsation generator, *Journal of Membrane Science* **142**, 173–189.
127. Li, H., Bertram, C. D. & Wiley, D. E. (1998) The mechanisms by which pulsatile flow affects crossflow microfiltration, *AIChE Journal* **44**, 1950–1961.
128. Hadzismajlovic, D. E. & Bertram, C. D. (1999) Flux enhancement in turbulent crossflow microfiltration of yeast using a collapsible-tube pulsation generator, *Journal of Membrane Science* **163**, 123–134.

SECTION B

INSTABILITY OF FLOW PAST COMPLIANT WALLS

Chapter 4

CONVECTIVE AND ABSOLUTE INSTABILITIES OF FLOW OVER COMPLIANT WALLS

CHRISTOPHER DAVIES
School of Mathematics, Cardiff University,
Senghennydd Road, PO Box 926, Cardiff CF24 4YH,
United Kingdom

Abstract

An introductory account is given of the absolute and convective instabilities that can occur in flow over a compliant wall. The model problem of flow in a plane channel between compliant walls is considered in order to provide a relatively simple illustration of the various physical mechanisms that can lead to instability. Estimates for the onset flow-velocities of surface-based instabilities are discussed in connection with the design of compliant walls for boundary-layer transition delay. It is shown that the use of a long wavelength approximation can give misleading predictions. Numerical simulation results are presented for the development of an absolute instability in Blasius boundary-layer flow over a compliant wall. Simulation results are also used to illustrate the suppression of absolute instability by a compliant surface, for the case of the three-dimensional von Karman boundary-layer flow over a rotating disk.

P.W. Carpenter and T.J. Pedley (eds.),
Flow past Highly Compliant Boundaries and in Collapsible Tubes, 69-93.

1. Introduction

Much of the challenge and interest of studying flow over a compliant wall stems from the fact that, because both the fluid and the compliant wall are wave-bearing media, there is a rich variety of possible modes of motion. A rough classification can be made by distinguishing between waves that would persist in the fluid in the absence of any wall motion and waves that would propagate along the compliant wall if there was no fluid flow. Thus it is possible to make reference to fluid-based and surface-based modes, respectively. However, the adoption of any such terminology can be somewhat artificial when, as is usual in cases of interest, there is strong interactive coupling between the wall motion and the fluid motion. The fluid-based character of waves that propagate in a flow over a rigid bounding surface can change, sometimes in a dramatic fashion, when the bounding surface is allowed to be compliant. Likewise, surface-based free waves that would propagate in a stable fashion along an isolated compliant wall can undergo significant alteration when a fluid flow is introduced.

A variety of distinct physical mechanisms can come into play to destabilize systems involving interactively coupled compliant wall and fluid motion if the compliant wall is made sufficiently flexible or the flow speed is taken to be large enough. Instabilities that arise from surface-based modes have been termed flow-induced surface instabilities by previous investigators. They include the so-called static divergence mode of instability, which can arise in a uniform potential flow, and travelling wave flutter instability, which can only exist when there is a shear in the flow. The introduction of a carefully selected degree of wall compliance can also lead to the stabilization, rather than the destabilization, of fluid-based modes of disturbance such as Tollmien-Schlichting waves. Thus, there are potential benefits to be reaped for flow control in engineering applications. In particular, it may be possible to obtain a delay in the onset of laminar-turbulent transition in a boundary-layer flow over a solid body, by simply exchanging a rigid bounding surface for a compliant one with well-chosen material properties.

Modal interactions amongst surface-based waves and between surface-based waves and fluid-based waves can give rise to relatively strong forms of instability. Typically, modal interactions lead to *absolute* instabilities, which are associated with temporally growing disturbances that may disrupt the basic state of the flow, by spreading both upstream and downstream from their point of excitation. By contrast, *convective* instabilities, such as travelling wave flutter, are usually less immediate in their effect, since they give rise to disturbances that only exhibit their growth spatially, as they convect away from the location where they were excited.

In the present paper we will not attempt to give a fully-comprehensive review of all the specific forms that instability can take when a fluid flows over a compliant surface. Instead we will try to illustrate some general features of the complex interactive motions that are possible, by concentrating our attention on parallel, or quasi-parallel, flows with a planar geometry. More detailed accounts of the instabilities that are found for other geometric configurations, such as pipe flow, are presented elsewhere in this book. We will also restrict our account of instabilities to two-dimensional forms of disturbance, except where we give a brief mention of some results for the three-dimensional flow over a rotating compliant disk.

To provide some focus for our discussion, we will address a number of interrelated problems that arise in optimizing the design of compliant surfaces for the purpose of laminar-turbulent transition delay. Compliant surfaces can only be used to delay transition in Blasius flow and other similar boundary-layer flows if the surface is flexible enough to inhibit the growth of fluid-based Tollmien-Schlichting waves, but not so soft that waves based in the compliant wall itself become unstable. If flow-induced surface instabilities are introduced, then they may usurp the role of Tollmien-Schlichting waves in initiating the transition to turbulence (Gaster 1987, Lucey & Carpenter 1995). Thus the practical design of compliant surfaces involves a process of optimization. It will be seen that this optimization turns out to be rather delicate and far from trivial.

We will also describe some results that were obtained using a direct numerical simulation method that was originally developed to investigate one particular facet of the optimization (Davies & Carpenter 1997a). However, the applications of the simulation method are by no means restricted to studies of the suppression of Tollmien-Schlichting waves. We shall report numerical simulations where there was evidence of absolute instability arising from a modal coalescence. The instability behaviour that was found in such simulations is likely to be relevant to physiological flow configurations, since it can persist down to low values of the Reynolds number, as well as being present at the relatively high Reynolds numbers that are more usually associated with laminar-turbulent transition.

2. Optimization of compliant surfaces for transition delay

As a first step in the optimization of compliant walls for the purpose of transition delay, it is necessary to obtain good estimates for the critical flow speeds associated with the onset of flow-induced surface instabilities. These estimates can be used to select values of the compliant-wall material properties that are viable for obtaining a reduction in Tollmien-Schlichting wave spatial growth rates, without any concomitant destabilization of surface-based waves. When the permissible compliancy properties have been thus delimited, a second selection can be made to characterize the compliant surfaces that give maximal stabilization of Tollmien-Schlichting waves.

Previous investigators have used potential flow theory and inviscid shear flow theory, respectively, to obtain estimates for the onset of the flow-induced surface instabilities known as divergence and travelling wave flutter. Any attempt to account for travelling wave flutter must incorporate the shear in the basic flow, because, as will be discussed more fully below, the instability mechanism involves the effects of a critical layer. However, divergence is destabilized by a mechanism that can operate in a potential flow, so it is possible to derive estimates for the onset of divergence without taking any account of shear flow effects. We shall show that, in fact, the neglect of the shear in the basic flow, and the associated lengthscale that this introduces, can lead to very poor and excessively conservative predictions for the onset of divergence instability. Moreover, potential theory yields its worst predictions for the onset of divergence precisely when the circumstances are such that a strong stabilization of Tollmien-Schlichting waves can be anticipated. The upshot of all this is that one of the selection constraints placed on the compliant-wall material properties can be relaxed, quite significantly, when shear flow effects are properly accounted for. Viable compliant surfaces can thus be made much softer than had previously been thought possible. In fact, it turns out to be feasible to design compliant surfaces so that, within localized streamwise regions at least, the Tollmien-Schlichting mode of instability is completely suppressed.

3. Onset of flow-induced surface instabilities

The essential features of the onset of flow-induced surface instabilities in flows over compliant surfaces can be identified by considering the somewhat idealized problem of symmetric disturbances in a plane channel flow between compliant walls. (A more detailed account of this problem can be found in Davies & Carpenter 1997b.) Because the basic flow takes a particularly simple parabolic form in this instance, it is possible to derive analytic results for critical flow velocities using inviscid shear layer theory without incurring too much labour. Similar results can be obtained, but generally in a less straightforward manner, for other flows involving planar geometries, such as Blasius flow over a compliant plate. (See, for example, Carpenter & Garrad 1986, Carpenter & Gajjar 1990.) We will also limit our attention to an elementary spring-backed plate model of the compliant surface. The various forms of instability that can be distinguished using such a simplified wall model provide a good starting point for categorizing the behaviour that is found when less tractable, but more realistic, wall models are employed.

Using the channel half-width as the reference lengthscale and the centre-line flow velocity to non-dimensionalize the velocity, the parabolic profile of plane-channel flow is given by $U = 1 - y^2$, where the undisturbed channel walls are located at $y = \pm 1$. Inviscid normal mode perturbations from this base-flow are governed by Rayleigh's equation,

$$(\alpha U - \omega)(D^2 - \alpha^2)\phi - \alpha U''\phi = 0, \tag{1}$$

where α is the wavenumber, ω is the temporal frequency and $\phi(y)$ is the profile of the disturbance streamfunction. We only consider perturbations that have the same parity as Tollmien-Schlichting waves upon reflection about the channel centre, *i.e.* the symmetry condition $\phi(y) = \phi(-y)$ is imposed. The behaviour of modes with the opposite parity has been considered, for example, by Larose & Grotberg (1997) and Huang (1998).

The spring-backed plate model of the compliant walls is specified by the following non-dimensional parameters: m the plate mass per unit area, B the flexural rigidity of the plate and K the spring stiffness. The phase speed c_0 of the free waves that would propagate along the compliant wall in the absence of any fluid is given in terms of these parameters by the expression

$$c_0^2 = \frac{1}{m}\left(B\alpha^2 + \frac{K}{\alpha^2}\right). \tag{2}$$

Defining a characteristic wavenumber and frequency by setting

$$\alpha_c = \left(\frac{K}{B}\right)^{1/4}, \quad \omega_c = \left(\frac{2K}{m}\right)^{1/2}, \tag{3}$$

the formula for the free wave speed can be rewritten as

$$c_0^2 = \frac{1}{2}\left(\frac{\omega_c}{\alpha_c}\right)^2 \left(\left(\frac{\alpha}{\alpha_c}\right)^2 + \left(\frac{\alpha_c}{\alpha}\right)^2\right), \tag{4}$$

from which it is readily seen that there is a minimum free wave speed

$$c_m = \frac{\omega_c}{\alpha_c}. \tag{5}$$

which occurs at the wavenumber α_c and frequency ω_c.

When wall damping is taken into account, temporally decaying free waves can propagate along the compliant walls with a complex phase speed $c = \omega/\alpha$ that satisfies the relation

$$m(c^2 - c_0^2) + i\frac{c}{\alpha}\,d = 0, \tag{6}$$

where d is the wall damping coefficient. For low levels of wall damping this has the approximate solutions

$$c = \pm c_0 - i\frac{d}{2m\alpha}, \tag{7}$$

which illustrates the fact that, in the absence of any fluid, there is a symmetry between the two possible directions of wave propagation. We shall see that this symmetry is broken when the plane channel fluid flow $U = 1 - y^2$ is introduced, since a definite distinction can then be made between the upstream and downstream directions.

Using the information obtained by matching the normal velocity of the compliant channel walls to that of the fluid, Rayleigh's equation can be solved for the disturbance in the fluid, and an expression derived for the perturbation pressure exerted by the fluid upon the walls. Substituting this expression for the wall pressure into the equation of motion for the compliant walls modifies the free wave relation (6) to yield a dispersion relation that can be used to determine the phase velocities of disturbance waves in the coupled wall and fluid system. This dispersion relation takes the general form

$$m(c^2 - c_0^2) + i\frac{c}{\alpha}\,d + \mathcal{P}(\alpha, c) = 0, \tag{8}$$

where $\alpha^2\mathcal{P}(\alpha, c)$ is the ratio between the perturbation wall pressure and the wall displacement. The function $\mathcal{P}(\alpha, c)$, which is obtained from the solution of Rayleigh's equation, is given by

$$\mathcal{P}(\alpha, c) = \frac{\int_0^1 (U-c)^2\,dy + \alpha^2 \int_0^1 (U-c)^2\,dy \int_0^y (U-c)^{-2}\,dy \int_0^y (U-c)^2\,dy + \mathcal{O}(\alpha^4)}{1 + \alpha^2 \int_0^1 (U-c)^{-2}\,dy \int_0^y (U-c)^2\,dy + \mathcal{O}(\alpha^4)}. \tag{9}$$

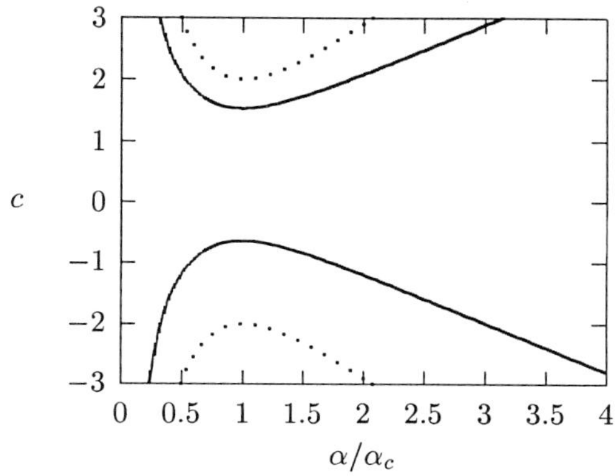

Figure 1. Phase velocities for disturbances in plane channel flow between compliant walls. $m = 0.5$, $c_m = 2$, $d = 0$. Dotted lines show the symmetrically disposed free wave solution branches. Solid lines show the asymmetric upstream and downstream solution branches in the presence of the fluid flow.

For long wavelength disturbances, this can be approximated by taking

$$\mathcal{P}(\alpha, c) \simeq \int_0^1 (U - c)^2 \, dy = c^2 - \frac{4}{3} c + \frac{8}{15}. \tag{10}$$

Substitution of the above approximation into the dispersion relation (8) gives a quadratic equation for the phase velocity c, which is readily solved. Provided that the minimum free wave speed c_m is not taken to be too small, the effect of the fluid flow is simply to break the symmetry between upstream and downstream propagating waves and to reduce their phase speeds. This is illustrated in Figure 1, which shows neutrally stable solutions of the dispersion relation (8) obtained using the longwave approximation (10) for a case where there is no wall damping.

The effects of taking physically softer compliant walls or increasing the dimensional speed of the mean flow can both be investigated by decreasing the value of the non-dimensional parameter c_m. As c_m is decreased, there are three significant staging posts at which a change in behaviour can be discerned. Firstly, if $c_m^2 \leq 8/(15m)$ then downstream propagating waves are found along part of the solution branch that originally consisted of only upstream propagating waves. This is illustrated in Figure 2. Secondly, if $c_m^2 \leq 1+1/(5m)$ then there is a portion of the downstream solution branch for which $0 < c \leq 1$, and so critical layers can come into play in the fluid at wall-normal locations $y = y_c$ where $U(y_c) = c$. It may be noted that, at least according to the longwave approximation, which of the above two possibilities occurs first, as c_m is decreased, depends only on the magnitude of the wall mass parameter m. Figure 2 shows the solution branches for a case where c_m and m have been chosen so that there are ranges of the wavenumber for which there are solutions with a critical layer, as well as solutions that have a

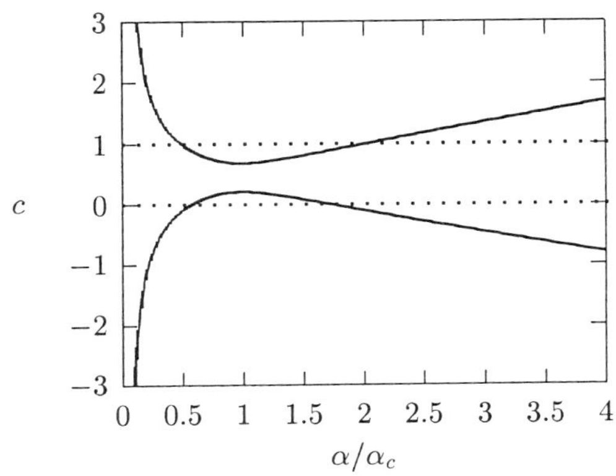

Figure 2. Phase velocities for disturbances in plane channel flow between compliant walls. $m = 0.5$, $c_m = 0.8$, $d = 0$. Portions of the solution branches that lie between the dotted lines at $c = 0$ and $c = 1$ are associated with waves that involve either a reversed direction of propagation (lower branch intersecting $c = 0$) or a critical layer (upper branch intersecting $c = 1$).

reversed direction of propagation. Finally, if c_m is decreased still further then the two solution branches can coalesce for a range of wavenumbers, so that solutions of the dispersion relation appear in the form of complex conjugate pairs.

Having identified three different changes that can occur in the behaviour of the solution branches as the quantity c_m is reduced, we will now describe how each of these changes is associated with the onset of a distinct mode of flow-induced surface instability. We first consider the change, Δc, in the phase velocity that occurs when a small amount of damping is incorporated in the wall model. In practice, of course, some measure of damping must be present in any model of a real compliant wall. Using the longwave approximation in the dispersion relation it is simple to show that

$$\Delta c \simeq -i\,\frac{d}{2\alpha m}\,\frac{c}{c_0}\left(\frac{dc}{d\alpha}\Big/\frac{dc_0}{d\alpha}\right). \tag{11}$$

It is also straightforward to confirm that the ratio $\frac{dc}{d\alpha}/\frac{dc_0}{d\alpha}$ is positive along the upper of the solution branches shown in each of Figures 1 and 2, but always negative along the lower solution branch. Thus it can be seen from (11) that wall damping is stabilizing for all solutions of the dispersion relation, unless the value of c_m is taken to be small enough so that downstream propagating waves emerge in the lower solution branch. Wall damping destabilizes such waves. The instability that arises in this way is usually referred to as static divergence because, at its onset, it is associated with waves that have a vanishing phase velocity. We shall return, in due course, to the issue of whether the approximate analysis outlined above gives an adequate account of the onset of divergence instability, particularly in circumstances where a stabilization of fluid-based waves is desired. For now, we just note that the non-dimensional version of the condition for the onset of divergence, $c_m^2 \leq 8/(15m)$, gives a condition on the dimensional centre-line velocity U_c of the fluid that can be written in the form $U_c \geq U_d$, where the critical velocity U_d is given by

$$U_d = \left(\frac{15}{8}\right)^{1/2} \left(\frac{2(B^* K^*)^{1/2}}{\rho h}\right)^{1/2} . \tag{12}$$

Here B^*, K^* denote the dimensional compliant-wall bending stiffness and spring stiffness parameters, ρ is the fluid density and h is the dimensional channel half-width. It may be observed that the critical velocity is independent of the wall mass. The divergence instability can thus be thought of as arising when the restoring structural forces of the compliant wall are overwhelmed by pressure forces exerted by the fluid. In the above account we showed that wall damping could facilitate such an imbalance of forces. It may be demonstrated, though, that essentially the same instability mechanism can be triggered by various other means, for example by the interplay of end effects for the case of finite-length panels (Lucey & Carpenter 1993).

We now turn to consider the stability effects of critical layers in the fluid. As was noted earlier, critical layers appear when the value of c_m is taken to be small enough for there to be waves along the upper solution branch that have phase velocities lying in the interval $0 < c \leq 1$. For the case of symmetric disturbances that we are considering for plane channel flow, it may be shown that critical layers contribute an $\mathcal{O}(\alpha^2)$ imaginary part to the quantity $\mathcal{P}(\alpha, c)$ that measures the ratio of the wall pressure to the wall displacement in the dispersion relation (8). Thus, retaining the long wavelength approximation for the real part of $\mathcal{P}(\alpha, c)$, we can incorporate critical layer effects by setting

$$\mathcal{P}(\alpha, c) \simeq \int_0^1 (U - c)^2 \, dy + i\alpha^2 f(c) , \tag{13}$$

where it may be shown, by a more detailed analysis than is included herein, that $f(c)$ is negative for all phase velocities $0 < c < 1$, but zero outside of this range (Davies & Carpenter 1997b). It is a simple matter to determine the change in the solutions of the dispersion relation that is produced when the longwave approximation is amended in such a fashion. The phase velocities are shifted by an amount Δc given by

$$\Delta c \simeq -i \frac{\alpha^2}{2m} \frac{f(c)}{c_0} \left(\frac{dc}{d\alpha} \Big/ \frac{dc_0}{d\alpha} \right) . \tag{14}$$

Recalling that the ratio $\frac{dc}{d\alpha} / \frac{dc_0}{d\alpha}$ is positive along the upper branch of solutions, it may be seen that any waves along the upper solution branch that have phase velocities in the range $0 < c < 1$ will be destabilized. The instability that arises in this way from critical layer effects is known as travelling wave flutter. The critical layer introduces a phase shift between the fluid perturbation pressure at the wall and the normal wall velocity that facilitates an irreversible energy transfer from the fluid to the wall. This instability mechanism was first identified by Miles in his attempts to model the generation of surface waves on water by the wind. Further details, and references dating back to the late 1950's, can be found in a recent review paper by Miles (1997).

In the same manner as for the divergence instability, the non-dimensional version of the onset condition for travelling wave flutter, $c_m^2 \leq 1 + 1/(5m)$, can be converted to a condition on the centre-line velocity of the fluid that takes the form $U_c \geq U_t$. The critical velocity U_t is given by

$$U_t = \left(\frac{2(B^* K^*)^{1/2}}{m^* + \rho h/5} \right)^{1/2}, \tag{15}$$

where m^* denotes the dimensional mass per unit area of the compliant walls. It is interesting to note that the parabolic profile of the base flow adds an effective fluid mass $\rho h/5$ to the wall mass m^*. Thus travelling wave flutter instability can persist even when the compliant wall mass is negligible. In fact, the derivation of the formula for the critical flow velocity may readily be amended to show that travelling wave flutter can still occur for the case of compliant walls that are modelled as massless tensioned membranes, rather than spring-backed plates. Altering the wall model simply leads to the replacement of the stiffness quantity $2(B^* K^*)^{1/2}$ by the longitudinal tension T^* in the expression for the critical velocity U_t.

If we compare the implications of the two results stated in (11) and (14) for the changes Δc in the complex phase velocity that are induced by wall damping and critical layers, respectively, it may be surmised that opposing stability effects are found for divergence and travelling wave flutter. Divergence is destabilized by wall damping but stabilized by critical layer effects, whereas travelling wave flutter is stabilized by wall damping but destabilized by the action of critical layers. In a similar fashion, it can be shown that incorporating viscous wall layers, to meet the no-slip condition for disturbances in the fluid, gives rise to opposing effects on the two modes of instability. Divergence is predicted to be destabilized by viscous wall layers, while travelling wave flutter is stabilized. In fact, the latter stabilizing effect can lead to a significant delay in the onset of travelling wave flutter instability, at least for our case of symmetric disturbances in plane channel flow. However, the overall influence of viscous wall layers on the onset of divergence turns out to be more difficult to discern, because the phase velocity must vanish. As the phase velocity approaches zero, there is a singularity in the simplest, but otherwise tractable, approximation that can be employed to modify the inviscid dispersion relation (8) to account for viscous wall layers. The same difficulty for vanishing phase velocities can be traced in the approximate analytic theory that was developed for external boundary-layer flows flow by Carpenter & Gajjar (1990).

The opposing effects on the stability of the divergence and travelling wave flutter modes due to wall damping, critical layers and viscous wall layers can be classified using a scheme that was first introduced and applied by Landahl (1962) and Benjamin (1963)[1]. Class A waves are destabilized (stabilized) by irreversible

[1] The same system of energy classification was developed, independently, to classify insta-

energy transfer out of (into) a system. Class B waves are more conventional in that they are stabilized (destabilized) by irreversible energy transfer out of (into) a system. Thus, according to the results that we have described so far, divergence is a Class A instability. Benjamin (1960) demonstrated that fluid-based Tollmien-Schlichting waves were destabilized by compliant wall damping, so they can also be categorized as Class A. Travelling wave flutter is a Class B instability, as may be discerned from the fact that compliant wall damping acts in its more usual manner to stabilize travelling wave flutter, and also from the fact that travelling wave flutter is destabilized when irreversible transfers of energy from the fluid to the compliant wall are set up by critical layers.

A third possible form of instability occurs when a mode of Class A waves coalesces with a mode of Class B waves. The resultant wave mode, described as being of Class C, is relatively indifferent to irreversible energy transfers. Instead, the instability arises from a process that involves conservative energy exchange, somewhat akin to the mechanism that is associated with the Kelvin-Helmholtz instability. We have already noted that, for disturbances in plane channel flow between compliant walls, the two solution branches that we identified could coalesce over a range of wavenumbers if the non-dimensional wall compliance parameter c_m was made sufficiently small. The relatively strong form of instability that can then arise is a Class C instability known as flutter. It is manifested by the appearance of complex conjugate pairs of solutions for the phase velocity.

Up to this point we have, for the purposes of exposition, concentrated on the behaviour of solutions to the inviscid dispersion relation (8) that can be obtained by applying a long wavelength approximation in conjunction with simple modifications to account for wall damping, critical layers and viscous wall layers. However, such an approach necessarily overlooks an important complication that will always arise unless the characteristic wavenumber of the compliant wall α_c is taken to be *very* small. Namely, the existence of fluid-based Tollmien-Schlicting waves with wavenumbers comparable to α_c. Naively, one might expect that, since Tollmien-Schlichting waves in flows over rigid surfaces are only destabilized by the action of viscosity, it might be possible to leave them out of any account of flow-induced surface instabilities that is based on inviscid shear layer theory. This supposition turns out be incorrect, as we shall now attempt to explain.

4. Disruption of the onset of divergence

The critical velocity formulae for the onset of surface-based instabilities that we noted in the previous section can be employed to constrain the choice of compliant wall properties for applications to transition delay. (Similar, but not identical, formulae can be obtained for Blasius flow over a compliant plate. These are stated later.) The aim is to maximize the flexibility of the compliant walls, with a view to stabilizing fluid-based Tollmien-Schlichting waves, without any consequent

---

bilities found in plasma physics (Briggs 1964). However, a more perspicuous terminology was adopted. Class A waves were described as being negative energy waves and Class B waves described as positive energy waves.

destabilization of surface-based waves. Thus, for a given flow-speed U_c we select the compliancy properties to make the wall as soft as possible by requiring that, nominally at least, there is marginal stability for both divergence and travelling wave flutter, *i.e.* we set $U_c = U_d = U_t$. If wall damping is ignored, these conditions determine a one parameter family of spring-backed-plate compliant-wall models that can be specified, most conveniently, by using the wavenumber α_c as the parameter. It should be recalled that α_c is the wavenumber at which the phase speed of surface-based free waves is minimized, and so defines a characteristic lengthscale for the compliant wall.

We would only expect a compliant wall to have a significant stability effect on fluid-based Tollmien-Schlichting waves if the characteristic lengthscale of the wall is matched to that of Tollmien-Schlichting waves. So for any compliant wall that is designed to delay laminar-turbulent transition, it will be necessary to have $\alpha_c \sim \alpha_{ts}$, where α_{ts} is a typical wavenumber for unstable Tollmien-Schlichting waves. For compliant walls with such a lengthscale matching, the approximations to the dispersion relation that we have previously described can yield very misleading results. It turns out that, in addition to the two solution branches that we have so far identified, a third solution branch must be considered in order to obtain accurate predictions. This branch of solutions corresponds to Tollmien-Schlichting waves. It can only be distinguished using an approximated dispersion relation if we replace the long wavelength approximation for the quantity $\mathcal{P}(\alpha, c)$, as stated in (10), by a more complicated Padé type of approximation

$$\mathcal{P}(\alpha, c) \simeq \frac{\int_0^1 (U-c)^2\,dy + \alpha^2 \int_0^1 (U-c)^2\,dy \int_0^y (U-c)^{-2}\,dy \int_0^y (U-c)^2\,dy}{1+\alpha^2 \int_0^1 (U-c)^{-2}\,dy \int_0^y (U-c)^2\,dy}\,. \tag{16}$$

This preserves a vestige of the fact that the quantity $\mathcal{P}(\alpha, c)$ is essentially the ratio of the fluid perturbation pressure at the wall to the wall displacement. If viscosity is fully accounted for, we know that the corresponding ratio would have singularities at values of α and c that represent Tollmien-Schlichting waves propagating in a fluid that is bounded by a rigid surface. For such waves the wall displacement vanishes, but the wall pressure does not. In the inviscid case, we can retain some trace of the singularities, and hence recover the Tollmien-Schlichting branch of solutions, by using the approximation (16) for $\mathcal{P}(\alpha, c)$ whenever $\alpha \sim \alpha_{ts}$. Thus, if the compliant wall is chosen so that $\alpha_c \sim \alpha_{ts}$, it is perhaps not too surprising to find that it is necessary to take account of the Tollmien-Schlichting branch of solutions, even when solving the inviscid dispersion relation to study the behaviour of surface-based waves.

If the approximation (16) is used instead of (10) then two major modifications are discovered. Firstly, the lower branch of solutions that was previously associated with static divergence instability is no longer able to cross the wavenumber axis at $c = 0$. There can be no reversal in the direction of propagation for the solutions along the lower branch. This means that, according to the inviscid dispersion relation with the approximation (16), there is no possibility of divergence instability. Secondly, there is no longer any modal coalescence between the upper and lower solution branches of surface-based waves that produces flutter instability.

This is replaced by a coalescence between the upper branch and the third branch of solutions comprised of Tollmien-Schlichting waves. The instability that results from this new form of modal coalescence was first identified by Sen & Arora (1988) who called it transitional. It only occurs after travelling wave flutter instability has set in. This is because the phase velocities along the solution branch of Tollmien-Schlichting waves remain smaller than the threshold $c = 1$, below which there are critical layers. When the compliant wall is taken to be so soft that there are phase velocities along the upper branch that are small enough for there to be a modal coalescence with Tollmien-Schlichting waves, the upper branch of surface-based waves must have already been destabilized by critical layer effects, giving rise to travelling wave flutter.

The brief account of the modifications to the onset of flow-induced surface instabilities that we have just given is drawn from an analysis that was conducted for the specific case of plane channel flow between compliant walls. Further details, including graphical illustrations of all the solution branches, can be found in the study by Davies & Carpenter (1997b). Rather than simply replicating this earlier work, we shall now show that a very similar pattern of behaviour is found for the case of Blasius flow over a compliant wall, which is more obviously relevant to laminar-turbulent transition delay and drag reduction applications.

For external boundary layers, potential flow and inviscid shear flow theory can be used to identify, respectively, the onset of divergence and the onset of travelling wave flutter (Carpenter & Garrad 1986, Carpenter & Gajjar 1990). Simple formulae for the critical velocities can be derived by following the same kind of procedure as we have described for the channel flow, *i.e.* by employing a long wavelength approximation in which no account is taken of the Tollmien-Schlichting branch of solutions. The predicted critical velocity for the onset of divergence is given by the expression

$$U_d^B = 2\left(\frac{B^* K^{*3}}{27\rho^4}\right)^{1/8}, \tag{17}$$

while the critical velocity for travelling wave flutter takes the form

$$U_t^B = U_t = \left(\frac{2(B^* K^*)^{1/2}}{m^* + \rho(1 - 1/H)\delta^*}\right)^{1/2}, \tag{18}$$

where H is the ratio of displacement to momentum thickness (=2.591 for Blasius flow) and δ^* is the displacement thickness. When the characteristic lengthscales of the compliant wall and the fluid are matched for an external boundary-layer flow, so that $\alpha_c \sim \alpha_{ts}$, the onset of divergence instability is again disrupted due to the presence of the Tollmien-Schlichting branch of solutions. We illustrate this for Blasius flow, by making a comparison between some numerical results that were obtained by solving the Orr-Sommerfeld equation and the solutions derived from Rayleigh's equation using a long wavelength approximation. The use of the Orr-Sommerfeld equation enables the full effects of viscosity to be incorporated, but this is less important, from the point of view studying of the onset of divergence, than the fact that the numerical solutions are obtained without any restriction being placed on wavenumber.

Figure 3(a) displays solution branches that were determined analytically, in the same kind of manner as was presented by Carpenter & Gajjar (1990), by using inviscid shear layer theory in conjunction with a long wavelength approximation. This means that no direct account has been taken of the Tollmien-Schlichting branch of solutions. It can be seen that there are two branches of surface-based waves and that the lower of these branches contains a range of wavenumbers for which the direction of propagation has been reversed. The flow velocity has been chosen in excess of the predicted critical velocity for divergence so that $U_d^B = 0.7\,U_\infty$, where U_∞ is the freestream velocity. We have also set $U_t^B = U_\infty$ to obtain marginal stability with respect to travelling wave flutter. Because the phase velocity threshold at $c = 0$ has been crossed by the lower branch of solutions, the inclusion of a small amount of wall damping leads to instability. Figure 3(b) shows how the behaviour is modified when, more correctly, the branch of solutions corresponding to Tollmien-Schlichting waves is interposed between the two branches of surface-based waves. The amended solution branches were obtained by solving the Orr-Sommerfeld equation, numerically, without making any assumption that the wavenumber is small. It can be seen that there is no longer any divergence instability, since the modified lower branch of surface-based waves has failed to reach the wavenumber axis. By contrast, it may be noted that the upper solution branch of surface-based waves has barely changed at all. Thus the simple formulae (18) can still be relied upon to predict the critical velocity for the onset of travelling wave flutter instability.

It is possible to give a more physically direct explanation of the disruption of the onset of divergence instability than we have done so far in ascribing it, rather formally, to the consequences of amending the inviscid dispersion relation to take account of Tollmien-Schlichting waves when $\alpha_c \sim \alpha_{ts}$. As noted previously, divergence can be viewed as an instability that arises when pressure forces from the fluid overwhelm the structural restoring forces of the compliant wall. As such it can exist in a purely potential flow. It can be shown that the boundary layer has the effect of diminishing the potential flow pressure forces that are exerted at the wall, thus delaying the onset of divergence. This sheltering effect on the wall pressure due to the boundary layer has been identified, for instance, in the results of inviscid shear layer numerical simulations that were conducted, using a discrete vortex method, by Lucey, Cafolla & Carpenter (1997). The effect only becomes significant when the thickness of the boundary layer is comparable to the intrinsic lengthscale of the compliant wall. Such comparability is achieved precisely when $\alpha_c \sim \alpha_{ts}$, since the boundary-layer thickness can be taken to characterize the lengthscale of unstable Tollmien-Schlichting waves.

If a more thoroughgoing investigation of the branch of solutions associated with divergence is conducted, by means of the numerical solution of the Orr-Sommerfeld stability equation, it can be demonstrated that it is still possible for divergence instability to occur, even when the lengthscales of compliant wall and unstable Tollmien-Schlichting waves are chosen to match each other. However, the onset of the instability is typically postponed to freestream velocities that are

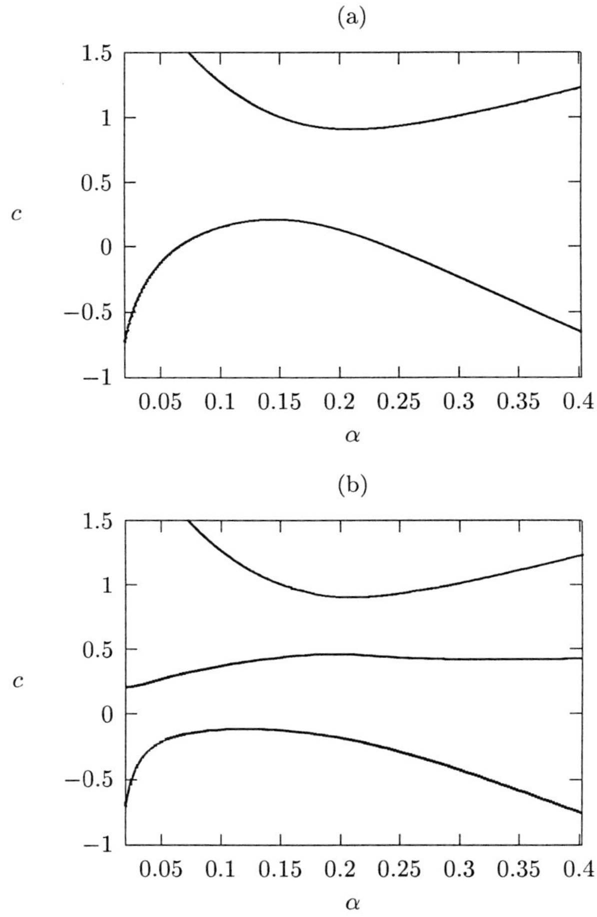

Figure 3. Phase velocities for disturbances in Blasius flow over a compliant spring-backed plate for a case where $\alpha_c \sim \alpha_{ts}$, $U_t^B = U_\infty$ and $U_d^B = 0.7\,U_\infty$. (a) analytic solutions obtained using inviscid theory and a long wavelength approximation, (b) numerical solutions of the Orr-Sommerfeld equation for $R = 2240$. In (a) the upper and lower solution branches correspond to surface-based waves. The Tollmien-Schlichting branch of solutions in (b) lies between the two branches of surface-based waves.

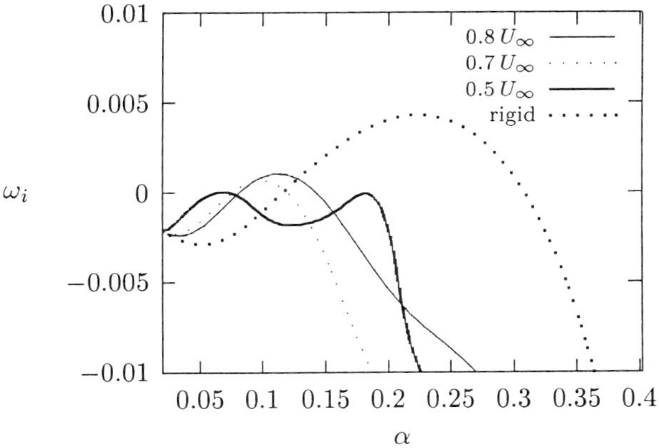

Figure 4. Temporal growth rates $\omega_i = \alpha c_i$ for disturbances in Blasius flow over a compliant plate for cases where $\alpha_c \sim \alpha_{ts}$, $U_t^B = U_\infty$ and (i) $U_d^B = 0.8\,U_\infty$, (ii) $U_d^B = 0.7\,U_\infty$ (iii) $U_d^B = 0.5\,U_\infty$. The remaining curve is for the case of a rigid wall. The solutions were obtained, numerically, from the Orr-Sommerfeld equation for $R = 2240$.

significantly larger than would have been predicted using the simple formula (17), which fails to take account of the disruptive effect of the Tollmien-Schlichting mode (or the equivalent boundary-layer sheltering effect). This means that in applications where the purpose of the compliant wall is to obtain a transition delay, much softer compliant materials can be employed than had previously been thought to be viable. Figure 4 illustrates the possibility of choosing the compliant-wall parameters so that, at a particular Reynolds number, the Tollmien-Schlichting instability is completely eliminated, without any further modes of instability being introduced. Plots are shown for the temporal growth rates of Tollmien-Schlichting waves, for three different compliant walls that are fashioned with increasing softness, subject to the constraint that there is always marginal stability with respect to travelling wave flutter. A plot of the growth rates for the case of a rigid wall is also included. For the softest compliant wall considered, the critical velocity for divergence predicted by the simple formula (17) has been exceeded by a factor of two. Nevertheless, no divergence instability was detected along the lower branch of surface-based waves. Thus the stabilization of the Tollmien-Schlichting branch has been achieved without any concomitant destabilization of either of the branches associated with surface-based waves.

As the results that we have just discussed should have made clear, the condition for the avoidance of divergence instability, $U_d^B > U_\infty$, places far too severe a restriction on the selection of the compliant-wall properties if the critical velocity U_d^B is calculated using the formula that was obtained on the basis of a long wavelength approximation. We noted that the putative critical velocity given by this formula could be exceeded by a factor of two without the appearance of any instability. However, further research is still required to derive an alternative formula, or some other simple recipe, that could be employed to give a more acceptable, and less conservative, constraint on the wall compliancy parameters that would ensure that divergence did not appear for a given freestream velocity.

5. Absolute instabilities

Thus far we have mostly been concerned with instabilities that are convective in nature. Physically, this means that an initially localized disturbance wavepacket will convect away from its point of excitation, but grow in amplitude as it does so. Thus the temporal behaviour of the disturbance at a fixed location in space is characterized by transient growth, followed by decay, as the disturbance first reaches, and then propagates past, the chosen spatial point. Absolute instabilities, by contrast, give rise to disturbances that display sustained temporal growth at any given spatial location. As such they tend to be much more dangerous than convective instabilities in their capacity to alter and disrupt the basic state of the flow.

The present paper is not the place to attempt any review of the extensive literature that has arisen in connection with the classification of instabilities as convective or absolute, for a wide variety of fluid flows. (The interested reader should refer, for example, to the review by Huerre & Monkewitz, 1990.) Instead, we will confine ourselves to noting the following two features (i) absolute instability is found when there is a modal coalescence between different branches of solutions (ii) the onset of absolute instability is associated with the emergence of waves with zero group velocity. However, neither the occurrence of a modal coalescence nor the appearance of a wave with zero group velocity is sufficient for there to be an absolute instability. Further conditions must be met. In practice, absolute instabilities can be identified by tracing the signature of pinch points in the complex wavenumber and complex frequency planes for normal modes of disturbance. (See Briggs 1964 and Kupfer, Bers & Ram 1987 for methods of doing this.) We will not dwell here on such technical issues, but simply provide some illustration of how particular modal coalescences can lead to absolute instability in a flow over a compliant wall.

The rich variety of the modes of motions that can occur in flows over compliant walls gives considerable scope for the generation of absolute instabilities. Typically, an absolute instability arises when there is a modal coalescence between a Class A mode and a Class B mode. The resultant instability is of Class C. In the present paper we will concentrate on the absolute instability that arises from a modal coalescence between travelling wave flutter (Class B) and the Tollmien-Schlichting branch of solutions (Class A) to give the transitional mode (Class C) that was first identified by Sen & Arora (1988). We have already discussed this modal coalescence in connection with the disruption of the onset of divergence instability, where it was noted that it could replace the coalescence between surface-based modes that would otherwise give rise to flutter instability. The situation is further illustrated in Figure 5, for the case of Blasius flow.

Figure 5(a) displays curves for the real and imaginary parts of the complex phase velocity that were obtained analytically by using inviscid shear layer theory in conjunction with a long wavelength approximation. The wall compliance parameters are chosen so that $\alpha_c \sim \alpha_{ts}$ and $U_d^B = U_t^B = 0.6\,U_\infty$. It can be seen that the upper and lower surface-based solution branches have merged, giving rise to complex conjugate phase velocities over a range of wavenumbers. In keeping with the approximation that has been used, there is no trace of the Tollmien-Schlichting branch of solutions. Figure 5(b) illustrates what really happens, for the same set of wall compliance parameters, at a particular Reynolds number $R = 5000$. The modified solution branches plotted in the figure were determined numerically from the Orr-Sommerfeld equation, without imposing any limitation on the wavenumber. It can be seen that there is now a third solution branch that lies between the two branches found before. This third branch can readily be identified as the representative of Tollmien-Schlichting waves. There is a strong modal interaction between the upper branch of surface-based waves and the Tollmien-Schlichting branch, giving rise to pairs of solutions for the phase velocity that are approximately complex conjugates. The magnitudes of the imaginary parts of these phase velocities are much larger than those which are usually associated with either Tollmien-Schlichting waves or travelling wave flutter. Thus a relatively strong form of instability is predicted. However, it should be noted that the mere fact that there is a modal coalesence is not sufficient to warrant any ascription of absolute instability. Further study of the behaviour of the normal mode solution branches in the complex wavenumber and complex frequency planes is required in order to establish this. The interested reader may wish to refer to the recent work by Lingwood & Peake (1999), and two earlier investigations that were conducted by Brazier-Smith & Scott (1984) and Crighton & Oswell (1991), for a detailed illustration of how the analysis of the solution branches should be performed, as well as an account of why the results of the analysis imply some rather counterintuitive physical behaviour. Lingwood & Peake studied the onset of absolute stability for boundary-layer flow configurations similar to those that we have considered, except that the spring-backed plate model of the compliant wall was replaced by an even simpler unsupported plate model. This meant that they were unable to consider the influence of any characteristic lengthscale for the compliant wall.

We will not attempt here to perform a solution branch analysis of the kind that was conducted in the studies that we have just mentioned. Instead, some direct numerical simulation results will be presented for the spatio-temporal evolution of an impulsively excited disturbance. It will be argued that these results provide convincing 'experimental' evidence for the absolute nature of the instability associated with the modal coalescence between the travelling wave flutter and the Tollmien-Schlichting branches. But before reporting the details of the numerical simulation we note, in passing, another feature that is displayed by the phase velocity plots in Figure 5(b). It may be observed that even though the compliant wall has been chosen to be soft enough for there to be a modal coalescence, there is no trace of divergence instability. The lower branch of surface-based waves has

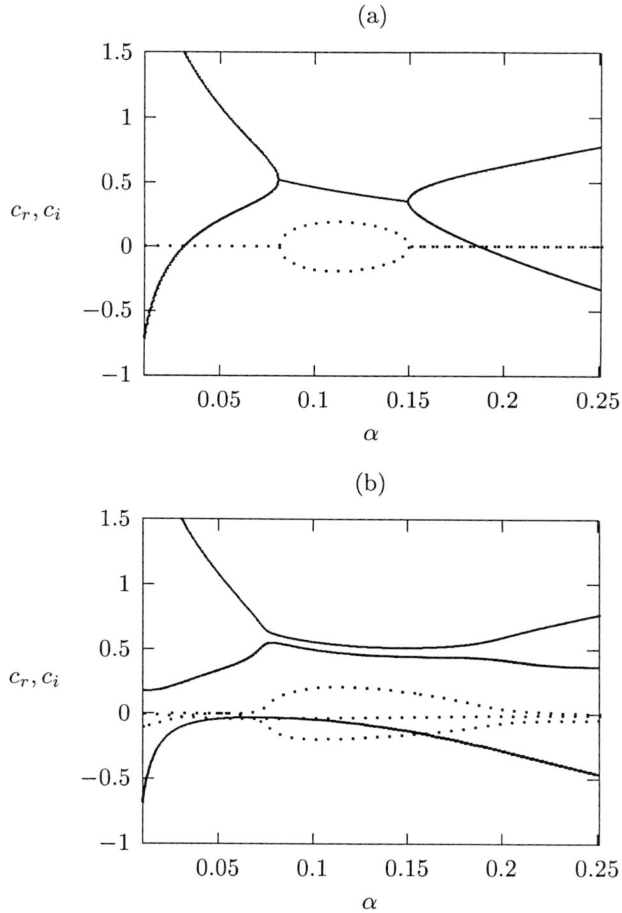

Figure 5. Complex phase velocities for disturbances in Blasius flow over a compliant plate for a case where $\alpha_c \sim \alpha_{ts}$ and $U_d^B = U_t^B = 0.6\, U_\infty$. (a) analytic solutions obtained using inviscid theory and a long wavelength approximation, (b) numerical solutions of the Orr-Sommerfeld equation for $R = 5000$. The solid lines represent the real parts of the phase velocities and the dotted lines represent the imaginary parts. In (a) the upper and lower solution branches corresponding to surface-based waves have merged over a range of wavenumbers. In (b) this merger is replaced by a coalescence between the Tollmien-Schlichting branch of solutions and the upper branch of surface-based waves.

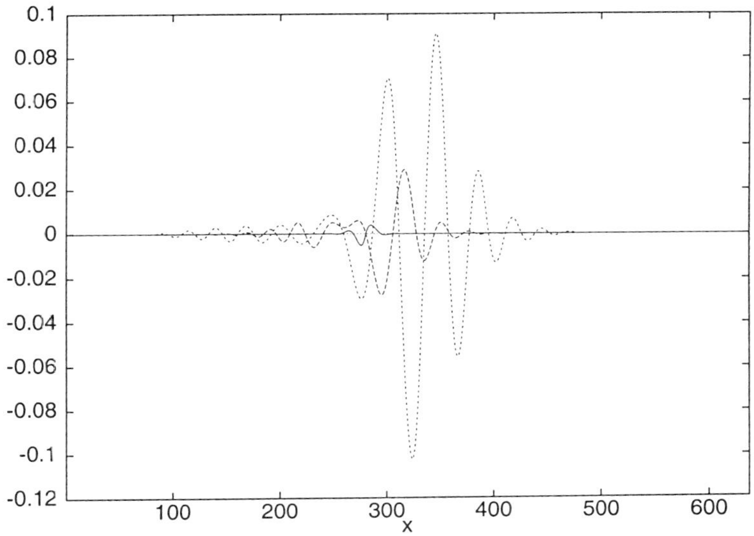

Figure 6. Snapshots of the wall displacement from a simulation of Blasius flow over a compliant panel at $R = 5000$. The wall compliance is such that $\alpha_c \sim \alpha_{ts}$ and $U_d^B = U_t^B = 0.6\, U_\infty$, just as for Figure 5. The impulsive forcing used to generate the disturbance was applied near $x = 280$. Three curves are shown, with increasing amplitudes, for times $t = \Delta T$, $4\Delta T$, $7\Delta T$, where ΔT is a fixed time interval.

failed, once more, to cross the wavenumber axis to give any reversal in the direction of propagation.

The numerical simulations that we shall describe were performed using a computationally efficient velocity-vorticity formulation of the linearized Navier-Stokes equations (Davies & Carpenter 1997a, 2001). A key feature of the discretization was that, in order to obtain numerical stability, it was necessary to, in effect, combine the wall-normal components of the fluid and the compliant surface momenta together as a single variable, and then impose a boundary condition on this variable in an implicit fashion. Thus far, the numerical simulation code has been implemented for two-dimensional linear disturbances developing in plane channel flow, parallelized boundary-layer flows, such as Blasius flow, and uniform shear flow. A three-dimensional version of the code has also been developed for simulating the evolution of disturbances in the von Karman boundary-layer flow over a rotating compliant disk. Some of the numerical results that were found for the rotating disk flow will be presented later. (Further discussion and information about flows over rotating compliant disks can be found in Carpenter & Thomas 2002.)

Figure 6 shows results obtained from a two-dimensional direct numerical simulation for disturbances developing along a very long compliant panel that is placed beneath a parallelized Blasius boundary layer. (This means that no account is taken of boundary layer growth.) Snapshots are displayed for the simulated wall

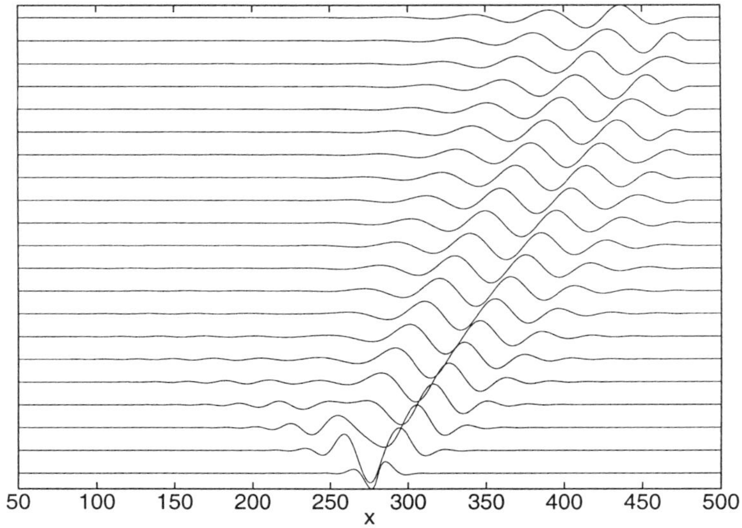

Figure 7. Scaled wall displacements from a simulation of Blasius flow over a compliant panel at $R = 5000$. The amplitude displayed in each of the curves has been reduced by a factor $\exp(-(\omega_i)_{\max} t)$, where the snapshots are for the times $t = \Delta T, 2\Delta T, 3\Delta T, \ldots$ with a fixed sampling interval ΔT. The wall parameters are the same as for the previous figure.

displacement at three selected time instants after a localized impulse had been applied at the centre of the panel. The wall compliance and flow parameters are the same as those that were considered in obtaining Figure 5(b), where we identified a modal coalescence between travelling wave flutter and the Tollmien-Schlichting branch of solutions. It can be seen that the disturbance propagates upstream and downstream from its point of generation, growing in both directions as it does so. A more detailed examination of the simulation data reveals that the disturbance grows continually with time at *all* locations along the length of the compliant panel. This certainly suggests that the instability has an absolute nature. Further evidence to confirm the existence of an absolute instability is displayed in Figure 7. Snapshots of the wall displacement are again plotted, but they are now presented one above another at regular time intervals. More importantly, they have been subjected to a systematic amplitude scaling. The amplitude used in the plot for each selected time instant t after the impulse has been reduced by a factor $\exp(-(\omega_i)_{\max} t)$, where $(\omega_i)_{\max} = (\alpha c_i)_{\max}$ is the maximum value of the temporal growth rate. This maximum growth rate was determined from the numerical solution of the Orr-Sommerfeld equation, independently of the results of the simulation. It can be seen that the temporal growth of the wavepacket maxima has been factored out in a very precise fashion by the use of the scaling based on $(\omega_i)_{\max}$. This is exactly the form of behaviour that would be expected for an absolute instability. (See, for example, Landau & Lifshitz (1987) for a more

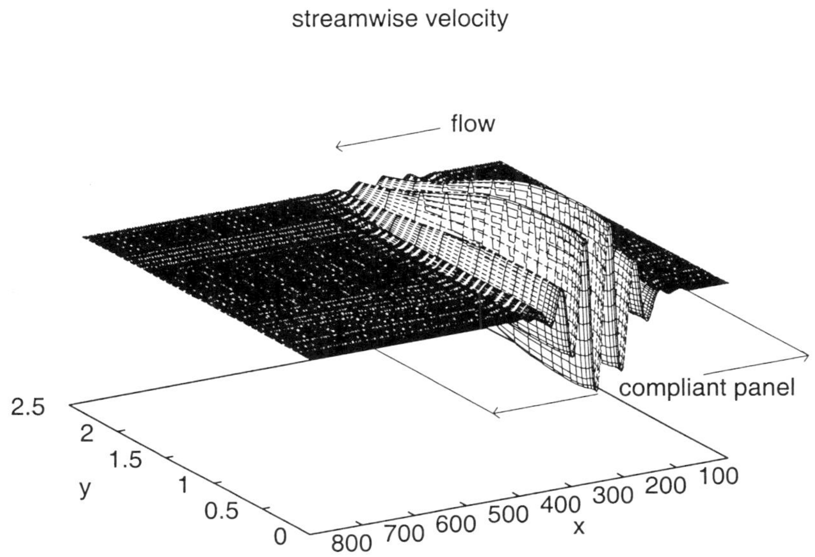

Figure 8. Instantaneous spatial variation of the streamwise perturbation velocity for an absolutely unstable mode of disturbance, as obtained from a numerical simulation of Blasius flow over a compliant panel at $R = 5000$. The wall compliance properties are the same as before.

detailed discussion of this point. It should be noted that, although an absolute instability can only arise when there is a mode with a vanishing group velocity, the maxima of any disturbance wavepacket will, in general, propagate at a non-zero velocity once the threshold for the onset of absolute instability has been exceeded.) The agreement between the temporal growth obtained in the simulation and that which is predicted using the maximum growth rate derived from the Orr-Sommerfeld equation is all the more striking because, over the time interval considered, the unscaled peak amplitude of the simulated disturbance has grown by many orders of magnitude. In Figure 8 we provide a final illustration of the wavepacket that is generated by the localized impulsive excitation. The spatial variation of the streamwise velocity perturbation field is plotted for a time instant at which the continual temporal growth associated with the absolute instability has increased the maximum wavepacket amplitude by several orders of magnitude from the values that were taken just after the impulse had been applied.

For the purposes of exposition we have focused our discussion of absolute instability on the modal coalescence that can arise between travelling wave flutter and the Tollmien-Schlichting branch of solutions. However, as was noted before, this is only one out of a number of possible modal coalescences that could lead to absolute instability. For example, Yeo, Khoo & Zhao (1996, 1999) have demonstrated that for Blasius boundary-layer flow and a variety of modified potential flows over a viscoelastic wall, the branch of solutions that leads to divergence can become associated with a form of absolute instability. Wiplier & Ehrenstein (2001) have also shown that for Blasius flow over a compliant wall modelled in a different fashion as a spring-backed plate, there can be a modal coalescence involving the Tollmien-Schlichting branch and the so-called evanescent wave solution branch. By conducting an appropriate analysis in the complex wavenumber and complex frequency planes, they were able to conclude that the resultant instability was absolute in nature. Evanescent waves are the counterparts, for the fully-coupled flow-structure problem, of the solutions to the free-wave relation (2) that are obtained by considering complex-valued spatial wavenumbers for temporal frequencies lower than the cut-off frequency $\omega_{\min} = (K/m)^{1/2}$. In the absence of any modal coalescence leading to absolute instability, evanescent waves play a rather limited role in the overall behaviour of disturbances. Typically, they only give rise to significant effects near to the ends of finite-length compliant panels (Davies & Carpenter 1997a). However, Wiplier & Ehrenstein were able to use a direct numerical simulation to show that these previously innocuous compliant-panel end-effects could trigger a globally growing instability, once the evanescent waves had been subjected to a modal coalescence.

We will end our discussion of absolute instabilities in flows over compliant walls by mentioning a case where surface compliance suppresses an absolute instability, rather than facilitating it. Cooper & Carpenter (1997) demonstrated that for the three-dimensional von Karman boundary layer that forms over a rotating disk, the introduction of a relatively small level of compliance in the disk surface could stave off an absolute instability for fluid-based disturbances that had previously been predicted by Lingwood (1995). We illustrate this stabilizing influence of surface compliance by presenting some results from a numerical simulation that was conducted for a disturbance evolving in the rotating-disk boundary layer when there was an annulus of compliant material embedded in the surface of an otherwise rigid disk (Davies & Carpenter 2000). Figure 9 shows the time evolution of a disturbance that was excited by applying a localized impulse at a position radially inward from the compliant annulus. By making a comparison with the amplitude envelopes that were found for an entirely solid disk surface, it can be seen that the compliant annulus is very effective in stabilizing the disturbance. It is particularly interesting to note that the disturbance is in fact stabilized at the two radial positions that are inboard from the inner boundary of the compliant annulus, where the disk surface remains rigid.

6. Concluding remarks

We have aimed to provide an *introduction* to the range of convective and absolute instabilities that can occur in flows over compliant walls. In keeping with this aim, we have not tried to be exhaustive or overly systematic in reviewing the previous work that has been conducted in this field. Instead, we have attempted to highlight some features of the intriguing complexity that can arise when the motion of a fluid is interactively coupled to the motion of a compliant wall. To

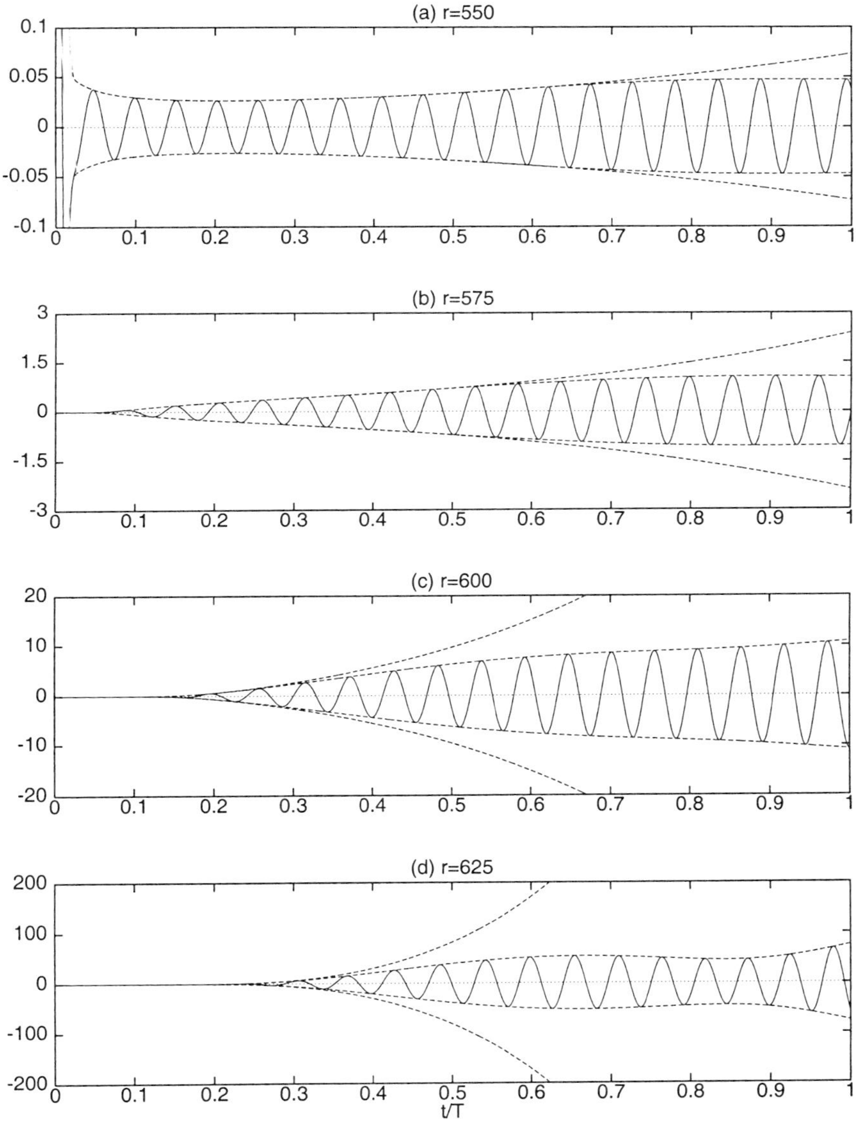

Figure 9. Time histories for the azimuthal component of the perturbation vorticity at the disk surface (solid lines) for a disturbance in the rotating disk boundary layer. The azimuthal mode number is $n = 75$. The disk surface is compliant for radii in the range $600 \leq r \leq 650$. An impulsive excitation was applied at a radius $r_e = 550$. Amplitude envelopes are shown (broken lines), both for the simulation with a compliant insert and for a corresponding simulation where the disk surface was entirely rigid. (a) $r = r_e = 550$; (b) $r = 575$; (c) $r = 600$; (d) $r = 625$. The surface compliance parameters are such that, notionally, there is marginal stability with respect to divergence and travelling wave flutter modes at the outermost radius of the compliant annulus.

achieve some measure of order in our account we have focused on issues that are pertinent to the transition-delaying applications of compliant surfaces, to the neglect of other applications such as flow-structure interactions in physiology. The other contributions to this volume (for example, Lucey & Peake 2002 and Yeo 2002) should be consulted to obtain a more comprehensive view of the roles that can be played by convective and absolute instabilities, for a wider range of applications that involve the coupled motion of a fluid-flow and a compliant structure.

References

BENJAMIN, T.B. 1960 Effects of a flexible surface on hydrodynamic stability. *J. Fluid Mech.* **9**, 513-532.

BENJAMIN, T.B. 1963 The threefold classification of unstable disturbances in flexible surface bounding inviscid flows. *J. Fluid Mech.* **16**, 436-450.

BRAZIER-SMITH, P & SCOTT, J. 1984 Stability of fluid flow in the presence of a compliant surface. *Wave Motion* **6**, 574-560.

BRIGGS, R.J. 1964 *Electron-Stream Interaction with Plasmas.* MIT Press.

CARPENTER, P.W. & GAJJAR, J.S.B. 1990 A general theory for two- and three-dimensional wall-mode instabilities in boundary layers over isotropic and anisotropic compliant walls. *Theor. Comput. Fluid Dyn.* **1**, 349-378.

CARPENTER, P.W. & GARRAD, A.D. 1986 The hydrodynamic stability of flow over Kramer-type compliant surfaces. Part 2. Flow-induced surface instabilities. *J. Fluid Mech.* **170**, 199-232.

CARPENTER, P.W. & THOMAS, P.J. 2002 Rotating flows over compliant walls. Chap. 8 of present volume.

COOPER, A.J. & CARPENTER, P.W. 1997 The stability of rotating-disc boundary-layer flow over a compliant wall. Part 2. Absolute instability. *J. Fluid Mech.* **350**, 261-270.

CRIGHTON, D. & OSWELL, J.E. (1991) Fluid loading with mean flow. I. Response of an elastic plate to localized excitation. *Philos. T. Roy. Soc.* A **335**, 557-592.

DAVIES, C. & CARPENTER, P.W. 1997a Numerical simulation of the evolution of Tollmien-Schlichting waves over finite compliant panels. *J. Fluid Mech.* **335**, 361-392.

DAVIES, C. & CARPENTER, P.W. 1997b Instabilities in a plane channel flow between compliant walls. *J. Fluid Mech.* **352**, 205-243.

DAVIES, C & CARPENTER, P.W. 2000 A novel velocity-vorticity formulation of the Navier-Stokes equations with applications to numerical simulation of boundary layer disturbance evolution, Part 2. *Fluid Dynamics Research Centre Technical Report* **00/1**, University of Warwick.

DAVIES, C. and CARPENTER, P.W. 2001 A novel velocity-vorticity formulation of the Navier-Stokes equations with applications to boundary layer disturbance evolution. *J. Comput. Phys.* **172**, 119-165.

GASTER, M. 1987 Is the dolphin a red herring? In *IUTAM Symp. on Turbulence Management and Relaminarisation, Bangalore, India* (ed. H.W. Liepmann & R. Narasimha), pp. 285-304. Springer.

HUERRE, P. & MONKEWITZ, P.A. 1990 Local and absolute instabilities in spatially developing flows. *Ann. Rev. Fluid Mech.* **22**, 473-537.

HUANG, L. 1998 Reversal of the Bernoulli effect and channel flutter. *J. Fluids Struct.* **12**, 131-151.

KUPFER, K., BERS, A. & RAM, A.K. 1987 The cusp map in the complex-frequency plane for absolute instabilities. *Phys. Fluids* **30**, 3075-3082.

LANDAHL, M.T. 1962 On the stability of a laminar incompressible boundary over a flexible surface. *J. Fluid Mech.* **13**, 609-632.

LANDAU, L.D. & , LIFSHITZ, E.M. 1987 *Fluid Mechanics*, 2nd edn. Pergamon.

LAROSE, P.G. & GROTBERG, J.B. 1997 Flutter and long-wave instabilities in compliant channels conveying developing flows. *J. Fluid Mech.* **331** 37-58.

LINGWOOD, R.J. 1995 Absolute instability of the boundary layer on a rotating disk. *J. Fluid Mech.* **299**, 17-33.

LINGWOOD, R.J. & PEAKE, N. 1999 On the causal behaviour of flow over an elastic wall. *J. Fluid Mech.* **396**, 319-344.
LUCEY, A.D., CAFOLLA, G.J. & CARPENTER, P.W. 1997 Numerical simulation of a boundary-layer flow interacting with a compliant boundary. *Lecture Notes in Physics* **490**, 406-411.
LUCEY, A.D. & CARPENTER, P.W. 1993 On the difference between the hydroelastic stability of infinite and very long compliant panels. *J. Sound Vib.* **163**, 176-181.
LUCEY, A.D. & CARPENTER, P.W. 1995 Boundary-layer instability over compliant walls: Comparison between theory and experiment. *Phys. Fluids* **7**, 2355-2363.
LUCEY, A.D. & PEAKE, N. 2002 Wave excitation on flexible walls in the presence of a fluid flow. Chap. 6 of present volume.
MILES, J. 1997 Generation of surface waves by wind. *Appl. Mech. Rev.* **50**, R5-R9.
SEN, P.K & ARORA, D.S. 1988. On the stability of laminar boundary-layer flow over a compliant surface. *J. Fluid Mech.* **197**, 201-240.
WIPLIER, O. & EHRENSTEIN, U. 2001 On the absolute instability in a boundary-layer flow with compliant coatings. *Eur. J. Mech. B - Fluids* **20**, 127-144.
YEO, K.S. 2002 Flow-induced waves on compliant surfaces subject to a turbulent boundary boundary layer. Chap. 11 of present volume.
YEO, K.S. KHOO, B.C. & ZHAO, H.Z. 1996 The absolute instability of boundary-layer flow over viscoelastic walls. *Theor. Comput. Fluid Dyn.* **8**, 237-252.
YEO, K.S. KHOO, B.C. & ZHAO, H.Z. 1999 The convective and absolute instability of fluid flow over viscoelastic walls. *J. Sound Vib.* **223**, 379-398.

Chapter 5

HYDRODYNAMIC STABILITY OF FLOW THROUGH COMPLIANT CHANNELS AND TUBES

V. KUMARAN
Department of Chemical Engineering
Indian Institute of Science
Bangalore 560 012
India

Abstract
The laminar flow in channels and tubes with compliant walls is destabilised by a variety of mechanisms, especially at moderate and low Reynolds number. There have been two types of classification schemes used for these instabilities. The first is the classification, adopted from boundary layer flow past compliant panels, into the Tollmien-Schlichting modes (Class A), the traveling-wave flutter (Class B) and the static divergence (Class C). The second is based on the Reynolds number regime, the flow structure, the scaling of the critical Reynolds number $(\rho V R/\mu)$ with the dimensionless parameter $\Sigma = (\rho G R^2/\mu^2)$, and the mechanism that destabilises the flow. Here, ρ and μ are the fluid density and viscosity, G is the shear modulus of the wall material, R is the cross-stream length scale and V is the maximum velocity. Linear stability analyses based on both of these classification schemes are discussed, with focus on flows which do not admit Tollmien-Schlichting waves, such as the Couette flow in a channel and the Poiseuille flow in a tube. The weakly non-linear studies carried out so far are briefly summarised.

P.W. Carpenter and T.J. Pedley (eds.),
Flow past Highly Compliant Boundaries and in Collapsible Tubes, 95-118.

1. Introduction

The stability of the flow past a compliant surface has been extensively studied due to its importance in marine and aerospace applications. Of particular interest is the possibility of drag reduction due to transition delay in the flow past compliant panels. This subject is discussed in more detail in this volume, and the consensus is that even though compliant walls do stabilise the 'Tollmien-Schlichting' (Class A) modes, which are responsible for destabilising the flow past rigid surfaces, there are two other types of 'solid based' instabilities, the 'traveling-wave flutter' (Class B) and the 'static divergence' (Class C), which destabilise the flow. In particular, the 'traveling-wave flutter' and the Tollmien-Schlichting mode could coalesce to form a more powerful instability. Thus, even though the compliant surface does stabilise the Tollmien-Schlichting instability, it may not fulfill the objective of achieving transition delay because it could trigger other instabilities which are not present in the flow past rigid surfaces.

Flow past compliant surfaces are also encountered in biological and biotechnological applications, where fluids are transported through tubes and channels made of soft materials such as tissue and polymer gels. Here, the characteristic length and velocity are much lower than in marine and aerospace applications, and the Reynolds number typically ranges from less than 1 to about 10^3. In these cases, it is necessary to maximise the rate of transport of nutrients and waste from the fluid to the walls of the channels, and so it is desirable to induce a transition even at low Reynolds numbers where the flow past a rigid wall is stable. For this, it is necessary to obtain a detailed understanding of the stability of the laminar flow past a soft material in both the low- and high-Reynolds-number regimes. Consequently, some of the methods of analysis used here have been different from those used for marine and aerospace applications.

1. In marine and aerospace applications, flow usually occurs past compliant bodies of finite length, such as airfoils, and there is the spatial development of the mean flow. In such cases, a spatial stability analysis is used, where the frequency of oscillations is assumed to be real, and the spatial growth of a disturbance is determined in the flow past airfoils or in boundary layers. When considering flow in channels and tubes where the length is large compared to the thickness, a 'temporal' analysis is usually carried out, where the temporal growth of a disturbance in a tube or channel of infinite length is determined.
2. The types of compliant surfaces aimed at marine and aerospace applications usually consist of one or two layers of viscoelastic elastomeric material. The dynamics of such walls is modeled by means of the Navier

equations, suitably modified to account for viscoelastic effects, but it is also often modeled approximately by spring-backed-plate membranes, where the pressure at the surface p is related to the normal displacement u by a constitutive relation of the form (11) discussed a little later. This type of a constitutive relation is suitable in the high-Reynolds-number regime, since the flow can be considered inviscid in the leading-order approximation, and the tangential velocity and stress at the surface do not significantly affect the flow. At low and moderate Reynolds numbers, such as those encountered in biological applications, it is also necessary to satisfy the tangential velocity and stress conditions at the interface. Models that have been used include those for a viscoelastic 'gel-wall', which is a solid continuum where the dynamics is described by a displacement field, as well as an infinitesimal 'membrane' which resists both tangential and normal motion.

3. The material properties of the compliant surface include an 'elastic' parameter (surface tension T, shear modulus G) which has dimensions of ML^aT^{-2}, where M, L and T are mass, length and time dimensions, and the parameter $a = 0$ for surface tension, $a = -1$ for the shear modulus. Consequently, there is an additional dimensionless group, $\Sigma = (\rho G R^2/\eta^2)$ or $\Sigma = (\rho T R/\eta^2)$, which influences the dynamics of the flow when elastic effects are significant. Therefore, the transition from stable to unstable modes is governed by two parameters, the Reynolds number and the dimensionless group Σ.
4. The Reynolds number regime (low or high) and the dependence of the Reynolds number on the parameter Σ provides one type of classification which is suitable for examining both the low- and high-Reynolds-number regimes, especially in cases such as the parabolic flow in a pipe and Couette flow in a channel where the Tollmien-Schlichting instability is absent. The classification into Tollmien-Schlichting modes, traveling-wave flutter and static divergence has also been used in this case.

The theoretical justification for the classification into class A, B and C waves was provided by (Benjamin, 1963) using the potential flow approximation for the flow in the fluid, and an equation similar to equation 11 to model the wall dynamics (see (Carpenter and Garrad, 1985) for a more complete description). This approximation is valid for boundary layer type flows at high Reynolds number, where the velocity far from the compliant surface is a constant, and the pressure variation in the 'friction layer' near the wall (where the velocity reduces to zero) is small. Though corrections to the pressure due to the variation of the mean velocity in the friction layer were calculated, these are still restricted to the limit where the thickness of the friction layer is small compared to the wavelength, and the flow outside the friction layer can be adequately treated by the potential flow approximation. Subsequent studies (LaRose and Grotberg, 1997; Gajjar *et al.*, 2002; Hamadiche and Gad-el-Hak, 2002) have used the same classification on the basis of the qualitative features of the instability, such as the relative magnitude of the wave speed and flow speed and the effect of an increase in wall dissipation, even at relatively low Reynolds numbers where the flow may not be adequately represented by the potential flow approximation.

The classification scheme of (Kumaran, 2000) was motivated by the recognition that the destabilising mechanism at low Reynolds number may be different from that in potential flow, and that there is an instability at high Reynolds number (wall-mode instability) where the velocity perturbations cannot be adequately approximated by the potential-flow approximation. This classification (discussed in Section 4) is based on the relationship between Re and the parameter Σ defined above, the parameter regime (low or high Reynolds number) at which these instabilities are observed, the structure of the unstable modes and the type of compliant surface for which these instabilities are observed. In this classification, there are four types of instabilities in addition to the Tollmien-Schlichting instability for the flow past rigid surfaces. The 'inviscid instability' in this classification scheme includes both the class B and class C waves of (Benjamin, 1963) which are present the absence of tangential motion in the wall, and where viscous effects are negligible in the bulk of the flow at high Reynolds number.

In the high-Reynolds-number regime, we restrict attention to flows such as the plane Couette flow and the Poiseuille flow in a tube, where the Tollmien-Schlichting instability is absent. The analyses of instabilities in channels where the Tollmien-Schlichting instability is present (Hains and Price, 1962; Green and Ellen, 1972; Davies and Carpenter, 1997) are similar to the stability of boundary layer flows past a flat surface, and these are discussed elsewhere in this volume. In addition, we also include studies that have examined the low- and moderate-Reynolds-number regime. The methods of analysis are briefly discussed in section 2, while the different wall models used are examined in section 3. The methods of classification of instabilities in the low- and high-Reynolds-number regimes are discussed in section 4. The non-linear studies carried out so far are briefly summarised in section 5.

2. Methodology

2.1. LINEAR STABILITY ANALYSIS

In most studies, the equations governing the fluid flow are the incompressible Navier-Stokes equations

$$\nabla.\mathbf{v} = 0, \tag{1}$$

$$\rho(\partial_t \mathbf{v} + \mathbf{v}.\nabla \mathbf{v}) = -\nabla p + \mu \nabla^2 \mathbf{v}, \tag{2}$$

where $\mathbf{v}$ and p are the velocity and pressure fields, ρ and μ are the fluid density and viscosity, and ∂_t represents the time derivative. However, there have been a variety of wall models used, which are discussed in the next section.

In the linear stability analysis, small perturbations $\mathbf{v}'$ and $\mathbf{u}'$ are placed on the base state velocity and displacement field. The base velocity profile for a plane Couette flow(figure 1(a)) is

$$\bar{v}_x = \frac{Vy}{R}, \ \bar{v}_y = 0, \tag{3}$$

and the perturbations are of the form

$$\begin{pmatrix} \mathbf{v}' \\ \mathbf{u}' \end{pmatrix} = \begin{pmatrix} \tilde{\mathbf{v}}(y) \\ \tilde{\mathbf{u}}(y) \end{pmatrix} \exp\left(\imath k x + s t\right), \tag{4}$$

where x and y are the flow and gradient directions respectively, k is the wave number in the flow direction, and $s = s_r + \imath s_i$ is the complex growth rate, such that s_r is the real growth rate, $s_i = -kc$ is the real frequency and c is the phase speed. For a tube (figure 1(b)), the base flow is a Hagen-Poiseuille flow,

$$\bar{v}_x = V\left(1 - \frac{r^2}{R^2}\right), \ \bar{v}_r = 0, \tag{5}$$

and the perturbations are of the form

$$\begin{pmatrix} \mathbf{v}' \\ \mathbf{u}' \end{pmatrix} = \begin{pmatrix} \tilde{\mathbf{v}}(r) \\ \tilde{\mathbf{u}}(r) \end{pmatrix} \exp\left(\imath k x + \imath n \theta + s t\right), \tag{6}$$

where r, θ and x are the radial, meridional and axial coordinates in a cylindrical coordinate system with axis along the center line of the tube. The above velocity and displacement fields are inserted into the conservation equations, which are then linearised. The solutions for the linear equations for the perturbations $\tilde{\mathbf{u}}$ and $\tilde{\mathbf{v}}$, which are consistent with the boundary conditions at the rigid surfaces, are determined as a function of

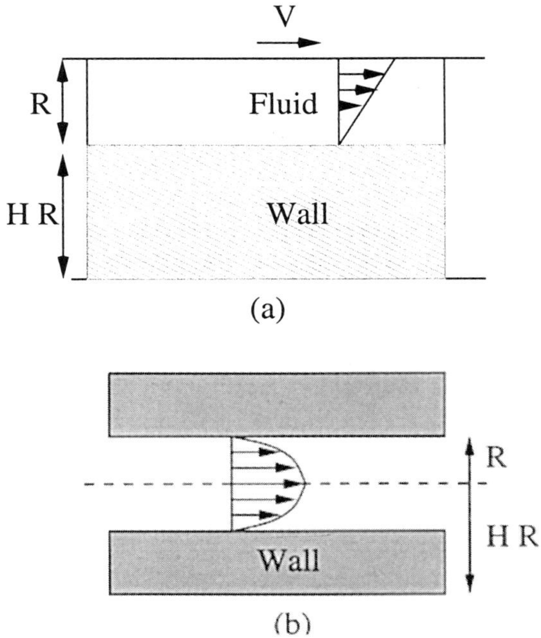

Figure 1. Couette flow in a channel (a) and Poiseuille flow in a tube (b).

wave number k and the growth rate s. These are inserted into the velocity and stress boundary conditions at the interface between the fluid and the wall material, and the dispersion matrix is determined. The determinant of this matrix is set equal to zero in order to determine the growth rate s. The perturbations are stable if the real part of the growth rate is negative, and they are unstable if the real part is positive. The transition from stable to unstable modes takes place when the real part of the growth rate passes through zero.

The velocity and displacement fields can be obtained using a simple integration scheme, such as the Runge-Kutta integrator (Shankar and Kumaran, 1999; Shankar and Kumaran, 2000) or the Tchebychev collocation scheme (Gajjar *et al.*, 2002). In general, there are multiple solutions for the growth rate, and iterative techniques such as a Newton-Raphson can be used for obtaining the solutions. The disadvantage with iterative techniques is that they require a good initial guess for the growth rate. To overcome this, a more sophisticated eigenvalue search technique was used by (Hamadiche and Gad-el-Hak, 2002), where the number of zeroes of a function (which is the determinant of the characteristic matrix in this case) within a domain in the complex plane is determined by carrying out a contour integration along the boundary of this domain. The values of the growth rate are then determined by evaluating the sums of the powers of the growth rate using the residue theorem.

2.2. WEAKLY NON-LINEAR ANALYSIS

The computation of the flow field after transition is a difficult problem due to the deformable nature of the boundary, but some qualitative features can be explored by carrying out a weakly non-linear analysis, where the leading-order effects of non-linearities in the vicinity of the neutral curve are incorporated. The objective of the weakly non-linear analysis is to derive the 'Landau equation', which is an amplitude equation of the form

$$\frac{dA}{dt} = s_{0r} A + s_1 |A^2| A, \tag{7}$$

where A is the amplitude of perturbations, s_{0r} is the real part of the growth rate from the linear analysis and s_1 is the 'Landau constant', which determines the nature of the bifurcation. When the Reynolds number is equal to the critical Reynolds number, s_{0r} is equal to zero. When the Reynolds number is slightly larger than the critical Reynolds number, the real part of the growth rate from the linear stability analysis can be expanded in a Taylor series

$$s_{0r} = (Re - Re_c) \left. \frac{ds_{0r}}{dRe} \right|_{Re=Re_c} + O(Re - Re_c)^2. \tag{8}$$

If s_1 is $O(1)$, a balance is achieved between the first and second terms on the right side of equation 7 if $|A| \sim (Re - Re_c)^{1/2}$. It is convenient to define a scaled amplitude $A_1 = (A/\epsilon)$ which is $O(1)$, where $\epsilon = (Re - Re_c)^{1/2}$. The Landau equation for the scaled amplitude becomes

$$\frac{dA_1}{d\tau} = \left. \frac{s_{0r}}{dRe} \right|_{Re=Re_c} A_1 + s_1 |A_1^2| A_1, \tag{9}$$

where $\tau = (\epsilon^2 t)$ is the 'slow time scale'.

The objective of the weakly non-linear analysis is to calculate the 'Landau constant' s_1. If s_1 is negative, a linearly unstable perturbation $s_{0r} > 0$ saturates at an amplitude $A = \sqrt{s_{0r}/|s_1|}$, and this type of bifurcation is termed 'supercritical'. Examples of supercritical bifurcations are the Rayleigh-Benard instability in fluids heated from below, and the Taylor-Couette instability in rotating fluids (Drazin and Reid, 1981). In this case, the amplitude of the perturbations grows continuously from zero after the flow becomes unstable. If s_1 is positive, modes that are stable in the linear analysis $s_{0r} < 0$ become unstable to finite-amplitude perturbations for $A > \sqrt{|s_{0r}|/s_1}$, and the bifurcation is termed 'subcritical'. In this case, the critical value of the Reynolds number is reduced by non-linear interactions, and there is a discontinuous change in the amplitude of perturbations when the flow becomes unstable. The Tollmien-Schlichting instability in a rigid-walled channel is known to be highly subcritical (Drazin and Reid, 1981), and there is a reduction of 65% in the critical Reynolds number when the velocity perturbation is 2.5% of the maximum velocity. Weakly non-linear studies of flow through compliant tubes and channels are carried out to determine how the nature of the Tollmien-Schlichting instability is affected by wall compliance, and to determine the nature of the other types of instabilities that are present at low and moderate Reynolds numbers.

Perturbations to the velocity and stress fields are expressed using the function $E(x,t) = \exp[\imath(kx + \omega t)]$ where k and ω are the wave number and frequency of the neutrally stable modes. An expansion of the following form is used for the dynamical variables,

$$\phi(x,y,t) = \sum_{m=0}^{\infty} \sum_{n=m}^{\infty} (\epsilon A_1(\tau))^n [E^m \tilde{\phi}^{(m,n)}(y) + E^{-m} \tilde{\phi}^{*(m,n)}(y)] \qquad (10)$$

where the superscript * denotes complex conjugate. These expansions are inserted into the governing equations and boundary conditions, and solved for progressively increasing values of (m,n) to obtain the higher corrections due to non-linearities. The linear stability analysis corresponds to $m = 1, n = 1$, while the non-linear correction to the mean flow is obtained for $m = 0, n = 2$. The non-linear correction to the growth rate of the first harmonic is recovered for $m = 2, n = 2$. In the $m = 1, n = 3$ problem, the linear operator acting on the dynamical variables is identical to that for the homogeneous $m = 1, n = 1$ problem, which corresponds to the linear stability analysis. The non-linearities due to the convective terms in the momentum equation and the boundary condition generate inhomogeneous terms in the governing equations for the $m = 1, n = 3$ problem. The Landau equation is generated by the solvability condition for the inhomogeneous terms in the $m = 1, n = 3$ problem.

There weakly non-linear analysis of the flow in a compliant channel differs from that for the flow in a rigid channel in two respects.

1. In a rigid channel, non-linearities arise due to the convective term in the Navier-Stokes momentum equation. In a compliant channel, additional non-linearities are generated due to the boundary conditions, when the

velocity and stress at the deformed interface are expanded about their values at the undeformed surface.

2. In a rigid channel, the governing equation for the linear stability analysis is a linear differential equation. Therefore, it is necessary to calculate the adjoint of the linear differential operator for determining the solvability condition for the $m = 1, n = 3$ problem. In the linear stability analysis for a compliant channel, the solutions for the linear differential equations for the fluid and wall consistent with the boundary conditions at the rigid surfaces are determined, and the values of the dynamical variables are inserted into the boundary conditions at the interface to obtain the dispersion matrix. The determinant of the dispersion matrix is set equal to zero to determine the growth rate in the linear stability analysis. Consequently, in the weakly non-linear analysis, it is sufficient to determine the adjoint of the dispersion matrix of the linear problem in order to obtain the solvability condition.

3. Types of surfaces

The wall models can be broadly classified into those that permit tangential motion, such as the gel-wall and membrane models, and those that do not permit tangential motion, such as the spring-backed-plate model and the kinematic boundary condition. As discussed a little later, the presence or absence of tangential motion has a significant effect on the stability of the flow.

3.1. SPRING-BACKED-PLATE MODEL

The 'plate-membrane' wall (figure 2) (LaRose and Grotberg, 1997; Davies and Carpenter, 1997) is a spring-backed-plate deformable only in the normal direction. The constitutive equation for normal displacements is usually of the form

$$(M\partial_t{}^2 + D_n\partial_t + B\partial_x{}^4 - T\partial_x{}^2 + K)u_y = n_i\tau'_{ij}n_j \,, \tag{11}$$

where $n_i\tau'_{ij}n_j$ is the normal fluid stress perturbation at the interface, n_j is the unit vector normal to the surface, K is the spring stiffness, D_n is the normal wall-damping coefficient, B is the flexural rigidity, T is the longitudinal tension per unit width, M is the mass per unit area of the membrane, x is the Cartesian coordinate along the wall, y is the direction normal to the wall, $\partial_x \equiv \partial/\partial x$, t is the dimensional time variable and $\partial_t \equiv \partial/\partial_t$. The tangential displacement at the membrane surface is usually considered to be zero in the plate/membrane model, though the study of (Carpenter and

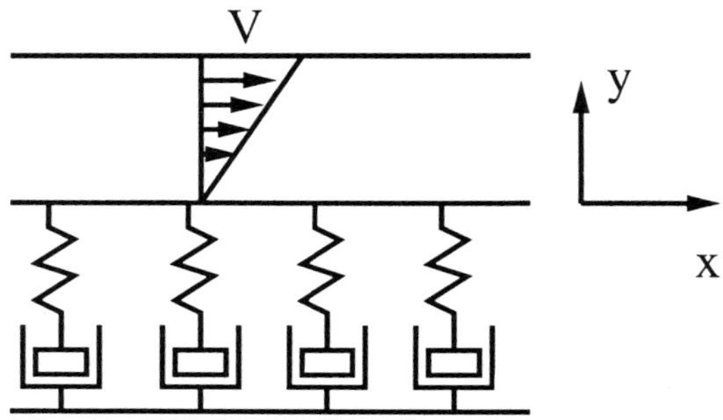

Figure 2. Spring-backed-plate model.

Morris, 1990) on boundary-layer stability used an 'anisotropic' wall model, where the ratio of normal and tangential displacements at the wall surface is considered to be a constant. In the study of (LaRose and Grotberg, 1997; Thaokar *et al.*, 2001; Shankar and Kumaran, 2002), the above normal-stress balance was augmented by another relation between the tangential stress and the tangential displacement,

$$(M\partial_t{}^2 + D_t\partial_t - E\partial_x{}^2)u_x = t_i\tau_{ij}n_j\,, \tag{12}$$

where u_x is the tangential displacement of the membrane surface, D_t is the tangential wall damping coefficient, $t_i\tau'_{ij}n_j$ is the tangential stress perturbation in the fluids, and E is the bending stiffness of the membrane.

The model equation 11 is used to model the properties of compliant walls which of multiple layers of viscoelastic elastomeric material. The material properties of these materials are the Young's modulus, the Poisson ratio and the actual thickness of the layers, and the density of the material. A comparison can be made with experiment only if the model parameters, K, B, M, D and T, are related to these material properties of the elastomeric materials. One of the applications is the flow of air in bronchial tubes, and the walls are modeled as a single layer, thin walled and isotropic tube, and the parameters are directly estimated from physiological measurements (LaRose and Grotberg, 1997). A more complicated surface, which is a dolphin-skin mimic called the Kramer-type surface, was analysed by (Carpenter and Garrad, 1985). In this case, the authors found that it is more difficult to estimate parameters since the spring stiffness depends on the type of measurement (single-point loading or distributed loading), while the damping coefficient is sensitive to the frequency of the perturbations.

3.2. KINEMATIC BOUNDARY CONDITIONS

These boundary conditions have been used by (Gajjar *et al.*, 2002) for the stability of the flow through compliant tubes, following the earlier work by (Sen and Arora, 1988) on the stability of boundary layer flows. Here, the boundary conditions used are the no-slip condition tangential to the surface, and the normal-velocity and normal-stress conditions in the direction normal to the surface. There is no tangential stress balance condition, since the tangential velocity is set equal to zero. In the formulation, the complex admittance $\tilde{Y} = -(\tilde{v}'_{yw}/\tilde{p}'_w)$ is defined as the ratio of the normal velocity and the pressure perturbation at the wall. For a spring-backed-plate type of model, where the normal stress is due to the pressure alone, the compliance is related to the parameters in the constitutive relation (11), and the Fourier transform of the compliance is

$$\tilde{Y} = -\frac{s}{Ms^2 + Tk^2 + s} = -\frac{\imath c}{Mk^2(c_0^2 - c^2 - (D\imath c/Mk))}, \tag{13}$$

where $c_0^2 = (T/M)$.

In the kinematic formulation, the stability analysis is not carried out for a specified set of wall parameters. Instead, the complex admittance is determined from the fluid side using a given value of the complex stream function evaluated at the wall. The value of the wall parameters are determined from the admittance using equation (13). The real and imaginary parts of equation (13) can be used to determine two of the parameters (usually D and T) in terms of the third parameter (M). However, there are constraints ($M > 0, D > 0, T > 0$) on these parameters for the wall to be physically realisable. These realisability conditions provide bounds on the parameters M, T and D, which have been derived by (Gajjar *et al.*, 2002).

The advantage of the kinematic boundary condition approach is that no assumption is made about the wall model, and so these can be used for treating large classes of wall models. However, it is difficult to extend this approach to cases where tangential motion is permitted, because it is also necessary to impose tangential velocity and stress conditions, and a single admittance would not be sufficient to completely describe the wall dynamics.

3.3. GEL-WALL MODEL

The gel-wall model has been used in various studies (Kumaran *et al.*, 1994; Kumaran, 1995a; Kumaran, 1995b; Kumaran, 1998a; Kumaran, 1998b; Kumaran, 1998c; Srivatsan and Kumaran, 1997; Shankar and Kumaran, 1999; Shankar and Kumaran, 2000; Shankar and Kumaran, 2001a; Shankar and Kumaran, 2001b). In this model (figure 3), the dynamics of the wall is represented by a displacement field $\mathbf{u}$, which is the displacement of material points from their steady-state positions due to fluctuations in the fluid stresses at the interface. The displacement field is governed by linear elasticity equations for an incompressible material

$$\nabla.\mathbf{u} = 0, \tag{14}$$

$$\rho\partial_t^2\mathbf{u} = -\nabla p + G\nabla^2\mathbf{u} + \mu_g\nabla^2(\partial_t\mathbf{u}), \tag{15}$$

where $\mathbf{u}$ is the displacement field which provides the displacement of material points from their steady state positions due to stresses exerted by the fluid. In the equation (15), the left side is the rate of change of momentum due to acceleration, and the density of the wall material is usually assumed to be equal to that of the fluid. The first term on the right is the pressure gradient required to enforce incompressibility. The second term on the right is the divergence of an elastic stress,

$$\sigma_e = G(\nabla\mathbf{u} + (\nabla\mathbf{u})^T), \tag{16}$$

where G is the coefficient of elasticity. The third term on the right side of equation (15) is the divergence of a viscous stress,

$$\sigma_v = \mu_g(\nabla\partial_t\mathbf{u} + (\nabla\partial_t\mathbf{u})^T), \tag{17}$$

where $\partial_t\mathbf{u}$ is the velocity field in the wall material in the linear approximation, and μ_g is the viscosity. The boundary conditions at the interface between the wall material and the rigid surface are zero-displacement conditions $\mathbf{u} = 0$, while the boundary conditions between a rigid surface and the fluid are the no-slip conditions $\mathbf{v} = \mathbf{v}_s$, where $\mathbf{v}_s$ is the velocity of the surface. Continuity of velocity and continuity of stress conditions are applied at the interface between the fluid and the wall material.

The shear modulus G, which can be directly measured, is a function of the frequency of oscillations, ω. However, in most theoretical studies, the shear modulus is assumed to be a constant. This assumption is valid for soft solids such as polymer gels, which exhibit a range of frequencies over which the modulus has a constant value (referred to as the 'plateau modulus'). The solid viscosity μ_g, can be related to the 'loss modulus' G'', which is also directly measured in experiments, by the relation $\mu_g = (G''/\omega)$. Both G'' and μ_g are functions of frequency, and it is necessary to determine the frequency dependence of these parameters in order to accurately compare theory and experiment. In the experiments of (Kumaran and Muralikrishnan, 2000; Muralikrishnan and Kumaran, 2002), the storage and loss modulus of the polymer gel were determined as a function of frequency

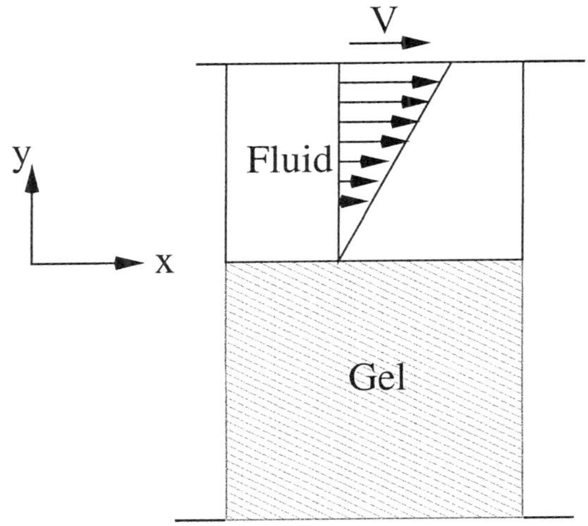

Figure 3. Gel-wall model.

in a rheometer. The storage modulus was found to be a constant over a frequency range extending over two decades, and this 'plateau' value was used for a comparison with theoretical results. An analytical form for the dependence of the loss modulus and gel viscosity on the frequency was obtained from measurements. This enabled a comparison between theory and experiments with no adjustable parameters.

3.4. MEMBRANE MODEL

This model has been used (Kumaran and Srivatsan, 1998; Thaokar *et al.*, 2001; Thaokar and Kumaran, 2002) to mimic the behaviour of membranes in biological systems. Membranes, which are composed of lipid monolayers or bilayers between two fluids, can be deformed by bending in the direction normal to the surface, as well as due to compression of the bilayers in plane, resulting in a change in the density (number of lipids per area) along the surface (figure 4). The amplitude of fluctuations is usually large compared to the membrane thickness, and so it is convenient to consider the membrane to be a surface of zero thickness. The material properties required to describe the membrane are the surface tension Γ, the bulk modulus K and shear modulus G for compression along the membrane surface. In addition, the effect of surface viscosity can also be incorporated using parameters η_{bm} and η_{sm} for the bulk and shear viscosities. Note that K, Γ and G are defined in units of (Force / length), and η_{bm} and η_{sm}

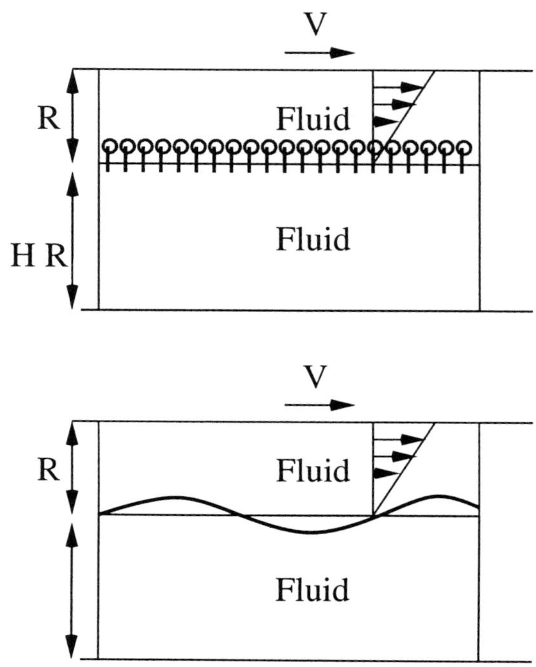

Figure 4. Membrane model.

are defined in terms of (Force × time / length). In addition to a normal stress which is the product of curvature and surface tension, there is an in-plane stress exerted by the membrane which is represented by a constitutive relation of the form

$$\sigma_{\alpha\beta} = (K + \eta_b \partial_t)\, \partial_\gamma u_\gamma\, \delta_{\alpha\beta} + (G + \eta_s\, \partial_t)(\partial_\alpha u_\beta + \partial_\beta u_\alpha), \tag{18}$$

where $\sigma_{\alpha\beta}$ is the stress (force per unit length) along the membrane surface, and u_α is the displacement vector, along the membrane surface, of material points from their equilibrium positions due to stresses exerted on the membrane. Note that Greek indices are used to denote two-dimensional vectors along the membrane surface. For two-dimensional disturbances, where surface displacement occurs only in the direction along the surface of the membrane (x direction), the constitutive relation reduces to

$$\sigma_{xx} = (K + \eta_{bm}\partial_t + 2G + 2\eta_{sm}\partial_t)\partial_x u_x. \tag{19}$$

The 'incompressible' viscoelastic membrane, where in-plane deformation of the membrane is not permitted, is a limiting case of the constitutive equation 19 in which the bulk modulus goes to infinity. In this case, the displacement in the direction tangential to the membrane is zero, since any finite displacement would cause an infinite stress in the tangential direction. The continuity of tangential velocity requires that the tangential velocity in the fluid should also be zero in the linear approximation.

It should be noted that the flow past a membrane cannot strictly be treated as invariant in the flow direction, because the shear stress exerted by the fluid causes a variation in the membrane tension along the surface. The tension can be considered a constant only if the variation in tension over length scales comparable to the wavelength of perturbations is small compared to the mean tension. As discussed in (Thaokar and Kumaran, 2002), this limits the applicability of the analysis to perturbations with wavelength large compared to the channel width. However, since the Reynolds number for transition from stable to unstable modes is lowest in the limit $k \to 0$, where k is the wave number of perturbations, the long wave approximation is appropriate for this case. The parameters in the model, such as the tension, bending modulus and the membrane viscosity, can be determined experimentally from the structure factor for fluctuations at the surface.

4. Classification

The instabilities in the flow through compliant channels have been classified into

1. the Tollmein-Schlichting instability, also referred to as the Class A instability, which is a modification of the instability in the flow through a rigid channel, and is amplified by an increase in wall dissipation;
2. the Class B or traveling-wave flutter instability, which is a downstream-traveling wave in the solid destabilised by fluid interaction, and is damped by an increase in the wall dissipation; and
3. the Class C or static divergence, which is similar to the Kelvin-Helmholtz instability and involves a unidirectional transfer of energy from the fluid to the solid, and which is not affected by a variation in the wall dissipation.

As mentioned in the introduction, we do not examine flows which admit the Tollmien-Schlichting instability, such as the parabolic flow in a channel (Hains and Price, 1962; Green and Ellen, 1972; Davies and Carpenter, 1997). However, it is useful to note that the conclusions of these studies indicate that the compliant surface has an effect qualitatively similar to that for boundary-layer flows past compliant surfaces. The presence of a compliant surface does damp the Tollmien-Schlichting (Class A) modes, but could give rise to other types of flow-induced surface instabilities, such as the static divergence and the traveling-wave flutter. In particular, the traveling-wave flutter could coalesce with the Tollmien-Schlichting instability to give rise to a powerful new instability.

In connection with an application to wheezing in the bronchial tubes in the lungs, many authors (Weaver and Paidoussis, 1977; Grotberg and Reiss, 1984; Grotberg and Reiss, 1984) have investigated the stability of a plug flow in a channel. We do not include this in the present review, since this is an idealised flow profile with zero strain rate in the channel. The stability of a developing flow near the entrance of a channel was studied by (LaRose and Grotberg, 1997) using a long-wavelength asymptotic analysis as well as a numerical analysis. In the long-wavelength limit, inertial effects are neglected in the leading approximation, and the system is neutrally stable. The first correction to the growth rate, due to inertial effects, becomes positive when the flow speed exceeds a critical value. This long-wave instability is similar to the low-Reynolds-number instability of (Kumaran and Srivatsan, 1998) discussed a little later. Numerical studies (LaRose and Grotberg, 1997) also revealed that there could be a finite-wavelength flutter-type instability at higher wave number.

The classification into Class A, B and C instabilities was used by (Hamadiche and Gad-el-Hak, 2002) for the instabilities in a viscoelastic tube. There is no Tollmien - Schlichting instability in the flow through a rigid tube, and so Class A instabilities are not expected in this case. The instabilities found by Hamadiche and Gad-el-Hak for axisymmetric and non-axisymmetric perturbations were classified into two types — those that converge to the least-damped rigid tube mode as the thickness of the viscoelastic wall goes to zero, and those whose damping rate approaches $(-\infty)$ as the wall thickness goes to zero. Both types of instabilities were damped by wall dissipation, indicating that these are Class B or Class C instabilities. The axisymmetric modes, which are a continuation of the least-damped modes in a rigid tube, had a significant group velocity, and were classified as a Class B instability. The non-axisymmetric modes whose growth rates approached $-\infty$ as the wall thickness approaches zero were classified as the static divergence instability belonging to Class C. The classification of the other two was less clear, since one of these had a negative group velocity, while the other had a positive group velocity for small wave number and a negative group velocity for large wave number.

A similar classification was used by (Gajjar *et al.*, 2002), who employed the the kinematic boundary condition technique (Sen and Arora, 1988) for the wall. This study found that modes in the Kelvin-Helmholtz (static

TABLE I. Classification scheme of Kumaran (2000).

Instability	Regime	Wall tangential motion	Mechanism	Flow structure
Viscous	$Re \ll 1$ $Re \propto \Sigma$	Required	Shear work at interface	Zero inertia
Low Re inertial	$Re \ll 1$ $Re \propto \Sigma^{1/2}$	Not required	Reynolds stress	Small wavenumber $k \to 0$
Inviscid	$Re \gg 1$ $Re \propto \Sigma^{1/2}$	Not required	Reynolds stress	Internal critical layer $O(Re^{-1/3})$
Wall-mode	$Re \gg 1$ $Re \propto \Sigma^{3/4}$	Required	Shear work at interface	Viscous layer at wall $O(Re^{-1/3})$

divergence), resonant (traveling-wave flutter) and transitional classes occur at all values of Reynolds and wave numbers, though the modes in the Tollmien-Schlichting class are highly damped at low Reynolds and wave numbers. The Kelvin-Helmholtz and resonant wave instabilities persist even in the low-Reynolds-number limit. They also derived a condition relating the admittance and phase velocity which ensures the presence of unstable modes. Using this, they derived two kinds of bounds on the mass, spring and damping coefficients which give rise to unstable modes. In the limit of low Reynolds number and wave number in a compliant tube, they found that the axisymmetric mode could be eliminated, but the non-axisymmetric mode with meridional wave number $n = 1$ continues to be unstable for all parameter values.

4.1. VISCOUS INSTABILITY

At zero Reynolds number, the flow dynamics depends on the dimensionless parameter $\Gamma = (V\mu/GR)$, which is the ratio of viscous forces in the fluid and elastic forces in the wall material, as well as the ratio of viscosities $\mu_r = (\mu_g/\mu)$, and the ratio of the thickness of the wall and the fluid layers. The stability of the flow in this limit was first analysed by (Kumaran *et al.*, 1994) for the Couette flow past a flexible surface, and subsequently by (Kumaran, 1995a) for the flow in a flexible tube. These studies indicated that there is a transition from stable to unstable modes when the parameter $\Gamma = (V\mu/GR)$ exceeds a critical value. More recently, a non-linear analysis was carried out by (Shankar and Kumaran, 2001b), and experimental studies (Kumaran and Muralikrishnan, 2000; Muralikrishnan and Kumaran, 2002) have also confirmed the existence of this instability. The instability is caused by a coupling between the mean flow and the fluctuations in the tangential-velocity boundary condition at the interface, due to the variation in the mean velocity with height. The tangential-velocity continuity condition is applied at the displaced interface, $y = u_y$, for a plane Couette flow,

$$v_x|_{y=u_y} = (\partial_t u_x)|_{y=u_y}, \tag{20}$$

where v_x is the tangential velocity at the surface, and u_x is the tangential displacement. A Taylor series expansion is used to obtain an approximation that is correct to first order for the perturbation in the tangential flow velocity at the wall, which permits the boundary condition (20) to be linearised as follows,

$$v'_x|_{y=0} + u'_y \frac{d\bar{v}}{dy}\bigg|_{y=0} = (\partial_t u'_x)|_{y=0}. \tag{21}$$

The second term on the left side of the velocity boundary condition (21) accounts for the variation in the velocity of the interface due to the change in surface height. This coupling between the mean flow and the fluctuations destabilises the flow when the dimensionless parameter $\Gamma > \Gamma_c$, where Γ_c is the critical value for the parameter Γ, and it can be easily inferred that the critical Reynolds number increases as $Re_c \propto \Sigma$. The physical mechanism for the instability is the transfer of energy from the mean flow to the fluctuations due to the shear work done by the mean flow at the interface. When this rate of energy transfer exceeds the rate of viscous dissipation of energy in the fluid and the wall material, the flow becomes unstable. This instability is encountered only for flows where tangential motion is permitted at the surface, such as in the flow past viscoelastic gels or compressible membranes (Thaokar and Kumaran, 2002), or the flow past a spring-backed-plate wall which permits tangential motion (Thaokar *et al.*, 2001). A salient feature of this instability is that it is damped by an increase in the dissipation in the wall. In the gel-wall model (Kumaran, 1995a; Kumaran, 1995b), for example, perturbations become unstable only when the ratio of viscosities of the fluid and wall is below a maximum value for a given ratio of fluid and wall thicknesses.

4.2. LOW-WAVE-NUMBER INSTABILITY

There is an inertially driven instability at low Reynolds number in the long wave limit, because in the limit $k \to 0$, the inertial terms in the momentum equation are proportional to kRe, while the damping of fluctuations due to the viscous terms scales as k^2. Therefore, the inertial terms in the momentum equation are significant in the limit $Re \gg k$ even when the Reynolds number is small.

The instability of the flow past a membrane between two fluids without tangential motion was analysed by (Kumaran and Srivatsan, 1998) in this limit, and it was shown that fluctuations could become unstable in the $k \rightarrow 0$ limit for certain values of the ratio of strain rates between the two fluids. The critical Reynolds number in this case has a scaling $Re_c \propto \Sigma^{1/2}$ in the low-Reynolds-number limit. A similar instability was also observed for the continuation of the wall modes in the low-Reynolds-number limit by (Shankar and Kumaran, 2001a). The mechanism of instability is the transport of energy from the mean flow to the fluctuations due to the Reynolds stresses in the fluid, and tangential motion of the interface is not necessary for inducing this instability. The long-wave instability of (LaRose and Grotberg, 1997) can also be classified as a low-wave-number instability, since an asymptotic expansion in the limit $k \rightarrow 0$ was used, and inertial terms were neglected at leading order in this limit. However, their numerical studies also indicated a second minimum for the neutral stability curve at finite wave number for this configuration, and this was classified as a 'flutter' instability by the authors.

4.3. INVISCID INSTABILITY

In the high-Reynolds-number limit, a balance between inertial and elastic stresses is achieved when the parameter $(\rho V^2/G)$ is $O(1)$, or when $Re \propto \Sigma^{1/2}$. In this limit, linear stability studies have shown that the flow in a tube exhibits interesting behaviour (Kumaran, 1996; Shankar and Kumaran, 1999; Shankar and Kumaran, 2000). The classical theorems of hydrodynamic stability of Rayleigh, Fjortoft and others were extended for this case by (Kumaran, 1996; Shankar and Kumaran, 2000). These studies indicate that for a flow with mean velocity $\bar{v}_x = 0$ at the wall, the possibility of an instability depends on the function

$$F(r) = \bar{v}_x \frac{d}{dr} \left(\frac{r}{n^2 + k^2 r^2} \frac{d\bar{v}_x}{dr} \right). \tag{22}$$

It can be shown that the flow could become unstable only if $F(r) < 0$ somewhere in the flow.

The function $F(r)$ is negative for the developing flow in a tube of constant cross section, and the flow in a converging tube. The stability of velocity profiles in a developing flow at various distances from the entrance of the tube, as well as those in a converging tube, to axisymmetric perturbations was analysed by (Shankar and Kumaran, 1999). In these flows, the inviscid stability equation (equivalent to the Rayleigh equation) has a singularity at a point where the wave speed is equal to the flow speed. In the 'critical layer' of thickness $Re^{-1/3}$ around this point, viscous stresses become significant and it is necessary to incorporate viscous effects. These were included using matched asymptotic expansions (Drazin and Reid, 1981), and the results showed that the flow does become unstable when the Reynolds number exceeds a transition value.

The function $F(r)$ is also negative for parabolic flows in a flexible tube when the perturbations are non-axisymmetric ($n > 0$ in equation 6)(Shankar and Kumaran, 2000). The behaviour of the Reynolds number for the transition from stable to unstable modes as a function of the wave number k is similar to that for non-parabolic flows, and shows a minimum at a finite value of the wave number k. The critical Reynolds number is a lowest for non-axisymmetric perturbations with azimuthal wave number $n = 1$. The mechanism of instability is the transfer of energy from the mean flow to the fluctuations due to the Reynolds stresses in the critical layer where the flow speed and wave speed are equal. Owing to this, the presence of tangential motion is not essential for this instability. The studies of (Shankar and Kumaran, 1999) and (Shankar and Kumaran, 2000) also show that there are multiple solutions for the growth rate in this parameter regime, and consequently there are is a sequence of neutral stability curves in the $Re - \Sigma$ plane. It should be noted that these modes are qualitatively different from those in a rigid channel or tube, since the neutral stability curve does not have the classical upper- and lower-branch structure, but the transition Reynolds number shows an increase proportional to k^{-1} at small k, and an increase proportional to k at large k.

The instabilities of (Hamadiche and Gad-el-Hak, 2002) at high Reynolds number also appear to be inviscid instabilities, though it must be noted that the maximum Reynolds number (1000) considered in this study is too low to infer the scaling of the Reynolds number with the dimensionless number Σ. The wave speed is of the same magnitude as the flow speed in this study, and the profiles for the disturbance velocity and stress show sharp variations, presumably at points where the wave speed is equal to the flow velocity, indicating the presence of critical layers. The authors find that only the non-axisymmetric modes are unstable in the high-Reynolds-number limit, and this finding is in agreement with the conclusions of (Shankar and Kumaran, 2000). However, at lower Reynolds numbers, both axisymmetric and non-axisymmetric perturbations become unstable.

4.4. WALL-MODE INSTABILITY

The wall-mode instability is observed in the high-Reynolds-number limit, where the viscous stresses become important in a wall layer of thickness $O(Re^{-1/3})$ smaller than the channel width or tube radius. The stability of the wall modes in the asymptotic regime $Re \propto \Sigma^{2/3}$ was analysed by (Kumaran, 1998a; Kumaran, 1998b), and it was found that the wall modes are always stable in this regime. However, the asymptotic analysis of (Shankar and Kumaran, 2001a; Shankar and Kumaran, 2002) showed that there is an instability in the asymptotic regime $Re \propto \Sigma^{3/4}$.

In contrast to the inviscid instability, the streamwise velocity in the wall layer for the wall modes is $O(Re^{-1/3})$ larger than that in the outer flow, which is inviscid in the leading approximation. Therefore, the velocity in the wall layer drives the outer flow, in contrast to the inviscid modes where the velocity in the outer flow is dominant. However, the force balances are rather subtle. The inertial and elastic stresses are of the same order of magnitude in the wall medium. The pressure forces due to the outer flow provide the dominant contribution to the normal force at the surface, even though the velocity in the outer flow is small compared to that in the wall layer. A balance of the pressure force due to the outer flow and the normal stresses in the wall material provides the scaling $Re \propto \Sigma^{3/4}$ for the critical Reynolds number. Thus, the critical Reynolds number in this case increases faster than that for the inviscid modes, and consequently these modes are more stable than the inviscid modes. However, this instability is likely to be observed in cases such as the Couette flow in a channel (figure 1(a)), where there is no inviscid instability. It should be noted that tangential motion is necessary for this instability, since the destabilising mechanism is the transfer of energy from the mean flow to the fluctuations due to the shear work done at the interface.

5. Non-linear analyses

The effect of a flexible surface on the nature of the Tollmien-Schlichting instability in a channel was analysed by (Rottenberry and Saffman, 1990), who found that the subcritical Tollmien-Schlichting instability in a rigid-walled channel was rendered supercritical if the walls are sufficiently compliant. However, the subsequent study of (Rottenberry, 1992) indicated that this effect was numerically small, and would not make a significant difference to the qualitative nature of the transition. The plane-channel flow was used by (Pierce, 1992) to illustrate the derivation of the Landau equation for general interfacial flows using symbolic computation.

The non-linear analysis of the viscous modes in the flow past a gel-wall model was carried out by (Shankar and Kumaran, 2001b). The results indicated that the type of bifurcation is sensitive to the boundary conditions between the gel and the rigid surface it is supported on. For 'grafted' gel boundary conditions, where the gel is fixed on to a rigid substrate, the bifurcation was found to be subcritical. However, for 'adsorbed' gel boundary conditions, where slip was permitted in the tangential direction between the gel and the rigid substrate, the bifurcation was supercritical for certain sets of parameter values. Even when the bifurcation is subcritical, it was found that the non-linear interactions did not result in a significant reduction in the critical Reynolds number, and there is a reduction of about 1% when the amplitude of the perturbation velocity is 2.5% of the mean velocity. This is in contrast to the reduction of the critical Reynolds number by 65% for the same perturbation amplitude for the Tollmien-Schlichting instability. The results that the bifurcation is subcritical is in qualitative agreement with experiments (Kumaran and Muralikrishnan, 2000; Muralikrishnan and Kumaran, 2002), where it was found that the flow could not be maintained in a supercritical state after transition. A non-linear for the flow past a membrane was carried out by (Thaokar and Kumaran, 2002), and it was found that the viscous instability is always supercritical in this case.

References

Benjamin, T. B. (1963) The threefold classification of unstable disturbances in flexible surfaces bounding inviscid flows. *J. Fluid Mech.*, **16**, 436-450.

Carpenter, P. W. and Garrad, A. D. (1985) The hydrodynamic stability of flows over Kramer - type compliant surfaces. Part 1. Tollmien - Schlichting instabilities. *J. Fluid Mech.*, **155**, 465-510.

Carpenter, P. W. and Garrad, A. D. (1986) The hydrodynamic stability of flows over Kramer - type compliant surfaces. Part 2. Flow-induced surface instabilities. *J. Fluid Mech.*, **170**, 199-232.

Carpenter, P. W. and Morris, P. J. (1990) Effect of anisotropic wall compliance on boundary layer stability, *J. Fluid Mech.* **218**, 171-223.

Davies, C. and Carpenter, P. W. (1997) Instabilities in a plane channel flow between compliant walls, *J. Fluid Mech.* **352**, 205-243.

Drazin, P. G. and Reid, W. H. (1981)*Hydrodynamic Stability*, Cambridge University Press, Cambridge.

Gajjar, J. S. B., Gibson, A. and Sen, P.K. (2002) Instabilities in compliant pipe and related flows, presentation at *IUTAM Symp. on Flow in Collapsible Tubes and past Other Highly Compliant Boundaries*, University of Warwick, England, 26 - 30 March.

Green, C. H., and Ellen, C. H. (1972) The stability of a plane Poiseuille flow between flexible walls, *J. Fluid Mech.* **51**, 403-416.

Grotberg, J. B. and Reiss, E. L. (1984) Subsonic flapping flutter, *J. Sound Vib.* **92**, 349-361.

Grotberg, J. B. and Shee, T. R. (1985) Compressible flow channel flutter, *J. Fluid Mech.* **159**, 175-193.

Hains, F. D. and Price, J. F. (1962) Effect of a flexible wall on the stability of Poiseuille flow, *Phys. Fluids* **5**, 365-365.

Hamadiche, M. and Gad-el-Hak, M (2002) Temporal stability of flow through viscoelastic tubes, in press, *J. Fluids Structures.*

Kumaran, V., Fredrickson, G. H. and Pincus, P. (1994) Flow induced instability at the interface between a fluid and a gel at low Reynolds number, *J. Phys. France II* **4**, 893-911.

Kumaran, V. (1995a) Stability of the viscous flow of a fluid through a flexible tube, *J. Fluid Mech.* **294** , 259-281.

Kumaran, V. (1995b) Stability of the flow of a fluid through a flexible tube at high Reynolds number, *J. Fluid Mech.* **302** , 117-139.

Kumaran, V. (1996) Stability of an inviscid flow in a flexible tube, *J. Fluid Mech.* **320**, 1-17.

Kumaran, V. (1998a) Stability of fluid flow in a flexible tube at intermediate Reynolds number, *J. Fluid Mech.* **357**, 123-140.

Kumaran, V. (1998b) Stability of wall modes in a flexible tube, *J. Fluid Mech.* **362**, 1-15.

Kumaran, V. (1998c) Asymptotic analysis of wall modes in a flexible tube, *Euro. Phys. J. B* **4**, 519-527.

Kumaran, V. (2000) Classification of instabilities in the flow past flexible surfaces, *Current Science* **79**, 766-773.

Kumaran, V. and Muralikrishnan, R. (2000) Spontaneous growth of fluctuations in the viscous flow of a fluid past a soft interface, *Phys. Rev. Lett.* **84**, 3310-3313.

Kumaran, V. and Srivatsan, L. (1998) Stability of fluid flow past a membrane, *Euro. Phys. J. B* **2**, 259-266.

LaRose, P. G. and Grotberg, J. B. (1997) Flutter and long wave instabilities in compliant channels conveying developing flows, *J. Fluid Mech.* **331**, 37-58.

Muralikrishnan, R. and Kumaran, V. (2002) Experimental study of the instability of fluid flow past a flexible surface, *Phys. Fluids* **14**, 775-780.

Pierce, R. (1992) The Ginzburg - Landau equation for interfacial instabilities. *Phys. Fluids A* **4**, 2486-2494.

Rottenberry, J. M. and Saffman, P. G. (1990) Effect of compliant boundaries on the weakly nonlinear shear waves in a channel flow. *SIAM J. Appl. Math.* **50**, 361-394.

Rottenberry, J. M. (1992) Finite-amplitude shear waves in a channel with compliant boundaries. *Phys. Fluids A* **4**, 270-276.

Sen, P. K., and Arora, D. S. (1988) On the stability of laminar boundary layer flow over a flat plate with a compliant surface, *J. Fluid Mech.* **197**, 201-240.

Shankar, V. and Kumaran, V. (1999) Stability of non-parabolic flows in a flexible tube, *J. Fluid Mech.* **395**, 211-236.

Shankar, V. and Kumaran, V. (2000) Stability of non-axisymmetric modes in a flexible tube, *J. Fluid Mech.* **407**, 291-314.

Shankar, V. and Kumaran, V. (2001a) Asymptotic analysis of wall modes in a flexible tube revisited, *Euro. Phys. J. B* **19**, 607-622.

Shankar, V. and Kumaran, V. (2001b) Weakly non - linear analysis of the viscous flow of a fluid past a flexible surface, *J. Fluid Mech.* **434**, 337-354.

Shankar, V. and Kumaran, V. (2002) Stability of wall modes in flow past a flexible surface, in press, *Phys. Fluids*

Srivatsan. L. and Kumaran, V. (1997) Flow induced instability of the interface between a fluid and a gel, *J. Phys. II (France)* , **7** , 947-963.
Thaokar, R. M., Shankar, V. and Kumaran, V. (2001) Effect of tangential interface motion on the viscous instability in the fluid flow past flexible surfaces, *Euro. Phys. J. B* **23**, 533-550.
Thaokar, R. M. and Kumaran, V. (2002) 'Stability of fluid flow past a membrane', in press, *J. Fluid Mech.*
Weaver, D. H. and Paidoussis, M. P. (1977) On collapse and flutter phenomena in thin tubes conveying fluids, *J. Sound Vib.* **50**, 117-132.

Chapter 6

WAVE EXCITATION ON FLEXIBLE WALLS IN THE PRESENCE OF A FLUID FLOW

A.D. LUCEY

Department of Mechanical Engineering, Curtin University of Technology, G.P.O. Box U1987, Perth, W.A. 6845, Australia.

N. PEAKE

Department of Applied Mathematics & Theoretical Physics University of Cambridge, Silver Street, Cambridge CB3 9EW U.K.

Abstract

Theoretical and computational investigations of wave excitation and evolution in fully-coupled flow-structure systems are discussed. A distinguishing feature of the body of work reviewed is its attempt to shed light on *how* and *where* waves come into being in response to some form of applied excitation. Identifying the properties of such waves and their contributions to the overall *system behaviour* is also a key objective of the studies reviewed. The model problem of waves initiated by line excitation applied to an elastic plate adjacent to a uniform flow (Brazier-Smith & Scott 1984 and Crighton & Oswell 1991) is first described. Thereafter, work that incorporates enhancements to the structural, geometric and fluid-flow models is presented. The results of these studies are related to the findings of the progenitor model problem in order to assess its general applicability in this branch of fluid-structure interaction.

P.W. Carpenter and T.J. Pedley (eds.),
Flow past Highly Compliant Boundaries and in Collapsible Tubes, 119-145.

1. Introduction

This article serves to review relatively recent work in which an essential component of the flow-structure system is a localised source of excitation. This work falls within the present volume's theme of 'flow over highly compliant boundaries'. With the exception of Lingwood & Peake (1999), potential flow is used throughout. Thus, the flow-structure applications envisaged here would most probably concern metal panels interacting with very high Reynolds-number flows in either marine or aerospace environments. The stiff, rubber-based, compliant coatings used by Choi *et al.* (1997) to reduce turbulent skin-friction in water present a further example of the parameter range covered herein. Significantly different, however, is the flow-structure combination required to achieve skin-friction reduction through the postponement of boundary-layer transition (Carpenter 1990). Nevertheless, the unstable waves discussed in this article may also be found on drag-reducing compliant surfaces, albeit modified by boundary-layer effects.

Traditional means for solving fully coupled flow-structure problems usually adopt a normal-mode boundary-value approach that yields an eigenproblem (for example, see Benjamin 1960, Landahl 1962, Carpenter & Garrad 1985, Lucey & Carpenter 1995). Such methods cannot address the matter of how disturbances come into being, nor do their results embrace spatial dependence other than through local mean-state conditions. The pioneering work of Brazier-Smith & Scott (1984) and Crighton & Oswell (1991), following the principles used in *Structural Acoustics*, incorporated a localised source of excitation and formally addressed the initial-value problem. The techniques they used are described in Section 2 below. This approach enabled them to identify which of the mathematical solutions present in the problem represented physically realisable waves and predict where such waves would be found - either upstream or downstream of the source of excitation. Moreover, they were able to determine rigorously whether waves were either neutrally stable, convectively unstable or absolutely unstable.

Brazier-Smith & Scott (1984) and Crighton & Oswell (1991) uncovered a number of unusual phenomena in their linearised studies of an elastic plate interacting with a uniform flow. These are described in Section 2 below. The main thrust of following work, described in this article, has sought to determine whether the findings for what is probably the simplest fully-coupled flow-structure model carry across to other, more refined, systems. In Section 2 different flexible-wall models are considered while Section 3 describes computational studies of the basic Crighton & Oswell problem for a flexible plate of finite streamwise extent with fixed ends. In Section 4, model-refinement entails the inclusion of flow shear representing a boundary layer. Nonlinear developments are presented in Section 5; these permit the evolution of linearly unstable waves present in the basic Crighton & Oswell problem to be studied. Finally, in Section 6, we briefly identify a number of topics that we believe merit study in order to progress understanding in the subject area.

2. Fundamental Model Problem

In this section we will set out the fundamental problem studied by Brazier-Smith & Scott (1984) and Crighton & Oswell (1991). Consider an infinitely long plate of thickness h, with bending stiffness $B \equiv Eh^3/12(1-\nu^2)$, where ν is the Poisson ratio and E is Young's modulus. The plate density is ρ_s, and its mass per unit area is therefore $m \equiv \rho_s h$. Above the plate there is an inviscid, incompressible fluid, of density ρ_f, which possesses a steady mean flow of speed U_f parallel to the plate. Below the plate there is a vacuum. The x axis is taken to be aligned with the (undisturbed) plate, and the y axis points into the fluid. In the main we will be considering two-dimensional motion, so that dependence on the spanwise coordinate z is suppressed. An unsteady line force, of strength $F(t)$, is located at the origin. The line force will be turned on at some initial time $t = 0$, so that $F(t) \equiv 0$ for $t < 0$. The plate motion is then driven by both the source and by the unsteady hydrodynamic pressure in the fluid, which is itself fully coupled to the plate motion. The plate transverse deflection, measured from its undisturbed position $y = 0$, is $\eta(x, t)$, and supposing that this remains small it follows that the motion can be described using linearised equations. Neglecting any gravitational effects, the plate equation is

$$m\frac{\partial^2 \eta}{\partial t^2} + B\frac{\partial^4 \eta}{\partial x^4} = F(t)\delta(x) - p(x, 0, t) \ , \tag{1}$$

where $p(x, y, t)$ is the unsteady fluid pressure. The fluid is incompressible and its motion assumed irrotational. Its dynamics can therefore be described by a velocity potential $\phi(x, y, t)$ that satisfies the Laplace equation

$$\nabla^2 \phi = 0 \ . \tag{2}$$

The fluid pressure is given by the linearised Bernoulli relation

$$p = -\rho_f \frac{\partial \phi}{\partial t} - \rho_f U_f \frac{\partial \phi}{\partial x} \ , \tag{3}$$

and finally the plate and fluid motions are coupled together by the boundary condition of continuous vertical velocity at the plate surface

$$\frac{\partial \phi}{\partial y} = \frac{\partial \eta}{\partial t} + U_f \frac{\partial \eta}{\partial x} \quad \text{on} \quad y = 0 \ . \tag{4}$$

These equations are nondimensionalised by introducing a lengthscale m/ρ_f and a timescale $m^{5/2}/\rho_f^2 B^{\frac{1}{2}}$, in which case $m = B = 1$ in the above equations, and the flow speed U_f is replaced by the nondimensional speed $U = U_f m^{3/2}/\rho_f B^{1/2}$.

The problem described by equations (1)-(4) is perhaps the simplest problem which could be envisaged in structural acoustics with mean flow, but even so it was shown by Brazier-Smith & Scott (1984) and Crighton & Oswell (1991) that its solution exhibits exceedingly complicated, and perhaps quite unexpected, behaviour. In order to obtain this solution it is crucial that the issue of causality is treated correctly, and this involves application of a rather subtle technique developed by Briggs (1964) and Bers (1983). First, equations (1)-(4) are Fourier transformed in time and in the streamwise coordinate x. The solution can then be written down in terms of the inversion integral

$$\eta(x,t) = \frac{1}{4\pi^2}\int_C \Psi(x,\omega)\bar{F}(\omega)\exp(-\mathrm{i}\omega t)\mathrm{d}\omega \ , \tag{5a}$$

$$\Psi(x,\omega) = \int_{-\infty}^{\infty} \frac{\exp(\mathrm{i}kx)}{D(k,\omega)}\mathrm{d}k \ . \tag{5b}$$

Here $\bar{F}(\omega)$ is the time transform of the forcing, and $D(k,\omega)$ is the so-called dispersion function. The result (5) is a very general one, and all that changes between different problems is the form of $D(k,\omega)$. For our problem of a plate with mean flow it is easy to show that

$$D(k,\omega) = \omega^2 - k^4 + \frac{(\omega - kU)^2}{|k|} \ , \tag{6}$$

where $|k| = k$ if $\mathrm{Re}(k) > 0$ and $|k| = -k$ if $\mathrm{Re}(k) < 0$. The most important point about the inversion integral is the choice of inversion contours. The contour in the (spatial) k plane is chosen to lie along the real axis, while the contour in the (temporal) ω plane is C, which we choose to be a straight line parallel to the real ω axis, but in the upper half of the ω plane and located above all singularities of the integrand in the ω integral. For $t < 0$ the ω integral is evaluated by closure in the upper half plane, and the choice of C then satisfies causality by yielding $\eta(x,t) = 0$ for $t < 0$. It should be noted that since the forcing is turned on at $t = 0$, the function $\bar{F}(\omega)$ is analytic in the upper half plane, so that the singularities of the ω integral in the upper half plane are identically those of $\Psi(x,\omega)$, and are not related to the forcing.

The next step in the Briggs and Bers procedure is to attempt to deform the ω contour C downwards towards the real axis. As this is done the inversion contour in the k plane must be deformed off the real axis in order to avoid any singularities in the k integrand from crossing it, thereby retaining an analytic solution. The singularities in the k integrand are any branch points of $D(k,\omega)$ and the poles corresponding to the spatial roots of $D(k,\omega) = 0$ for ω on C. Brazier-Smith & Scott (1984) analysed the motion of these poles as C is lowered, and found that, provided U is sufficiently large, two poles which originate on opposite sides of the real k axis coalesce before C has been deformed all the way down onto the real ω axis. This is a so-called 'pinch point', and occurs, say, at $k = k_0$ for $\omega = \omega_0$. The significance of this is that once the pinch has occurred the spatial inversion contour can no longer be deformed to avoid a pole crossing, so that the downward deformation of the temporal inversion contour C must be halted. The pinch point in the k plane is a double root zero of $D(k,\omega)$, so that $\omega = \omega_0$ is a branch point of $\Psi(x,\omega)$ in the ω plane. Brazier-Smith & Scott show that once C is then deformed round the corresponding ω branch cut, the solution of the problem at late times (i.e. as $t \to \infty$) is dominated by a contribution from the pinch point, i.e. by a mode proportional to $\exp(-\mathrm{i}\omega_0 t + \mathrm{i}k_0 x)$. Since $\mathrm{Im}(\omega_0) > 0$, it follows that this mode grows exponentially in time at all x. This is ***absolute instability***, and means that disturbances will grow in situ, rather than, as is more normal, being convected away downstream. (This point of course corresponds to the fact that the group velocity at the pinch point is zero). It also means that the system response selects its own frequency, $\mathrm{Re}(\omega_0)$, rather than a frequency forced externally by the driver. The critical minimum value of dimensionless U for absolute instability is shown to be $U_c \approx 0.0742$ by Brazier-Smith & Scott, while Crighton & Oswell derive an exact closed-form expression for U_c by noting that the threshold for absolute instability occurs when three spatial zeros of $D(k, \omega)$ coalesce. For steel in water this threshold takes the rather high value of 14.5 ms^{-1}, and for aluminium the even higher value of 43.1 ms^{-1}.

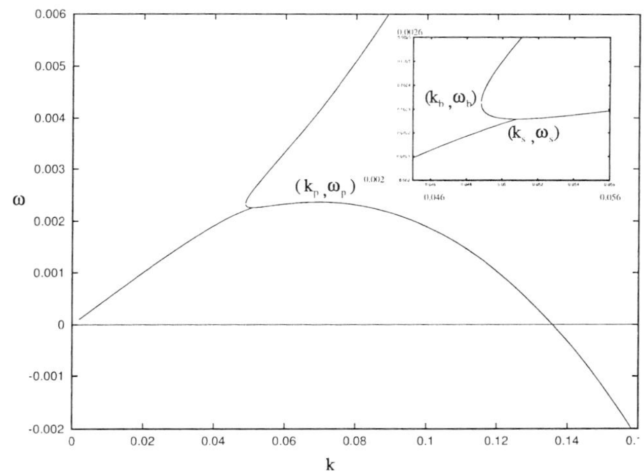

Figure 1. Plot of the roots $k(\omega)$ of the dispersion relation $D(k, \omega) = 0$. The solid lines are neutral modes, and the dashed line represents the real part of the convective instability. Behaviour for $k < 0$ can be inferred from the symmetry $D(k, \omega) = D(-k, -\omega)$. From Peake (2001).

If $U \leq U_c$ it follows that the no pinch occurs for $\mathrm{Im}(\omega) > 0$, and the contour C can be deformed all the way down onto the real axis. One can then consider the spatial stability problem of determining the k zeros of $D(k, \omega)$ for real ω. The results of this calculation are shown in figure 1. For $0 < \omega < \omega_s(U)$ there are two real values of k roots (one negative value of k, together with the positive value shown), and a pair of complex conjugate roots with $\mathrm{Re}(k) > 0$. As the temporal integration contour C is lowered towards the real axis, both these complex roots originate in the upper half of the complex k plane and move downwards, and one of them crosses the real k axis and comes to rest in the lower half of the k plane when C reaches the real ω axis. To avoid a pole crossing the spatial integration contour has therefore been deformed below this root, so that the root of the complex-conjugate pair which possesses negative imaginary part is located in $x > 0$, and grows downstream. This is a *convective instability*

For $\omega_s < \omega < \omega_p$ there are four neutral modes, but within this range the interval $\omega_s < \omega < \omega_b$ is especially interesting. Essentially, the remarkable feature here concerns the anomalous spatial location of one of these four neutral modes. A standard procedure, due to Rayleigh and Lighthill (see Lighthill 1960), is to determine the spatial location of a neutral mode by insisting that its group velocity, $\partial\omega/\partial k$, should be directed *away* from the driver at $x = 0$. Inspection of figure 1 then suggests that in this ω range there is just one neutral mode downstream, while there are three neutral modes (including one with $k < 0$) upstream. However, this conclusion is incorrect. If one considers instead how these four modes move in the complex k plane as C is lowered, then it turns out that the mode shown in figure 1 with $k_b < k < k_s$ originates in the upper half of the k plane, and moves downwards and crosses the real k axis as C is lowered. Before C reaches the real ω axis, the locus of the mode turns round in the k plane, and comes to rest on the real k axis from below as C reaches the real ω axis. Hence, this mode has a negative group velocity (because it approached the real k axis from below), and according to the Rayleigh-Lighthill criterion should therefore be located in $x < 0$. However, because it originated in the upper half of the complex k plane and crossed the real k axis from above, the more rigorous Briggs-Bers method tells us that this mode is in fact located downstream of the driver, in $x > 0$, and therefore has its group velocity pointing *towards* the driver. In the frequency range $\omega_s < \omega < \omega_b$ there are therefore two neutral modes upstream and two downstream. In the range $\omega_b < \omega < \omega_p$ all the neutral modes behave in the usual way, with group velocity pointing *away* from the driver, so that there are again two modes upstream and two downstream. For $\omega > \omega_p$ there are only two neutral modes (one with $k < 0$), which as ω increases match onto the usual bending waves on an elastic plate in still fluid. Two evanescent modes, emanating from the turning point at $k = k_p$, also exist, but are not shown in figure 1.

A much fuller account of the behaviour described in the previous two paragraphs is given by Crighton & Oswell (1991), including analytical and asymptotic analysis of the various modes in the limit of small U (note that this is a realistic limit, since the dimensionless speed threshold for absolute instability is a small number). Another issue which receives a great deal of attention from Crighton & Oswell is the energetics of the system, and it turns out that some unusual behaviour can be observed here as well. In fact, the first to consider the energy of neutral modes on an elastic plate with flow were Landahl (1962) and Benjamin (1963). The wave energy, or activation energy, E of a neutral mode is defined to be the energy required to create the mode, starting from rest at time $t = -\infty$. For the plate with uniform flow, Cairns (1979) was able to show that for a wave of unit amplitude

$$E = \frac{\omega}{4}\frac{\partial D}{\partial \omega} \quad . \tag{7}$$

For the neutral modes in figure 1 it is easy to show that $E > 0$ for those modes on the upper branch of the neutral curve with $\omega > \omega_b$, and for those modes with $\omega < 0$. These are termed Positive Energy Waves (PEW), or Class B Waves by Benjamin (1963), and behave conventionally in that if dissipation is present in the system, for instance via the introduction of dashpots on the plate, they are damped. However, those modes on the lower branch from $\omega = \omega_b$ round to $\omega = 0$ have $E < 0$. These are Negative Energy Waves (NEW), or Class A Waves, and behave anomalously in that when dissipation is added they are destabilised, since the dissipation acts to reduce the system energy and therefore increases the amplitude of the NEW. The presence of NEWs has important implications for the direction of energy flow into the driver. For a plate with zero mean flow, the real part of the line drive admittance, $\mathrm{Re}(A_0)$, is always positive, so that energy is flowing away from the driver into the fluid and the structure. However, Crighton & Oswell (1991) show that with nonzero mean flow $\mathrm{Re}(A_0) < 0$ throughout the frequency range $0 < \omega < \omega_p$ where NEWs exist. The net energy flux carried by the various modes is directed towards the driver, and this has arisen because those NEWs with group velocity pointing *away* from the driver are effectively transporting positive energy *towards* the driver. This observation also explains why the Rayleigh-Lighthill group velocity criterion is inapplicable to the present problem, because in the derivation of that criterion it is assumed that energy flux and group velocity always point in the same direction.

Abrahams & Wickham (2001) have also considered the same problem as Brazier-Smith & Scott and Crighton & Oswell, but this time using a different mathematical approach, in which the governing equations (1)-(4) are Fourier transformed in x only, and the resulting ODEs in t then solved explicitly. In the long-time limit $t \to \infty$, their results are fully consistent with the results of Crighton & Oswell. However, it should be noted that for a moving observer (i.e. x/t fixed with $t \to \infty$), a case not considered by Crighton & Oswell, transient waves associated with turning on the driver at $t = 0$ are important. Abrahams & Wickham consider a number of other modifications to the Crighton & Oswell model, which will be described in the next section

It is now clear that the problem of an infinite elastic plate in uniform mean flow admits some very unusual and unexpected behaviour. Since their underlying model is clearly rather simplistic, however, the question remains as to whether or not this behaviour is physically realisable. Modifications to the model to include more physically realistic geometry and flow features will be discussed in the following sections.

2. Linear Structural Extensions

We will now present a number of modifications which have been made to the Brazier-Smith & Scott and Crighton & Oswell model of the plate. In this section we will restrict attention to linear plate equations.

2.1 MODIFIED PLATE PROPERTIES

De Langre (2001) has considered the effects of pre-tensioning the plate. Mathematically, this involves introducing an extra term $-T\partial^2\eta/\partial x^2$ onto the left hand side of equation (1), so that the dispersion relation (6) has an extra term Tk^2, where T is a nondimensional tension. De Langre calculates the boundaries of both convective and absolute instability, and his results are shown in figure 2. For $T = 0$ the results of Crighton & Oswell are recovered. Note that convective instability occurs down to $U = 0$ when $T = 0$, but that as T is increased the plate becomes stable over an increasingly wide range of U. For T large de Langre's results approach the corresponding stability boundaries for a membrane, as calculated by Kelbert & Sazonov (1996). The crucial point here is that introduction of even a small amount of pretension significantly increases the values of flow speed U required for both convective and absolute instability to occur.

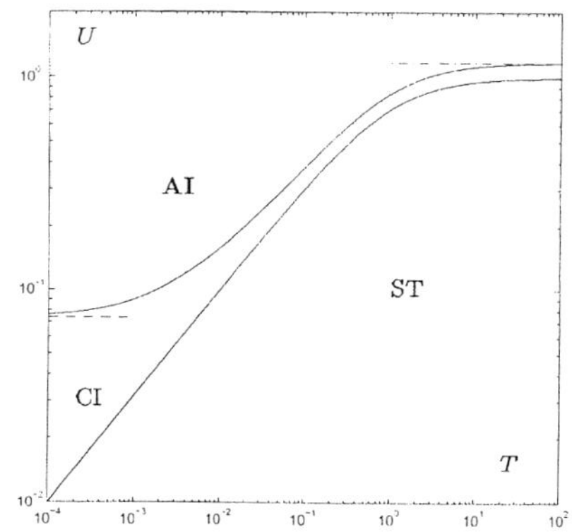

Figure 2. Regions of absolute instability (AI), convective instability (CI) and stability (ST) for a pre-tensioned plate. The dashed line is the absolute instability boundary for a plate with zero tension, and the dash-dot line is the absolute instability boundary for a membrane. Reproduced from de Langre (2001), with permission.

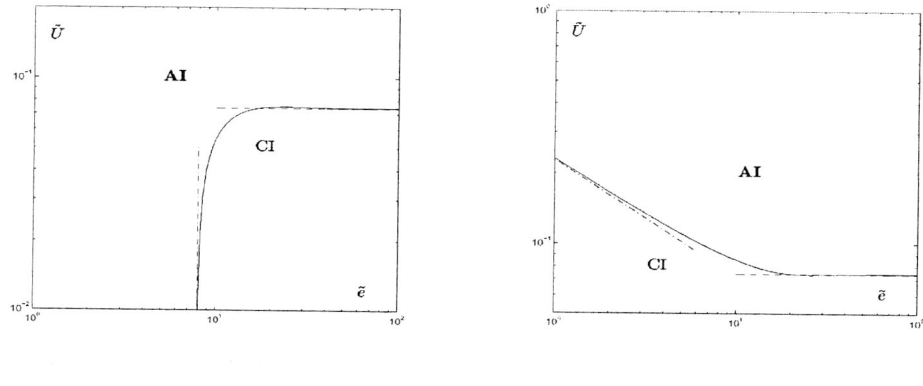

Figure 3. Regions of absolute instability (AI) and convective instability (CI) for channel flow, with e the channel width; (a) sinuous mode; (b) varicose mode. The dashed lines are the absolute instability boundary for a single plate, found by Brazier-Smith & Scott (1984) and Crighton & Oswell (1991), and the dash-dot lines are the absolute instability boundaries in the thin channel limit. Reproduced from de Langre (2001), with permission.

A second modification considered by de Langre is the case of two parallel elastic plates with uniform mean flow between them, as a two-dimensional model of pipe flow. For sinuous modes it is shown that the dispersion relation becomes

$$\omega^2 - k^4 + \frac{\tanh(ke)}{|k|}(\omega - Uk)^2 = 0 \ , \tag{8}$$

where e is the (dimensionless) plate separation (so that in particular equation (6) is regained if $e \to \infty$). For varicose disturbances, the tanh in (8) is replaced by coth. Results for the stability boundaries in both cases are shown in figure 3, and it is now clear that the single-plate results are only regained for quite large plate separation. Note also that as U is increased the transition to absolute instability occurs through the sinuous mode first. For very thin channels, de Langre is able to show that the absolute instability boundary occurs at flow speed $U = 0.228/\sqrt{e}$, confirming that the close proximity of the two plates has a strongly stabilising effect.

Abrahams & Wickham (2001) consider the effects of fluid compressibility, by replacing the Laplace equation (2) with the linear convected wave equation. They also consider the use of an improved model of the plate motion, in which the thin plate is replaced by an elastodynamic layer of isotropic material, whose motion is described by the usual two displacement potentials, which themselves satisfy wave equations with primary and secondary wave speeds. However, for weak compressibility (*i.e.* low freestream Mach number, as must always be the case in practice underwater) and for a thin layer under light fluid loading, respectively, Abrahams & Wickham show that these modifications do not qualitatively alter the Crighton & Oswell results, and equations (1)-(4) are therefore described as being 'structurally stable' to these perturbations.

Abrahams & Wickham argue that one additional perturbation is potentially significant. Following the argument of the previous section, it can be seen that the neutral turning point (k_p, ω_p) is a NEW, and moreover it is also a pinch point. When weak dissipation is introduced, this point is displaced (by a small amount) into the upper half of the ω plane, and leads to absolute instability. Abrahams & Wickham therefore note that with the inclusion of arbitrarily small, but nonzero, dissipation the plate displacement will eventually grow in time for all x irrespective of the forcing or the (nonzero) flow speed.

2.2 PLATE CURVATURE

Peake (1997) sought to generalise the Crighton & Oswell geometry to include the effects of plate curvature by considering a thin cylindrical shell, with uniform axial mean flow either inside or outside, and driven by a ring line force with fixed tangential mode order. Using the Donnell thin-shell equations for the three components of plate displacement, a rather complicated, but closed-form, dispersion relation is derived. This dispersion relation depends on both the fluid-solid density ratio, $\rho \equiv \rho_f/\rho_s$ and on a dimensionless cylinder radius $a = r\rho/h$, where r is the dimensional cylinder radius and h is the wall thickness. In the limit of large a, it turns out that the complicated dispersion relation reduces to

$$\omega^2 - k^4 + \frac{(\omega - Uk)^2}{|k|} - \frac{12(1-\nu^2)}{\rho^2 a^2} = 0 \ , \tag{9}$$

which is very similar to the dispersion relation (6) for the flat plate, except that we now have an additional term which is independent of both k and ω. Interestingly, this is exactly the dispersion relation one would obtain for a flat plate with a spring foundation, so that in this large radius limit the hoop stress introduced by plate curvature is equivalent to spring stiffness.

The stability boundaries, for both the full dispersion relation and for the approximation (9) are derived in Peake (1997). The minimum value of U required for absolute instability is shown to satisfy the implicit relation

$$\rho^2 a^2 = \frac{(1-\nu^2)5^{5/3}2^{8/3}}{U^{8/3}[U^{2/3} - U_0^{2/3}]}, \tag{10}$$

with U_0 the flat-plate value. As $a \to \infty$ the two-dimensional result is regained, but for finite values of a it turns out that exceedingly large flow speeds are required for absolute instability. For instance, if $a = 100$ (which for steel in water corresponds to a cylinder radius of 780 times the plate thickness, a large value by any standards), then we would require $U \geq 0.87$, which corresponds to the unrealistically high dimensional flow speed of 170ms^{-1}. This suggests that the introduction of plate curvature has a very strong effect in inhibiting the onset of absolute instability.

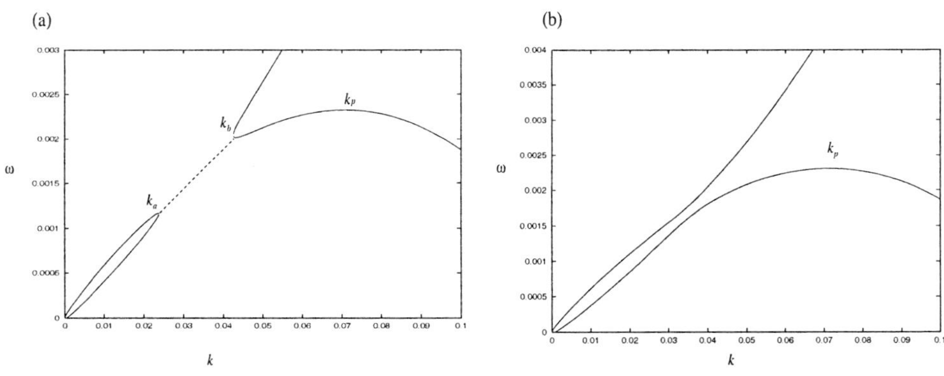

Figure 4. Plot of the roots $k(\omega)$ of the dispersion relation for a cylindrical shell with uniform mean flow of speed $U = 0.05$ outside, with radius (a) $a = 25000$ and (b) $a = 20000$. The solid lines are neutral modes, and the dotted line represents the real part of the convective instability.

When the dispersion diagram for real ω is plotted, a similar picture to that seen by Crighton & Oswell emerges, provided that a is sufficiently large - see figure 4. One new feature, however, is the appearance of a second loop of neutral modes at low frequency, so that the range of convective instability is restricted between the two branch points $k = k_{a,b}$. As a is decreased, the branch points move together, and coalesce at some critical value of a, which from equation (10) can be approximated by

$$a_1(U) = \left[\frac{48(1-\nu^2)}{\rho^2 U^4}\right]^{\frac{1}{2}} . \tag{11}$$

For $a < a_1(U)$ no convective instability is present, and the system is stable - again see figure 4. Moreover, the neutral mode with group velocity directed towards the driver is associated with the existence of the branch point $k = k_b$, and cease to exist for $a < a_1(U)$. For $U = 0.05$ it turns out that $a_1 = 21370$, which means that the convective instability, and the anomalous neutral mode, are only present when the cylinder radius of curvature exceeds about 166,00 plate thicknesses, a degree of flatness which may well be unattainable in engineering practice!

Peake (1997) also shows that NEWs exist on the cylindrical shell, this time for all values of a in excess of $a_2(U)$, with

$$a_2(U) \approx \frac{2^{7/3}(1-\nu^2)^{\frac{1}{2}}}{\rho U^{4/3}} . \tag{12}$$

The way that NEWs are eliminated from the system is that the turning point $k = k_p$ shown in figure 4 moves to progressively lower frequencies as a is reduced, until it crosses $\omega = 0$ at $a = a_2$. After this point all of the lower neutral branch lies in $\omega < 0$, and by (7) therefore corresponds to positive wave energy. For $U = 0.05$ it is found that $a_2 = 2040$, a much smaller value than a_1, indicating that NEWs are present down to more realistic plate curvatures.

3. Computational Studies

3.1 ELASTIC PLATE

The theoretical studies reviewed above assume that the flexible surface is of infinite extent so as to make the the problem tractable. Thus, the long-time response of the system has a frequency identical to that of the applied line-excitation, ω_F (for example, see Crighton & Oswell 1991 and Peake 1997). This is because waves at other frequencies, excited by the start-up procedure and having non-zero group velocity, will convect away to infinity in the form of a wave-packet in the long-time limit. Furthermore, the energy of transients associated with both the upstream- and downstream-propagating PEW's of figure 1 would quickly be eliminated if structural damping were present. In contrast, transients associated with the NEW-part of the lower branch would be destabilised by damping. On this branch Abrahams & Wickham (2001) (and Peake 1997 for the corresponding cylindrical geometry) have recognised that the point (k_p, ω_p) on the dispersion diagram of figure 1 represents an absolutely unstable wave. This wave would persist locally and could strongly contribute to the overall system response. Equally important may be the rôle of the convectively unstable transient waves, as noted by Abrahams & Wickham (2001), in a system of finite length. At finite time these will impact the downstream end of the flexible surface to launch upstream propagating waves that will then remain a part of the overall system response. Lastly, for ω_F lying within Crighton & Oswell's frequency range of convectively unstable waves (existing downstream of the driver), the matter of their fate when they arrive at the trailing edge of a finite panel is an important question. The effect of finiteness on the basic Crighton & Oswell* problem was therefore studied by Lucey (1998) using numerical simulations and the principal findings are reported below.

The computational model used to solve the system equations, (1)-(4), deploys a higher-order, source-sink, boundary-element method to solve the Laplace equation, (2), subject to the linearised no-flux boundary condition, (4). The velocity potential and its derivatives are therefore found for a set of discrete interfacial points which then yield the perturbation pressure through equation (3). This pressure, along with the line-excitation, is then used to force the flexible-wall dynamics couched in finite-difference form. A semi-implicit time-stepping algorithm then updates the wall and fluid displacement, velocity and acceleration at each of the interfacial points in the discretised spatial domain. A detailed description of the method is given in Lucey & Carpenter (1992).

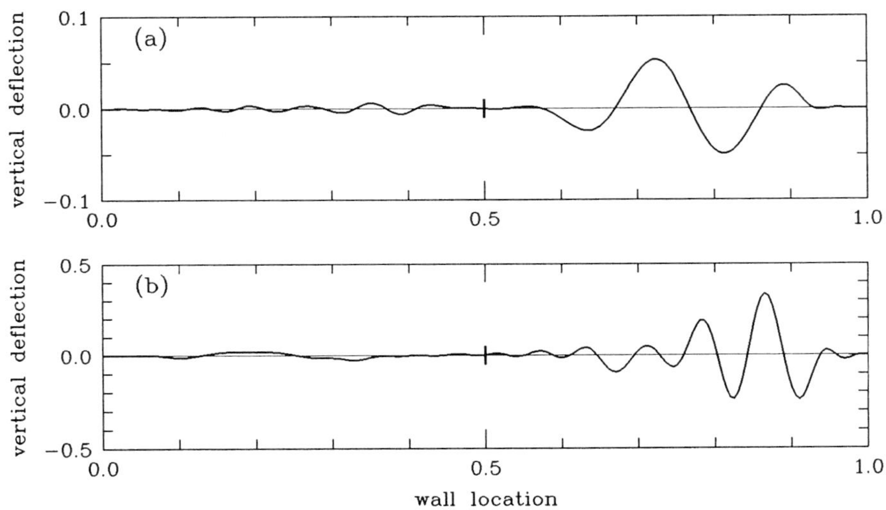

Figure 5. Instantaneous wall profiles for $U = 0.05$ (fluid flow is from left to right) for low-frequency line excitation at the plate mid-point (marked by vertical bar) at (a) $t = 4\Delta T$, and (b) $t = 8\Delta T$. Adapted from Lucey (1998).

Figure 5 shows instantaneous wall profiles at two times, $4\Delta T$ and $8\Delta T$ (where ΔT represents 1340 units of the nondimensional timescale) after start-up for a plate undergoing oscillatory line excitation at $\omega_F = 0.001492$ at its mid-point. Throughout this subsection, $U = 0.05$ and the flow direction is from left to right in the figures. The plate length, L, is 738 units of the nondimensional lengthscale and hinged end conditions are modelled. A further feature of the plate model is

* The 'inclusion of finiteness' can also be regarded as a *perturbation* to the Crighton & Oswell model in the terminology of Abrahams & Wickham (2001) (see Section 2.1 above).

that structural (or dash-pot) damping has been included adjacent to the ends to suppress reflections of modified flexural waves and, more importantly, to provide an exit for wall energy in a finite system which includes a potential source of energy input, this being the continuous line excitation. However, omitting such damping and/or using built-in end conditions does not qualitatively change the wall-flow system behaviour. The excitation frequency for the numerical experiment of figure 5 lies in the range $0 < \omega < \omega_S$. Theory predicts the existence of a convectively unstable wave downstream of the driver and this can be seen at $t = 4\Delta T$ (figure 5a). We will refer to this as a Type I wave. Upstream of the driver is seen a wave on the NEW branch of the dispersion curve, figure 1; energy records show that this is also unstable. This would be the case for a plate of infinite length *only* if it were dissipative. The wall profile $t = 8\Delta T$ (figure 5b) clearly finds a different unstable wave occupying the downstream half of the plate. (Note the different vertical scales used in figures 5a and 5b.) This wave originates with the impact of the original convectively unstable wave on the trailing edge of the plate and it has exactly the same wave properties as the wave located upstream of the driver at the earlier times represented by figure 5a. It will be referred to as a Type II wave.

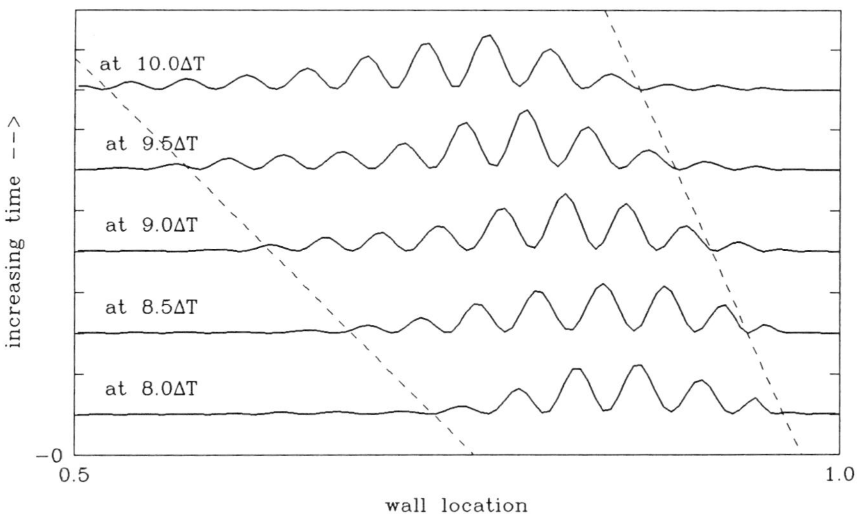

Figure 6. Wall-energy-density evolution for the Type II wave seen in Figure 5b. (Note that only the downstream half of the domain is shown.) Reproduced from Lucey (1998).

Both theory and numerical simulation predict that the convectively unstable Type I wave seen downstream of the excitation in figure 5a has downstream-directed group velocity and plate-energy flux. The wave which has effectively replaced it at $8\Delta T$ is theoretically predicted to be a neutrally-stable NEW and have upstream-directed group velocity. This combination means that the fluid-structure wave has a downstream-directed energy flux. In applications where the plate is of finite length, one is typically interested in both the total energy and the energy flux *within* the elastic plate. Thus, in figure 6 we show the evolution of the plate's total (strain and kinetic) energy density for times immediately after $8\Delta T$ when the Type II wave prevails in the downstream half of the domain. Two features are evident: (i) the plate energy grows in time, albeit slowly as compared to its growth when supporting the Type I wave and, (ii) propagation is upstream directed.

At times later than those illustrated in figure 5, there is progressive amplification of wall deflection until the present linear model is no longer appropriate. The increase of wall energy occurs through the combination of Type I and Type II waves present in the system. Just as the Type I wave launched the Type II wave of figure 5b from the trailing edge, the upstream propagating Type II triggers a Type I wave when it reaches the leading edge of the plate. Thus the energy-density of the finite plate increases at all spatial locations through a combination of convective instabilities. The oscillatory line-excitation becomes increasingly insignificant with the passage of time and accompanying deflection growth. As time passes, it is the ends of the panel which become the principal initiators of waves in the finite system.

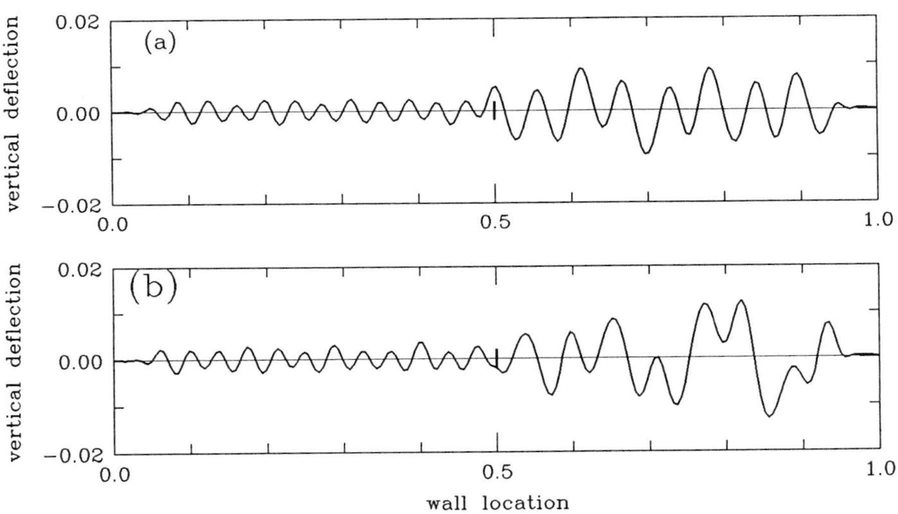

Figure 7. Instantaneous wall profiles for $U = 0.05$ for high-frequency line excitation at the plate mid-point (marked by vertical bar) for (a) $t = 2\Delta T$, and (b) $t = 4\Delta T$. Adapted from Lucey (1998)

It is perhaps unsurprising that the results above lead to wholescale destabilisation of the plate at $U = 0.05$, because the frequency of the oscillatory excitation lies within the range of convectively unstable behaviour predicted by theory. A 'global' theoretical approach can be applied to the system using standard techniques of hydroelasticity (for example see Weaver & Unny 1971 and Ellen 1973). In the absence of applied excitation disturbances are written as an expansion of orthogonal modes satisfying the end conditions and a Galerkin method is then used to solve the system equations. For an unsupported flexible plate with hinged ends, the nondimensional flow speed, U_D, above which the plate *may* be unstable is $6.33(m/\rho_f L)^{\frac{3}{2}}$, the deformation at U_D being proportional to $\sin(\pi x/L)$, *i.e.* a standing wave with wavenumber $\pi m/\rho_f L$. For the data used in the present simulations, this gives $U_D = 0.000297$ and for a plate of infinite length there exists the possibility of instability at any non-zero flow speed. What the simulations summarised above show is *how* unstable behaviour develops from a localised source of initial excitation.

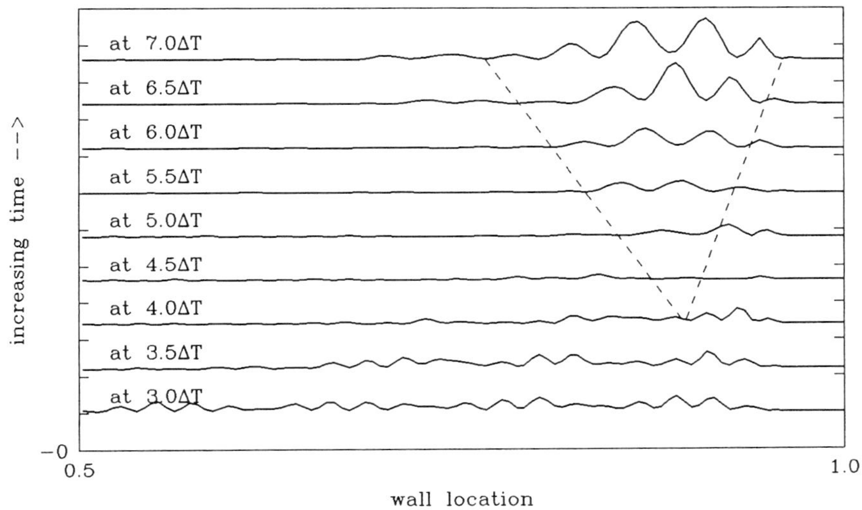

Figure 8. Wall-energy-density evolution for the response of figure 7 but having switched off the driver at $3\Delta T$. (Note that only the downstream half of the domain is shown.) Reproduced from Lucey (1998).

In order to assess the importance of transients in the finite system, the numerical experiment that gave figures 5 and 6 is repeated now with high-frequency excitation, $\omega_F = 0.01343$. Figure 7 shows instantaneous wall profiles at $2\Delta T$ and $4\Delta T$. At the earlier time, the dominant response conforms to theoretical prediction, the response frequency (lying in the range $\omega > \omega_p$) matches that of the excitation and a pair of neutrally stable PEW's propagate outwards from the line excitation. However, downstream of the excitation at $4\Delta T$, there is evidence of an additional, and amplifying, wave superimposed upon the modified flexural response. Spectral analysis in both frequency and wavenumber domains, locates this wave at (k_p, ω_p) on the dispersion curve of figure 1. It is possible to 'isolate' this part of the response by switching off the exciter that sustains the modified flexural response a short time after start-up; the results of this are shown in figure 8 where the total plate energy density is plotted in space-time form. The emergence of the (k_p, ω_p)-instability is clearly evident and its propagation is in both upstream and downstream directions, precisely what one would expect of an absolute instability. This result – the selection and amplification of a transient, absolutely unstable, (k_p, ω_p)-wave from the start-up procedure – is theoretically predicted by

Abrahams & Wickham (2001) (and Peake 1997 for the cylindrical geometry) for an infinitely long plate with dissipation. However, for a plate of finite length, this phenomenon can occur in the absence of plate damping.

At times later than those shown in figure 7, the continued growth of the (k_p, ω_p)-instability impacts on the trailing edge whence it launches a convectively unstable Type II wave with a frequency slightly greater than ω_p. This marks the beginning of a sequence of events qualitatively identical to that which caused wholescale destabilisation of the panel at low frequency, namely repeated excitation and amplification of convectively unstable waves. Again, the finite plate is ultimately destabilised at this flow speed (*i.e.* $U > U_D$) but, for high-frequency excitation, the route to 'global' destabilisation includes a period of absolutely unstable behaviour.

3.2 ELASTIC PLATE WITH A SPRING FOUNDATION

Having recognised that very long elastic plates will ultimately succumb to instability at exceedingly low flow speeds, we now consider the behaviour of flexible plate for which the structural stiffness is supplemented by a uniformly distributed spring foundation. The spring coefficient per unit area of plate is $K^* = (\rho_f^4 B/m^4)K$ Nm^{-3}, where K is its nondimensional value. The dispersion relation is then

$$\omega^2 - k^4 + \frac{(\omega - Uk)^2}{|k|} - K = 0 \ , \tag{13}$$

which takes a similar form to (9), that of Peake's (1997) thin shell. We therefore obtain a similar set of dispersion curves except that, for fixed U, the control parameter is K instead of a. In the following work, however, we fix the wall properties by choosing K, and investigate the flexible-wall response at different flow speeds. To demarcate régimes of response, we first identify two important flow speeds, U_D and U_F, respectively known as the divergence-onset and modal-coalescence flutter-onset critical speeds in hydroelasticity.

The lower of the two, U_D, is the value for which the lower branch touches the $\omega = 0$ axis of the dispersion diagram, having a_2 as its equivalent in Peake's (1997) shell problem – and see Section 2.2 above. Using (13), we find that

$$U_D = (2)(3^{-\frac{3}{8}})K^{\frac{3}{8}} \qquad \text{with} \qquad k_D = (K/3)^{\frac{1}{4}} \ , \tag{14a, b}$$

which, for $K = 0.000146$ used in the simulations below*, gives $U_D = 0.0483$ with the associated critical wave having the wavenumber $k_D = 0.0835$. The above results hold strictly only for a wall length, L, that is very long compared to the wavelength of the critical wave, *i.e.* for $(Lk/2\pi) \to \infty$. However, Carpenter & Garrad (1986) have shown that even for $(Lk/2\pi)$ as low as 4, (14) gives good results when compared to thoretical predictions based on the analysis of a wall of finite-length. We also note that U_D is independent of the flexible-wall length in contrast to the corresponding result for an unsupported flexible plate.

---

* All other details of the system are identical to those of the unsupported plate described in Section 3.1 above.

At the higher flow speed, U_F, the lower and upper branches of the dispersion curve coalesce to give a pair of complex conjugate roots to (11). In the corresponding shell problem, this occurs at $a = a_1$. Solving (11), coalescence occurs for

$$U^2 = \frac{1}{k^2}(|k| + 1)(k^4 + K) \ . \tag{15}$$

Minimising the above U yields the following approximate *critical* values for flutter onset:

$$U_F \approx \sqrt{2} K^{\frac{1}{4}} (K^{\frac{1}{4}} + 1)^{\frac{1}{2}} \qquad \text{with} \qquad k_F \approx K^{\frac{1}{4}} \ . \tag{16a, b}$$

For the present value of K, these expressions give $U_F = 0.164$ and $k_F = 0.110$ – direct numerical evaluation using (13) gives corresponding values of 0.164 and 0.107.

In summary, then, by fixing K, we expect to categorise wave behaviour using the three flow-speed régimes: (i) $0 < U < U_D$, wherein the typical response comprises a pair, one upstream-and one downstream-propagating, neutrally-stable PEW's, (ii) $U_D < U < U_F$, giving a dispersion diagram with a structure characterised by that of figure 4b and, (iii) $U > U_F$, yielding the structure seen in figure 4a. Here, we present typical results of numerical simulations in régimes (ii) and (iii).

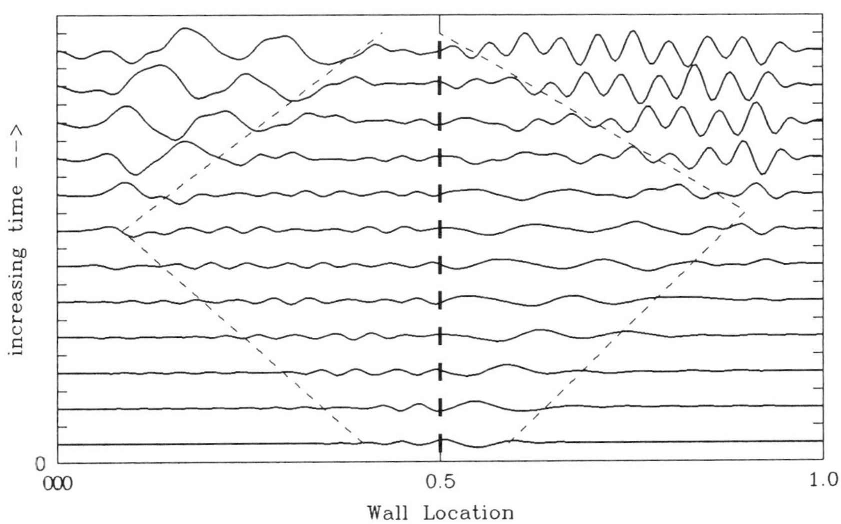

Figure 9. Space-time plot of instantaneous wall profiles for low-frequency excitation of a plate-spring flexible wall with a flow speed in the range $U_D < U < U_F$.

Low-frequency oscillatory excitation, $\omega_F = 0.00412$, and a flow speed $U = 0.0992$, lying in the range $U_D < U < U_F$, generates the wave behaviour seen in the space-time plot figure 9. The first instantaneous wall profile is plotted at $t = 485$ units of the nondimensional timescale and the following profiles are separated by 485 units. To aid interpretation, rays of wave propagation have been sketched in using broken lines. The principal finding is that the flexible wall is destabilised by the flow. Results not presented here show that the driver actually *absorbs* energy for the duration of the simulation; this accords with the analysis of the shell problem for which it has been shown that the lower branch of the dispersion curve above the $\omega = 0$ axis in figure 4b corresponds to NEW's. The destabilisation of the wall is most strikingly seen at later times when a pair of amplifying waves, propagating in opposite directions, occupies the domain. Their properties identify them as the Type I and Type II waves discussed in Section 3.1 above. At earlier times, careful examination of figure 9 shows that a similar Type II wave (plus modified flexural waves) has propagated upstream outwards from the line excitation. Concurrently, a Type I has propagated downstream from the line excitation. Like the unsupported elastic plate, destabilisation of the entire wall proceeds through a combination of convectively unstable waves. On the dispersion curve (figure 4b) these waves can be identified with two points on the $\omega > 0$ part of the lower branch, one on each side of the turning point and the direction of their wall energy-density propagation is given by the sign of $\partial\omega/\partial k$, the group velocity.

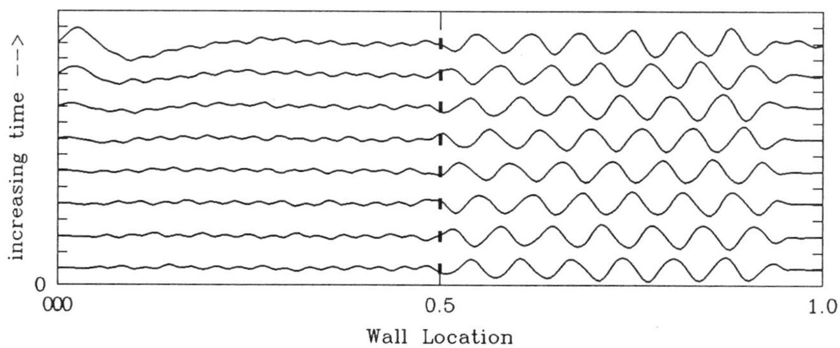

Figure 10. Space-time plot of instantaneous wall profiles for later times after start-up; high-frequency excitation of a plate-spring flexible wall with a flow speed in the range $U_D < U < U_F$.

When high-frequency excitation, $w_F = 0.0165$ (*i.e.* substantially above ω_p, the lower branch turning point) is used, results such as those presented in figure 10 are found; the first profile plotted is found at $t = 5460$ units of the nondimensional timescale while subsequent profiles are separated by 607 units. At earlier times, and throughout the period chronicled by figure 10, there exists a wave downstream of the driver which can be associated with the upper branch of the dispersion curve of figure 4b. However, this wave can be shown to be convectively unstable - although only mildly so - because the rate at which its energy is dissipated by the wall damping included just near the trailing edge exceed the power input by the driver. Upstream of the driver, a PEW associated with the $\omega < 0$ part of the lower branch propagates energy in the upstream direction. At later times, though, a low-frequency, strongly unstable, wave is seen to emerge from the leading edge of the wall. This is is identified as a Type I wave which eventually dominates the wall response as its sweeps downstream. It therefore seems that for $U_D < U < U_F$, a finite-length wall will ultimately become unstable. By contrast with the destabilisation of an unsupported plate undergoing high-frequency excitation, there seems to be no rôle for absolute instability in the process.

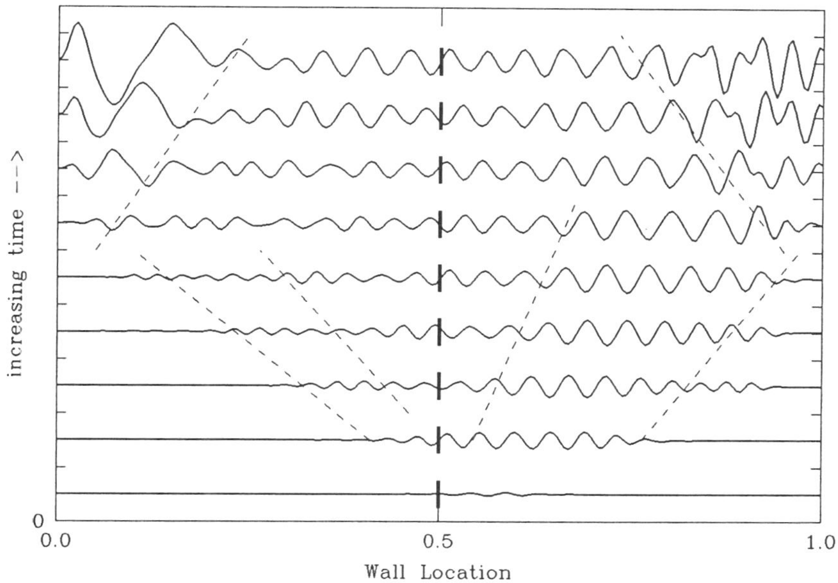

Figure 11. Space-time plot of instantaneous wall profiles for a plate-spring flexible subjected to high flow speed, $U > U_F$.

We now consider a high-speed flow, $U = 0.198$, lying in the range $U > U_F$, for which the structure of the dispersion relation is represented by figure 4a. Figure 11 shows the results obtained for oscillatory line-excitation with a frequency $\omega_F = 0.0247$. This frequency lies within the range of modal-coalescence response. The first plot is at $t = 485$ while subsequent plots are separated by 243 units of the nondimensional timescale. Clearly, the response is now very complicated. At early times, two wavepackets have been highlighted in figure 11 by rays of propagation. The downstream-located wavepacket is amplified as it propagates towards the trailing edge. The dominant frequency of this wavepacket is significantly higher than ω_F; this and its measured wavenumber locate it on the upper branch of the high-k loop in figure 4a. It is therefore surprising that the wave is found to be unstable in the present simulations. Upstream of the driver at early times, we tentatively identify a wavepacket which appears to be neutrally stable. Its importance lies in the fact that, upon arrival at the leading edge, it launches a strongly unstable wave which is readily identifiable with the downstream-propagating modal-coalescence flutter, *i.e.* lying on the complex branch that connects the two loops in figure 4a. It is also surprising that the modal-coalescence flutter is first excited by the wall leading edge as opposed to the line excitation just after $t = 0$. Finally, at later times, a new, unstable wave is seen to appear near the trailing edge, propagating in the upstream direction. This wave can be identified with the lower branch of the high-k loop of the dispersion curve and has the characteristics of a Type I wave. Other numerical experiments show that the sequence of wave-excitations seen in Figure 11 is almost identically reproduced even if the forcing frequency is reduced by a factor of one-third. These results suggests that the mechanism by which the flexible wall is destabilised for $U > U_F$ is dominated by the effect of transients.

4. Shear Flow

The work described so far has been concerned with uniform flow, but in practice the steady flow over the plate will have a boundary-layer structure, with shear flow in some (thin) layer separating the plate from the uniform freestream. This problem has been studied before, Carpenter & Garrad (1985,1986), but in the context of the Crighton & Oswell analysis full details are given by Lingwood & Peake (1999). The steady velocity parallel to the plate is of the form $Uf(y/\epsilon)$ where $f(0) = 0$ to satisfy the no-slip condition on the plate, $f \to 1$ as $y \to \infty$ to match onto a uniform outer flow of speed U, and ϵ here is the nondimensional boundary layer thickness. The plate equation is unchanged, but the unsteady fluid motion must account for the steady shear (assuming inviscid perturbations, this leads to Rayleigh's equation), and also the boundary condition on $y = 0$ is simply that the fluid normal velocity must equal $\partial\eta/\partial t$, since the convective term now vanishes on the plate.

Full details of the solution of this system are given in Lingwood & Peake (1999), and only a short discussion is necessary here:

(1) For the simplest case of a piecewise-linear profile (e.g. $f = y/\epsilon$ for $y < \epsilon, f = 1$ for $y > \epsilon$), the results as $\epsilon \to 0$ agree with what is seen with uniform flow. For instance, the introduction of a thin shear layer reduces the minimum flow speed required for absolute instability, but for $\epsilon \to 0$ the uniform flow result is regained.

(2) For a steady profile with curvature, and hence a critical layer, such as the Blasius boundary layer, the limit $\epsilon \to 0$ seems to be much less clear. In particular, it appears that there are more pinch points than observed for uniform flow, and that these pinch points interact with each other to change the topology of $D(k, \omega)$. The absolute instability boundary, for instance, moves to exceedingly high values of U as $\epsilon \to 0$. Recently, Miles (2001) has suggested that in actuality the absolute instability boundary does actually return to the flat plate value for very small ϵ, but this point cannot be seen in the numerics reported by Lingwood & Peake.

(3) Modes with group velocity pointing towards the driver are still present for a Blasius boundary layer. For instance, it is possible to find a mode with $\text{Im}(k) < 0$ when $\text{Im}(\omega) = 0$, which moves downwards and then upwards in the k plane as $\text{Im}(\omega)$ is increased. This is therefore a convective instability located downstream of the driver, whereas considerations based on the direction of the group velocity would imply incorrectly that the mode is located upstream and is damped!

5. Nonlinear Effects

In this section we will briefly describe the extension of the linear model presented in section 2 for an infinite plate with mean flow to include the effects of nonlinearity - for full details see Peake (2001). The first point to make is that for the regime in which the effects of mean flow are of most interest, i.e. for $\omega \leq \omega_p$, see figure 1, it is clear that $\omega \ll 1$. One measure of the level of fluid loading on the plate is the ratio of the mass of fluid within one bending wavelength of the plate, per unit length, to the mass of the plate, per unit length, and in our nondimensional units this ratio is proportional to $\omega^{-\frac{1}{2}}$, which is a large number. This means that (Sorokin 2000) nonlinearity must be included in the plate equation, the boundary condition and the fluid motion. Peake (2001) takes the plate equation to be

$$\eta_{tt} + \frac{\partial^2}{\partial x^2}\left[\frac{\eta_{xx}}{(1+\eta_x^2)^{\frac{3}{2}}}\right] - \frac{\tau}{s}\left(\int_0^s \eta_{x'}^2 \mathrm{d}x'\right)\eta_{xx} + \epsilon c_d \eta_t = -p(x, y=\eta, t) \ . \quad (17)$$

Here $\tau = 6(1-\nu^2)(\rho_s/\rho_f)^2$ is the coefficient of the nonlinear tension term induced by the plate bending, and s is the wavelength of a single wave (precise conditions for this term to be valid are described in Peake 2001, but essentially we need to consider just a single spatial wavelength). The final term on the left hand side of (17) corresponds to the introduction of (small) dashpot dissipation, which will be important when we consider the nonlinear evolution of NEWs. The pressure in the fluid is given by the full Bernoulli equation

$$p = -\phi_t - U\phi_x - \frac{1}{2}\left(\phi_x^2 + \phi_y^2\right) \ , \quad (18)$$

while the velocity potential $\phi(x, y, t)$ still satisfies Laplace's equation for incompressible flow. The boundary condition is

$$\phi_y = \eta_t + (U + \phi_x)\,\eta_x \quad (19)$$

applied on the actual plate surface $y = \eta(x, t)$. Equations (17,18,19) are to be compared with their linear counterparts (1,3,4) respectively.

Considering first the evolution of a NEW, we introduce a slow time scale $T = \epsilon t$, and write

$$\eta(x,t) = \epsilon^{\frac{1}{2}}\left[A\mathcal{E} + \mathrm{c.c}\right] + \epsilon\left[\eta_1^{(2)}A^2\mathcal{E}^2 + \eta_1^{(1)}\mathcal{E} + \mathrm{c.c.} + \eta_1^{(0)}\right] + \dots \ , \quad (19)$$

with similar expansions for the velocity potential and pressure, where

$$\mathcal{E} = \exp(-\mathrm{i}\omega t + \mathrm{i}kx)$$

represents the carrier wave (a NEW), with $D(k, \omega) = 0$, and $A = A(T)$ is the slowly-evolving amplitude of the carrier. After a considerable amount of algebra it then turns out that A satisfies the evolution equation

$$\frac{\mathrm{d}A}{\mathrm{d}T} + \frac{c_d\omega^2}{4E}A + \mathrm{i}\beta A|A|^2 = 0 \ , \quad (20)$$

where β is a negative real number. Note that since we are considering a NEW, so that $E < 0$, this equation predicts that $|A(T)|$ grows exponentially in time, and that the nonlinearities in the governing equations have only had the effect of modifying the phase. Further efforts to balance this exponential growth with a quintic nonlinearity are described in Peake (2001), and produce a similar result. In order to produce a saturated nonlinear state, as one would envisage in static divergence for example, it is necessary to change these scalings by noting that the nonlinear tension term is typically more important than is suggested above, because the coefficient τ is proportional to the square of the solid-fluid density ratio, and is therefore large. Choosing the preferred scaling $\tau = \epsilon^{-2}\bar{\tau}$, with $\bar{\tau} = O(1)$, this time the wave frequency will evolve in slow-time, so we need to write

$$\mathcal{E} = \exp\left(\mathrm{i}kx - \frac{\mathrm{i}}{\epsilon}\int^{T} \omega(T')\mathrm{d}T'\right) \quad . \tag{21}$$

It then turns out that $\omega(T)$ and $A(T)$ are coupled at leading order through the amplitude-dependent dispersion relation

$$\omega^2 - k^4 + \frac{(\omega - Uk)^2}{|k|} - 2k^4\bar{\tau}|A|^2 = 0 \ , \tag{22}$$

so that the effect of the induced nonlinear tension is to increase the plate stiffness.

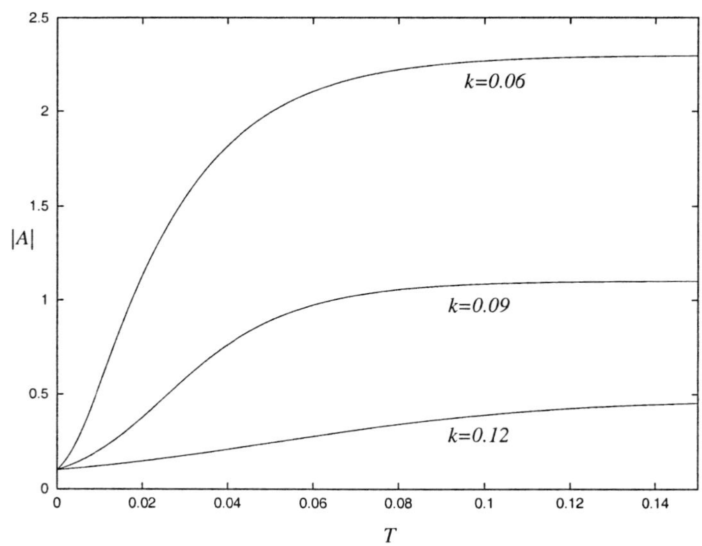

Figure 12. Evolution of the NEW amplitude, for $U = 0.05$, and $|A(0)| = 0.1$. From Peake (2001).

The evolution of the amplitude is given at the next order by the equation

$$\frac{\mathrm{d}|A|}{\mathrm{d}T} + \frac{c_d\omega^2}{4E_{NL}}|A| = 0 \ , \tag{23}$$

where now E_{NL} is a nonlinear analogue of the linear wave energy E (see Peake 2001, equation 3.16). The crucial point, however, is that for a NEW we have $E_{NL} < 0$ as well, so that from equation (23) the amplitude of a NEW will initially grow, but this causes ω to decrease according to (22), so that in the limit $T \to \infty$ a saturated state of constant $|A|$ (see figure 12) is reached, in which $\omega = 0$.

Numerical simulations by Lucey *et al.* (1997), for a flexible wall of finite length, show agreement with this prediction. The computational modelling described in Section 3 has been extended so that the boundary elements now lie *upon* the deformed interface. The Laplace equation is solved subject to (19) imposed at $y = \eta$ and the pressure obtained using (18). Because the flexible wall is of finite length, L, and held at both its ends, the induced tension – see term 3 on the left-hand side of (17) – is found by replacing s with L.

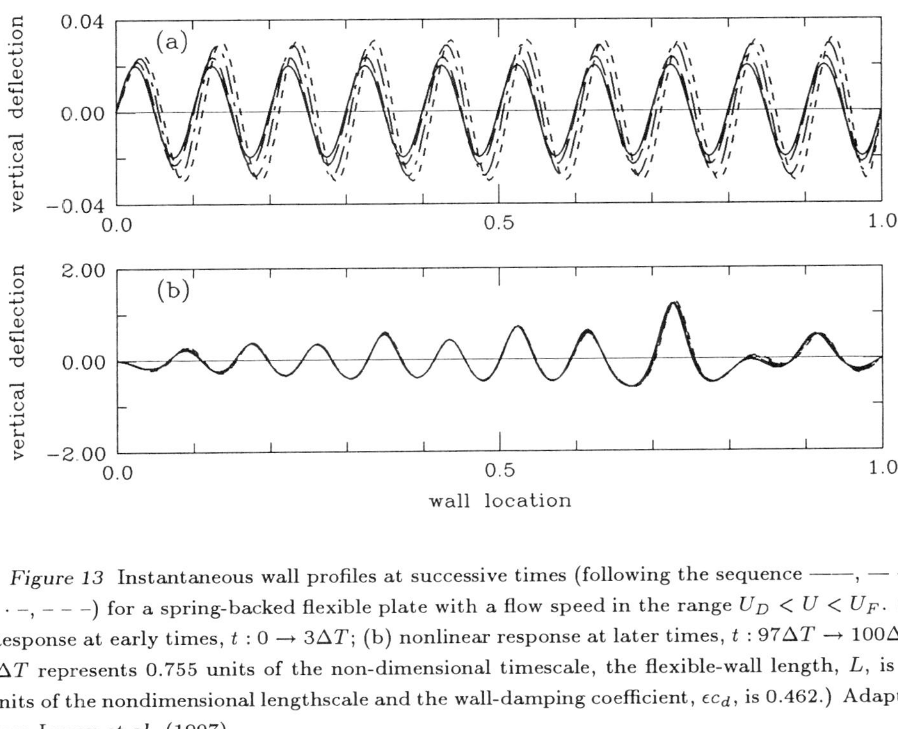

Figure 13 Instantaneous wall profiles at successive times (following the sequence ——, — —, – · –, – – –) for a spring-backed flexible plate with a flow speed in the range $U_D < U < U_F$. (a) Response at early times, $t : 0 \to 3\Delta T$; (b) nonlinear response at later times, $t : 97\Delta T \to 100\Delta T$. ($\Delta T$ represents 0.755 units of the non-dimensional timescale, the flexible-wall length, L, is 63 units of the nondimensional lengthscale and the wall-damping coefficient, ϵc_d, is 0.462.) Adapted from Lucey *et al.* (1997).

Figure 13 illustrates the evolution of an unstable wave for a damped spring-backed flexible plate, i.e. including an additional wall term $K\eta$ in (17). For this flexible wall, $K = 3.4$ has been chosen. Thus, using (14a), $U_D = 2.10$ and minimising (15) numerically* gives $U_F = 2.89$. In generating figure 13, $U = 2.43$ has been used; hence $U_D < U < U_F$ and the appropriate dispersion-relation

* Note that the chosen properties could represent a rubber-based compliant wall for which the control parameters, U and K are no longer the small quantities they were for the metal plates. Even so, (16a) continues to give a good estimate of U_F.

structure for small-amplitude waves is that seen in figure 4b. After releasing (at $t = 0$) an imposed small-amplitude deflection with a wavenumber greater than that of the dispersion-relation lower-branch turning point, amplitude growth of the chosen NEW is seen in figure 13a. However, at the much later times represented by figure 13b, a state of saturation has been reached – note the different vertical scales in the two figures. The flexible wall has settled into a nearly static state of divergence and the deformation now possesses a profile more characteristic of nonlinear waves. Thus, the action of damping has permitted the flexible wall to shift from an (unstable) equilibrium of zero deflection into the state of nonzero (weakly nonlinear) static deflection.

Nonlinear analysis can also be completed when damping is absent. For instance, suppose that the nonlinear tension term can be removed (relaxed to zero along the infinite plate), then we can consider neutral waves of different wavelengths, and in particular we can allow slow spatial, as well as a slow temporal, modulation. Introducing the slow space scale $X = \epsilon x$ and time scale $T = \epsilon^2 t$, and supposing now that the amplitude of a wave on the lower branch of the neutral curve takes the form $A(X,T)$, then by suitable expansion of equations (17-19) it can be shown that A satisfies the Nonlinear Schrodinger (NLS) equation

$$\frac{D_{kk}}{2}\frac{\partial^2 A}{\partial X^2} - \mathrm{i}D_\omega \frac{\partial A}{\partial T} = lA|A|^2 \ , \tag{24}$$

where l is a positive real number. On the lower branch of the neutral curve we have $D_\omega < 0$, while D_{kk} changes sign between $k = k_b$ and $k = k_p$; close to $k = k_b$, $D_{kk} > 0$, while close to $k = k_p$, $D_{kk} < 0$. The behaviour of (24) is well-known; in the case $D_{kk} > 0$ we have the minus, or defocusing, form of the NLS equation, and all initial data with compact support will decay in time due to diffraction; but for $D_{kk} < 0$ we have the plus form of the NLS equation, and initial data will evolve into the famous envelope solitons of Zakharov & Shabat (1972). The specific one-soliton solution is

$$A(X,T) = a\left|\frac{D_{kk}}{l}\right|^{\frac{1}{2}} \operatorname{sech}\left[a\left(X + \frac{bD_{kk}}{D_\omega}T\right)\right] \exp\left[\mathrm{i}bX - \frac{\mathrm{i}D_{kk}}{2D_\omega}(a^2 - b^2)T\right] \ ,$$

for any a and b. For the parameters of interest in fluid-loading problems, the amplitude of this soliton is large compared to its width, meaning that it could possibly be visible in numerical or experimental studies. Further extensions of this analysis to account for three-dimensional motion, in which the plate has a significant spanwise curvature, may also be possible, and one could envisage a very rich variety of soliton behaviour and interactions in this case.

6. Closing Remarks and the Way Forward

Many of the phenomena first brought to light by Brazier-Smith & Scott (1984) and Crighton & Oswell (1991) have been shown to be robust to modifications to the structural, geometric and fluid-flow elements of the original simple flow-structure model. However, the inclusion of dissipation in the structure (Abrahams & Wickham 2001 and Peake 1997, 2001) has been shown to render unstable all waves on the lower, NEW, branch of the dispersion curve. While most of these waves are convective instabilities, the point (k_p, ω_p) represents an absolutely unstable wave which will be excited in any start-up procedure, or by other 'non-uniformities' present in a real environment, and will ultimately dominate the system response. Convective lower-branch instability and absolute instability at (k_p, ω_p) also feature in the computational study of finite panels with fixed ends; here, however, structural damping is not required for their realisation. This calls into question, their categorisation as NEW's in the finite system. Wave reflections and the excitation of convective instabilities at the panel ends are also seen to be very important in the computational work. Clearly, then, there is a need for reconciliation between the studies of infinitely long flexible walls and those that model finite plates. As a first approach to modelling fixed panel ends, Abrahams & Wickham (2001) have suggested using a spatially-varying value of the flexural rigidity, $B(x)$ to give a continuous change from flexible to rigid parts of the flow boundary. For certain choices of the function $B(x)$, the analytical methods discussed in this article could be carried through.

While Peake (2001) has made significant progress on nonlinear aspects of the problem at hand, there remains a need to study the evolution of the strongly unstable convective instability (denoted Type I in the computations of Section 3) that arises from coalescence of the upper- and lower-branch roots of the dispersion relation (for elastic plates, thin shells and spring-backed elastic plates); in particular, the waves on this branch with the highest growth rates. As suggested by Abrahams & Wickham (2001), in a moving frame of reference with x/t fixed, a wavepacket of transients containing response in the range $0 < \omega < \omega_b$ would amplify rapidly. For an elastic plate of finite length, this would have consequences on the overall system response. Two-dimensional analyses appear to suffice for nonlinear modelling (Lucey *et al.* 1997, Peake 2001) of the divergence-type (or NEW-type) instabilities because the saturated waves, seen in the experiments of Gad-el-Hak *et al.* 1984, have a quasi-two-dimensional structure. However, a fully three-dimensional model will, most probably, need to be developed to study the nonlinear evolution of the coalescence instability.

Finally, there is a pressing need for experiments to be carried out in this area of study. Most aero/hydro-elasticians would agree that a given flexible panel subjected to a fluid flow at high enough speed becomes destructively unstable. What careful experiments could show is *how* it becomes unstable. The understanding generated by the theoretical and computational work reviewed in this article would provide clear guidance for the design of a suitable experimental programme.

7. References

Abrahams I.D. & Wickham G.R. 2001 On transient oscillations of plates in moving fluids. *Wave Motion*, **33**(1), 7–24.

Benjamin, T.B. 1960 Effects of a flexible boundary on hydrodynamic stability. *J. Fluid Mech.* **9**, 513–532.

Benjamin, T.B. 1963 The threefold classification of unstable disturbances in flexible surfaces bounding inviscid flows. *J. Fluid Mech.* **16**, 436–450.

Bers, A. 1983 Space-time evolution of plasma instabilities - absolute and convective. In *Handbook of Plasma Physics* , ed M.N. Rosenbluth, R. Z. Sagdeev. 1:451–517. North-Holland.

Briggs, R.J. 1964 *Electron-stream Interaction with Plasmas* . Monograph No. 29. MIT Press.

Brazier-Smith, P.R. & Scott, J.F. 1984 Stability of fluid flow in the presence of a compliant surface. *Wave Motion* **6**, 547–560.

Cairns. 1979 The role of negative energy waves in some instabilities of parallel flows. *J. Fluid Mech.* **92**, 1–14.

Carpenter, P.W. 1990 Status of transition delay using compliant walls. In *Viscous Drag Reduction in Boundary Layers*, AIAA, New York, 79–113.

Carpenter, P.W. & Garrad, A.D. 1985 The hydrodynamic stability of flow over Kramer-type compliant surfaces. Part 1. Tollmien-Schlichting instabilities. *J. Fluid Mech.* **155**, 465–510.

Carpenter, P.W. & Garrad, A.D. 1986 The hydrodynamic stability of flow over Kramer-type compliant surfaces. Part 2. Flow-induced surface instabilities. *J. Fluid Mech.* **170**, 199–232.

Choi, K.-S., Yang, X., Clayton, B.R., Glover, E.J., Altar, M., Semenov, B.N. & Kulik, V.M. 1997 Turbulent drag reduction using compliant surfaces. *Proc. Roy. Soc. A* **453**, 2229–2240.

Crighton, D.G. & Oswell, J.E. 1991 Fluid loading with mean flow. I. Response of an elastic plate to localized excitation. *Phil. Trans. R. Soc. Lond. A* **335**, 557–592.

Ellen, C.H. 1973 The stability of simply supported rectangular surfaces in unifrom subsonic flow. *Trans. ASME* E**95**, 68–72.

Gad-el-Hak, M., Blackwelder, R.F. & Riley, J.F. 1984 On the interaction of complaint coatings with boundary-layer flows. *J. Fluid Mech.* **140**, 257-280.

Kelbert, M & Sazonov, I. 1996 *Pulses and other wave processes in fluids.* Modern Approaches in Geophysics. Kluwer Academic.

Landahl, M.T. 1962 On the stability of a laminar incompressible boundary layer over a flexible surface. *J. Fluid Mech.* **13**, 609–632.

de Langre, E. 2001 Absolutely unstable waves in inviscid hydorelastic systems *J. Sound Vib.*, submitted.

Lighthill, M.J. 1960 Studies on magnetohydrodynamic waves and other anisotropic wave motions. . *Phil. Trans. R. Soc. Lond.*. **252**, 397-430.

Lingwood, R.J. & Peake, N. 1999 On the causal behaviour of flow over a compliant wall. *J. Fluid. Mech.*, **396**, 319–344.

Lucey, A.D. 1998 The excitation of waves on a flexible panel in a uniform flow. *Phil. Trans. R. Soc. Lond.* **356**, 2999–3039.

Lucey, A.D. & Carpenter 1992 A numerical simulation of the interaction of a compliant wall and an inviscid flow. *J. Fluid Mech.*, **234**, 121–146.

Lucey, A.D. & Carpenter 1995 Boundary layer instability over compliant walls: comparison between experiment and theory. *Phys. Fluids*, **7**, 2355–2363.

Lucey, A.D., Cafolla, G.J., Carpenter, P.W. & Yang, M. 1997 The nonlinear hydroelastic behaviour of flexible walls. *J. Fluids & Structures*, **11**, 717-744.

Miles, J. 2001 Stability of inviscid shear flow over a flexible boundary. *J. Fluid Mech.*, **434**, 371–378.

Peake, N. 1997 On the behaviour of a fluid-loaded cylindrical shell with mean flow. *J. Fluid. Mech.*, **338**, 387-410.

Peake, N. 2001 Nonlinear stability of a fluid-loaded elastic plate with mean flow. *J. Fluid Mech.*, **434**, 101-118.

Sorokin, S.V. 2000 Nonlinear oscillations of a baffled elastic plate in heavy fluid loading conditions. *J. Sound Vib.* **232**(3), 619–643.

Weaver, D.S. & Unny T.S. 1971 The hydroelastic instability of a flat plate. *Trans. ASME* E**37**, 823-827.

Zakharov, V.E. & Shabat, A.B. 1972 Exact theory of two-dimensional self-focusing and one-dimensional self-modulating waves in nonlinear media. *Sov. Phys. J.E.T.P*, **34**, 62–69.

Chapter 7

PROPAGATION OF WAVES ACROSS JUNCTIONS BETWEEN RIGID AND COMPLIANT WALLS

PETER W. CARPENTER
School of Engineering, University of Warwick
Coventry, CV4 7AL, U.K.

PRADEEP K. SEN
Department of Applied Mechanics
Indian Institute of Technology Delhi, New Delhi 110011, India

Abstract

This chapter reviews the generic problem of the propagation of vortical waves across junctions between one wave-bearing medium and another. Two simpler examples are considered first where the waves are not vortical, namely waves across junctions in elastic tubes and those between two spring-supported elastic plates. It is assumed that the eigensolutions are known for the corresponding spatially homogeneous problems. The task is how to determine the amplitudes of the reflected and transmitted waves given the amplitude of the incident wave. In general, there may be more than one incident, reflected or transmitted wave. It is shown how this sort of problem may be solved in terms of the homogeneous eigensolutions by drawing an analogy between the junction and a wave-driver. The particular problem studied is that of a Tollmien-Schlichting wave, propagating along a rigid-walled channel flow, that is incident on a section of the channel where the walls consist of compliant panels. It is shown how the wave system over the compliant panels and the amplitude of the Tollmien-Schlichting wave leaving the compliant section may be determined in terms of the incident wave. The techniques reviewed for this problem are considered to be generic.

P.W. Carpenter and T.J. Pedley (eds.),
Flow past Highly Compliant Boundaries and in Collapsible Tubes, 147-166.

1. Introduction

In this chapter we will discuss the propagation of waves across a junction between one wave-bearing system and another. The type of system we have in mind consists of a fluid flow in a flexible tube or channel, or over a flexible wall. More specifically, we consider the situation where the properties of the waves in two or more such wave-bearing systems are known when the systems are spatially homogeneous (i.e., infinitely long). The basic problem then becomes: how do we relate the amplitudes of the known (spatially homogeneous) wave states in the two wave-bearing systems when they are joined? For example, one could envisage an incident wave of known amplitude and phase propagating downstream in a wave-bearing system and suddenly meeting a junction with a different wave-bearing system. The problem at its simplest would then become: how do we determine the amplitudes and phases of the reflected and transmitted waves? In general, however, two or more eigenstates (wave types) may be involved in each of the two (spatially homogeneous) wave-bearing systems.

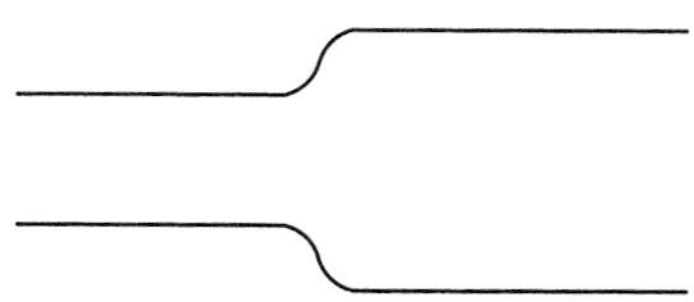

Figure 1. A flexible-walled channel or tube with a sudden change in cross-sectional area.

A simple example of such a problem is the propagation of longitudinal waves along a fluid in an elastic tube that abruptly changes area (Fig. 1). The theory for this case is well-known; see, for example, Lighthill (1978). And it will be discussed briefly in Section 2. In the simple version of this theory the flow is inviscid and the variation of pressure and velocity across the cross section of the tube is neglected. The problem becomes much more challenging when viscous effects are included and when some or all of the waves are vortical. For most applications waves of this sort are involved and viscous effects are important. A model problem for which viscous effects are dominant was introduced by Davies and Carpenter (1997). It is illustrated in Figure 2. This shows the flow in a two-dimensional channel with rigid-walled inlet and outlet sections that are separated by a section with compliant walls. A growing Tollmien-Schlichting (TS) wave is propagating along the flow in the inlet section. The basic problem is: given the amplitude and phase of this inlet TS wave, how do we calculate the wave system over the compliant section, and what will be the amplitude of the TS wave when it leaves the compliant section and enters the outlet section? This will be considered in some detail in Sections 5 and 6 where a general approach to such problems will be described.

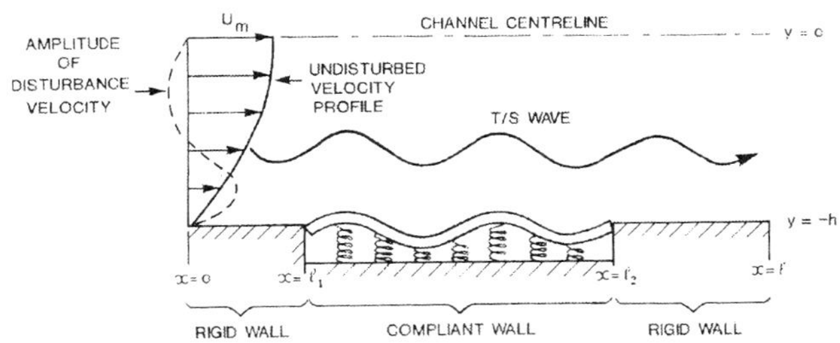

Figure 2. A schematic sketch of the problem under study.

The study of Davies and Carpenter was mainly motivated by an interest in using a series of short compliant panels to suppress the growth of T/S waves and thereby maintain laminar flow at very high Reynolds numbers. But their problem is plainly related to many biomedical applications; for example, the use of stents to reinforce blood vessels, and pulsatile flows through a stenosis. There are also evident links with the study of collapsible tubes (see Chap. 2 by Heil and Jensen in the present volume), particularly the numerical studies of Luo and Pedley (1995, 1996) and Pedrizzetti (1998). These studies are nonlinear, however, whereas the general problem considered here is linear. In other branches of engineering, similar problems are found in a variety of applications (see, for example, Howe 1998, Paidoussis 1998, and Nguyen *et al.* 1994).

In §3, we briefly consider a purely elastic system with no fluid flow. A wave propagates along an elastic plate supported on a spring foundation, similar to the compliant wall depicted in Figure 2, and is incident on a junction with an elastic plate on a spring foundation having a different stiffness. This model problem is studied because of the insight it provides for dealing with the more general class of problems like the one illustrated in Figure 2.

2. Junctions in elastic tubes

If inviscid flow is assumed with uniform velocity and pressure across the tube cross section, the propagation of longitudinal waves along an elastic tube of constant cross-sectional area, A, with stationary undisturbed flow is governed by the classic wave equation. Accordingly, if a wave is propagating in the positive x direction in such a system, the intramural (or excess) pressure and flow velocity are given by

$$p = f\big(t - \frac{x}{c}\big), \qquad u = \frac{1}{\rho c} f\big(t - \frac{x}{c}\big), \tag{1a, b}$$

where ρ is the fluid density and the wave speed is given by

$$c^2 = \frac{A_0}{\rho}\left[\frac{dp}{dA}\right]_{p=0} \equiv \frac{1}{\rho D}, \tag{2}$$

where A_0 corresponds to $p = 0$. D is known as the *distensibility*; it depends on the elastic properties of the tube and its use provides us with a simple way of modelling the effects of these properties.

Let us now consider the cases illustrated in Figure 1 where there is a sudden increase or decrease in the tube's cross section. Plainly, eq. (2) shows that this will result in a sudden change in wave speed. Much the same effect could be achieved by a sudden change in distensibility. When the incident wave in the left-hand section reaches the junction, part of it will be refected and part will be transmitted. The formal solutions in the left-hand and right-hand portions of the tubes joined at $x = 0$ can be written as

$$p = \underbrace{f\left(t - \frac{x}{c_1}\right)}_{\text{incident wave}} + \underbrace{g\left(t + \frac{x}{c_1}\right)}_{\text{reflected wave}}, \quad u = \frac{1}{\rho c_1}\left[f\left(t - \frac{x}{c_1}\right) - g\left(t + \frac{x}{c_1}\right)\right] : \quad x < 0 \tag{3a}$$

$$p = \underbrace{h\left(t - \frac{x}{c_2}\right)}_{\text{transmitted wave}}, \quad u = \frac{1}{\rho c_2}\left[h\left(t - \frac{x}{c_2}\right)\right] : \quad x > 0. \tag{3b}$$

The fundamental problem now becomes: given function f, what are g and h? To solve this problem we need two conditions linking the three fundamental solutions at the junction. In the classic approach to such problems, it is assumed that any near-field effects in the immediate neighbourhood of the junction can be neglected. Indeed, this is implicit in the general forms of the solutions given in eqs. (3a,b) since f, g and h are all solutions of the classic wave equation as applied to spatially homogeneous, infinitely long domains. If near-field effects are neglected then continuity of pressure at the junction demands that

$$f(t) + g(t) = h(t). \tag{4}$$

Before deriving a second condition we introduce the admittance, defined as

$$\text{Admittance}: \quad Y = \frac{\text{Volume flow in propagation direction}}{\text{Intramural pressure}}. \tag{5}$$

Admittance is the reciprocal of impedance, a concept that comes from an analogy with electric circuits. It is more general than, but closely linked to, distensibility.

The second condition to be satisfied at the junction can be derived by requiring continuity to be satisfied, namely that the volume flow entering the junction region from the left should equal that leaving from the right, i.e.,

$$A_1 u_1 = A_2 u_2 \quad \text{or} \quad \underbrace{\frac{A_1}{\rho_1 c_1}}_{Y_1}[f(t) - g(t)] = \underbrace{\frac{A_2}{\rho_2 c_2}}_{Y_2} h(t). \tag{6a}$$

Substituting eqs (3a) and (3b) into eq. (6a) gives

$$Y_1\big(f(t) - g(t)\big) = Y_2 h(t). \tag{6b}$$

Equations (4) and (6b) can be solved to give

$$\frac{g(t)}{f(t)} = \frac{Y_1 - Y_2}{Y_1 + Y_2}, \qquad \frac{h(t)}{f(t)} = \frac{2Y_1}{Y_1 + Y_2}. \tag{7a, b}$$

It should be noted that this approach neglects the rate of change of volume within the junction region (a near-field effect), assuming it to be higher-order. For this reason some may consider that a condition based on conservation of energy might be preferable. For any travelling wave the rate of energy transfer in the propagation direction is the product of the volume flow and the intramural pressure, so conservation of energy demands that

$$A_1 u_1 p_1 = A_2 u_2 p_2, \qquad i.e. \qquad \frac{A_1}{\rho_1 c_1}\big(f(t) - g(t)\big)\big(f(t) + g(t)\big) = \frac{A_2}{\rho_2 c_2}\big(h(t)\big)^2. \tag{8}$$

It follows immediately from eq. (4) that eq. (8) reduces to eq. (6b), leading to the same results as before.

Some basic results are evident from the results given in eqs. (7a,b). For example, even though setting $Y_1 = Y_2$ does not imply that the left-hand and right-hand elastic-tube systems are identical (for example, it is quite possible to have $A_1 \neq A_2$ and $D_1 \neq D_2$ but still allow $Y_1 = Y_2$), there is no reflected wave when this condition is satisfied. This is almost true even when the two admittances are slightly different. For instance, suppose that $Y_2 = Y_1(1 + \epsilon)$ where $|\epsilon| \ll 1$ then eq. (7a) implies that the energy reflected is

$$\left(\frac{g(t)}{f(t)}\right)^2 \simeq \frac{\epsilon^2}{4}.$$

We note that the reflected wave has the same sign as the incident wave when $Y_1 > Y_2$ (i.e., when passing from a more to a less compliant region). In the opposite case when the downstream pipe section is the more compliant a negative reflection is required. (Essentially this is because for the more compliant downstream region the same intramural pressure produces a greater volume flow in the direction of propagation, requiring the reflected wave to augment the upstream volume flow.)

3. Elastic wave incident on a junction between two plates

Two wave-bearing systems in the form of semi-infinite plates supported on spring foundations (similar to the compliant wall depicted in Fig. 2) are joined at $x = 0$ by a pinned joint. The two systems can either be identical or not (having different spring stiffnesses, say). A driver at $x = -\infty$, say, creates an elastic wave that propagates along the left-hand plate in the positive x direction. What will happen when the wave is incident on the junction? How much of the wave will be reflected and how much transmitted? The advantage of this problem is that it can be solved exactly. This analytical solution can then be used to assess certain general approximate approaches to such problems.

The equation of motion for an unloaded elastic plate supported on a spring foundation is given by

$$m\frac{\partial^2\eta}{\partial t^2} + d\frac{\partial\eta}{\partial t} + B\frac{\partial^4\eta}{\partial x^4} + K\eta = -p_w, \tag{9}$$

where η is the vertical displacement of the plate; x is the co-ordinate along the undisturbed plate surface; t is time; m, d, and B are respectively the areal density, damping, and flexural rigidity of the plate; and K is the spring stiffness of the continuous foundation. In general, p_w would be the fluid pressure acting on the plate; here it is equivalently zero.

Suppose that the plate is driven at frequency β. Providing this frequency is above cut-off, i.e.,

$$\beta > \sqrt{\frac{K}{m}}, \tag{10}$$

waves will propagate along the plate. In general, when $d = 0$, there are four fundamental solutions taking the forms:

$$\exp(i\alpha x - i\beta t): \qquad \alpha = \pm\tilde{\alpha}, \pm i\tilde{\alpha}; \qquad \tilde{\alpha} \equiv \left|\left(\frac{m\beta^2 - K}{B}\right)\right|^{1/4}. \tag{11}$$

Thus there are two travelling-wave solutions, one travelling to the right, the other to the left; and two evanescent waves, decaying for $x < 0$ and $x > 0$ respectively.

We will consider two such wave systems joined at $x = 0$, such that $m_1 = m_2$, $B_1 = B_2$, $d_1 = d_2 = 0$, but, in general, $K_1 \neq K_2$ (where suffices 1 and 2 denote properties to left and right of the junction respectively). For such a case the fundamental solutions (11) allow us to write down the forms of the solution for both $x < 0$ and $x > 0$. Thus

$$\eta = \underbrace{\hat{\eta}_i e^{i(\tilde{\alpha}_1 x - \beta t)}}_{\text{incident wave}} + \underbrace{\hat{\eta}_r e^{i(-\tilde{\alpha}_1 x - \beta t)}}_{\text{reflected wave}} + \underbrace{\hat{\eta}_{e1} e^{\tilde{\alpha}_1 x} e^{-i\beta t}}_{\text{evanescent wave}} : \qquad x < 0 \tag{12a}$$

$$\eta = \underbrace{\hat{\eta}_t e^{i(\tilde{\alpha}_2 x - \beta t)}}_{\text{transmitted wave}} + \underbrace{\hat{\eta}_{e2} e^{-\tilde{\alpha}_2 x} e^{-i\beta t}}_{\text{evanescent wave}} : \qquad x > 0. \tag{12b}$$

For a pinned joint we require that the bending moment be zero at the junction, force being transmitted through the junction by the shear force that must therefore be continuous there. Together with the requirement of continuity of displacement, these requirements give four conditions to be imposed at the junction, namely

$$\eta_1 = \eta_2, \quad \frac{\partial^2\eta_1}{\partial x^2} = \frac{\partial^2\eta_2}{\partial x^2} = 0, \quad \frac{\partial^3\eta_1}{\partial x^3} = \frac{\partial^3\eta_2}{\partial x^3} \quad \text{at} \quad x = 0. \tag{13a, b, c, d}$$

Equations (13b,c) correspond to the requirement of zero bending moment, and eq. (13d) follows from the requirement of continuity of shear force. The slope is, of course, discontinuous at the junction.

The four conditions (13a,b,c,d) allow the amplitudes of the reflected and transmitted waves to be determined (as well as those of the evanescent waves), giving

$$\hat{\eta}_r = -\frac{\lambda^3 - i}{1+\lambda^3}\hat{\eta}_i, \qquad \hat{\eta}_t = \frac{1+i}{1+\lambda^3}\hat{\eta}_i; \qquad \lambda \equiv \frac{\tilde{\alpha}_2}{\tilde{\alpha}_1}. \tag{14))}$$

If the near field represented by the evanescent modes is neglected, then only conditions (13a,b,c) can be satisfied (or some other combination of the four conditions) to give

$$\hat{\eta}_r = -\hat{\eta}_i; \qquad \hat{\eta}_t = 0. \tag{15}$$

Qualitatively, this result is similar to the exact solution for the reflected wave, but this simplified approach predicts that no wave is transmitted. Plainly, then, it fails as a reasonable approximation. Accordingly, it would appear that one cannot predict the correct behaviour in the far field without taking into account the near-field solution (the evanescent modes). However, one can model the correct far-field response by regarding the system as consisting of the incident wave superposed on a wave system created by a driver of frequency β located at $x = 0$. Thus a driver of amplitude

$$\hat{\eta}_d = -\hat{\eta}_i\frac{\lambda^3 - i}{1+\lambda^3} \tag{16}$$

produces the waves

$$\eta_{d1} = -\frac{\lambda^3 - i}{1+\lambda^3}\hat{\eta}_i e^{i(-\tilde{\alpha}_1 x - \beta t)} : \quad x < 0, \qquad \eta_{d2} = -\frac{\lambda^3 - i}{1+\lambda^3}\hat{\eta}_i e^{i(\tilde{\alpha}_2 x - \beta t)} : \quad x > 0.$$

Superposition of these waves on the incident wave gives the desired wave solutions on both sides of the junction. Evidently, then, if a way could be found to determine $\hat{\eta}_d$ from first principles, it may well be possible, after all, to determine the correct far-field response without knowing the near-field solution.

4. Definition and discussion of the main mathematical problem

The mathematical problem is illustrated in Fig. 2. It shows a Tollmien-Schlichting wave propagating along a fully developed, plane channel flow. Part of the rigid wall of the channel from $x = \ell_1$ to $x = \ell_2$ is replaced by a compliant panel. In essence the problem is to calculate the amplitude of the Tollmien-Schlichting wave slightly downstream of the compliant panel (i.e., at $x = \ell_2+$), given its amplitude slightly upstream of the panel at $x = \ell_1-$. The response of the panel is also of interest. This problem has been solved by numerical simulation by Davies & Carpenter (1997). Very similar results have been obtained more recently for the Blasius flat-plate boundary layer (see Carpenter *et al.* 2000, 2001)

Unless otherwise stated all quantities are made non-dimensional with reference to the channel half-width h, the flow speed U_m along the centreline, and the fluid density ρ and kinematic viscosity ν. Attention is restricted to disturbance velocities with normal and streamwise components v and u that are symmetrical and antisymmetrical respectively. Disturbances of this type determine the flow stability in the case of rigid walls (Drazin & Reid 1981). They are also analogous to the the form of the disturbances found in the Blasius boundary layer.

Here only small-amplitude disturbances will be considered, so that the two-dimensional, linearized, Navier-Stokes equations in the form of the transport equation for the vorticity disturbance, ω, can serve as the governing equation:

$$\frac{\partial \omega}{\partial t} + U\frac{\partial \omega}{\partial x} + U''v = \frac{1}{R}\nabla^2\omega. \tag{18}$$

where $U = 1 - y^2$ is the velocity in the undisturbed channel flow, $U'' = -2$ is its second derivative with respect to y, and Reynolds number $R = U_m h/\nu$. In order to reduce the number of dependent variables to unity, a perturbation streamfunction, ψ, is introduced, such that

$$u = \frac{\partial \psi}{\partial y}, \qquad v = -\frac{\partial \psi}{\partial x}, \qquad \omega = \nabla^2\psi. \qquad (19a, b, c)$$

The boundary conditions at the wall are as follows:

$$u = v = 0 \qquad \text{for} \quad x < \ell_1 \quad \text{and} \quad x > \ell_2 \qquad (20a)$$

$$u = -\eta U_w', \qquad v = \frac{\partial \eta}{\partial t} \qquad \text{for} \quad \ell_1 < x < \ell_2 \qquad (20a)$$

where η is the vertical displacement of the compliant wall (horizontal displacements will be neglected throughout).

The plate-spring model of Carpenter & Garrad (1985) is used for the dynamics of the compliant wall; accordingly, it is governed by the eq. (9) with the pressure disturbance p_w acting on the wall included on the right-hand side. All of the terms in eq. (9) are now made non-dimensional. It is inconvenient to use U_m as a reference quantity for wall parameters, since they will change when it is altered. Consequently, kinematic viscosity is used instead. This has the effect of introducing the factors involving R in definitions below. Using an asterisk to denote a dimensional quantity we have

$$m \equiv \frac{m}{\rho h}, \qquad \frac{d^*}{\rho U_m} \equiv \frac{1}{R}\underbrace{\frac{d^*\rho h}{\nu}}_{d}, \qquad \frac{B^*}{\rho h^3 U_m^2} \equiv \frac{1}{R^2}\underbrace{\frac{B^*}{\rho h \nu^2}}_{B} \qquad \frac{K^* h}{\rho U_m^2} \equiv \frac{1}{R^2}\underbrace{\frac{K^* h^3}{\rho \nu^2}}_{K}. \tag{21}$$

Conditions are also required at the panel ends $x = \ell_1$ and $x = \ell_2$. Two common cases would be to assume 'built-in' ends (requiring zero displacement and slope) or hinged conditions. In most respects the latter are more demanding, so these are what we will use here, namely

$$\eta = \frac{\partial^2 \eta}{\partial x^2} = 0 \qquad \text{at} \qquad x = \ell_1 \qquad \text{and} \qquad x = \ell_2. \tag{22}$$

The pressure disturbance acting on the wall is obtained by integrating the linearized y-momentum equation of the Navier-Stokes equations:

$$p_w = -\int_0^1 \left(\frac{\partial v}{\partial t} + U\frac{\partial v}{\partial x} + \frac{1}{R}\frac{\partial \omega}{\partial x} \right) dy \tag{23}$$

At entry ($x = 0$) a velocity and vorticity disturbance in the form of a TS wave is assumed, so that

$$v = \mathcal{R}\{\hat{v}_{os} \exp(i\alpha x - i\beta t)\}, \qquad \omega = \mathcal{R}\{\hat{\omega}_{os}(y) \exp(i\alpha x - i\beta t)\}, \tag{24a, b}$$

where $\hat{v}_{os}$ and $\hat{\omega}_{os}$ are complex eigenfunctions of the Orr-Sommerfeld equation, $\mathcal{R}$ denotes the real part, and α and β are respectively the complex wave-number and real frequency of the TS wave.

Two main features were revealed by the numerical simulations of Davies & Carpenter (1997). In the first place they found that the response of the compliant panel to the incident TS wave depended very strongly on whether or not its frequency, β, was above the cut-off frequency of the panel. When β is below cut-off, only the modified TS wave was seen to propagate along the compliant panel. A typical simulation of this type is shown in Figs. 3 and 4. When the cut-off frequency of the panel is exceeded its response is much more complex. Figure 5 shows a typical simulation of this type. The second main feature is closely connected to the first. It was shown by Fourier analysis that the response of the panel above cut-off was almost completely determined by a superposition of three eigenmodes of the coupled (spatially homogeneous) Orr-Sommerfeld/compliant-wall system. One mode was the modified TS mode. The second one was a damped mode with a similar wave-length that propagates in the upstream direction from the trailing edge of the compliant panel. And the third mode was a near-neutral, slightly damped, long-wave mode that propagates in the downstream direction from the leading edge of the compliant panel. This suggests that it should be possible to find a solution to the problem solely in terms of eigenmodes of (spatially homogeneous) Orr-Sommerfeld/compliant-wall system, provided a way could be found of linking their amplitudes to the amplitude of the incident (rigid-wall) TS wave. It is this problem that is the subject of the remaining part of this chapter.

5. Approaches to the problem

When solving the (spatially homogeneous) coupled Orr-Sommerfeld/compliant-wall eigenproblem most authors avoid the difficulties of simultaneously solving the two coupled equations of motion by introducing one or other of two concepts from acoustics and electric circuits. These are complex compliance, Z, first used by Benjamin (1960) and complex admittance Y (cf. eq. 5) introduced by Landahl (1962). The definitions are:

$$Z \equiv \frac{\text{wall displacement}}{\text{pressure}} \qquad \text{and} \qquad Y \equiv -\frac{\text{normal velocity}}{\text{pressure}}$$

One way to use the concepts is to equate the admittances for the wall and fluid at $y = 0$, i.e.,

$$Y_f = Y_w \qquad \text{at} \qquad y = 0, \tag{25}$$

and to use this as a boundary condition on the Orr-Sommerfeld equation (for example, see Carpenter & Garrad 1985 and Carpenter & Morris 1990). Y_f can be determined by using eqn (23) or some other equation for the pressure; and for the plate-spring wall model, the following expressions for complex compliance and admittance can be obtained from eq. (9) using eq. (21):

$$Z_w = \frac{1}{m\beta^2 + id\beta/R - (B\alpha^4 + K)/R^2}, \qquad Y_w = i\beta Z. \tag{26a, b}$$

Two papers (Manuilovich 2001 and Sen *et al.* 2001, see also Carpenter *et al.* 2002) were presented at the Symposium that described analytical approaches to the problem specified in the previous section. Before describing these studies, it is worth discussing two approaches that do not work. It is natural to think of trying to develop an approach similar to the one described in §2 for analysing what happens when a longitudinal wave is incident on a junction in an elastic tube (see Fig. 1). In this case it is possible to determine the reflected and transmitted waves by satisfying conservation of energy (or volume flow) and requiring continuity of pressure at the junction. Accordingly, one may well imagine that the present problem could be dealt with by enclosing the junction in a control volume that extends slight upstream and downstream of it, and equating the kinetic-energy fluxes entering and leaving the control volume. But it is clear from the results presented in Davies & Carpenter (1997) that this approach is not valid. Principally this is because there is strong irreversible energy transfer (especially energy production by the Reynolds stresses) in the neighbourhood of the junction. Notwithstanding this evidence, let us assume for the moment that mechanical energy is approximately conserved within the control volume. If this were so, the kinetic energy of the flow disturbance would initially drop over the compliant wall, owing to a portion of the energy being transferred to the wall. In fact, the opposite happens (see Fig. 6 of Davies & Carpenter 1997) and there is a substantial upwards jump in the kinetic-energy integral as the wave passes from the rigid to the compliant wall. Energy-conservation principles do work, as would be expected, in the equivalent conservative system. For example, when there is a wall wave propagating along a compliant wall of rapidly changing properties immersed in a potential flow (see Lucey *et al.* 2002).

The other approach that might initially appear feasible is to regard the rigid-wall eigenfunction as an initial condition at the start of the compliant wall. One could then envisage determining the compliant-wall solution in the form of a series of compliant-wall eigenfunctions. Mathematically, this approach fails because, although the two-dimensional, rigid-wall eigenfunctions are orthonormal (Henningson & Schmid 1992), the compliant-wall ones are not. Physically, this approach ignores the rapid changes that take place as the incident TS wave approaches the junction on the rigid-wall side. In fact, the recent numerical simulations of Wiplier and Ehrenstein (2000) use this type of initial condition. When making a comparison with some of their results using the methods of Davies & Carpenter, we have found substantial differences between our results and theirs. (It is not suggested, however, that this in any way invalidates the study of Wiplier & Ehrenstein.)

We turn now to approaches that do work. The fundamental concepts of Manuilovich (1995, 2001) and Sen *et al.* (2001) are quite similar. They both use Fourier Transform techniques and draw an analogy between the effect of the junction on the TS wave and the receptivity problem of how a wave is created by a driver of some sort. Manuilovich drawing on earlier work (Manuilovich 1992) thinks of the junction being analogous to the situation where a controlled wall motion creates a TS wave. In a quite similar fashion, Sen regards the effect of the junction as being equivalent to a vibrating ribbon. As we have seen in the simpler problem discussed in §3, there is a sense in which the near-field behaviour at a junction can be modelled by a driver.

Manuilovich does not compare the results of his analysis directly with those of Davies & Carpenter (1997). Indeed, his specific method is rather limited in scope; and because he works with pressure as his main variable, he formulates the problem for compressible flow and presents results corresponding to a Mach number of 0.8. Accordingly, only a brief outline of his theory is given here. But a more detailed description of the theory of Sen *et al.* is presented in §6.

Manuilovich writes the small-amplitude displacement of the compliant wall in the form:

$$\eta = \mathcal{R}\{a_\eta w(x) \mathrm{e}^{-i\beta t}\}. \tag{27}$$

As an initial problem he determines the wall motion required to cancel an incident TS wave. In this case the form of $w(x)$ is specified *a priori* and the problem is to determine the required amplitude a_η and phase of the wall motion. He then draws an analogy between this wave–cancellation problem and the junction problem. He introduces a real compliance of the form:

$$Z = -\sigma S(x) \tag{28}$$

where σ is a constant real parameter and $S(x)$ is a shape function taking the form:

$$\begin{aligned} S(x) &= 1 \qquad |x| \le \ell \\ S(x) &= 1 - 2\Big(\frac{|x|-\ell}{a}\Big)^2 \qquad \ell \le |x| \le \ell + \frac{a}{2} \\ S(x) &= 2\Big(\frac{|x|-\ell}{a} - 1\Big)^2 \qquad \ell + \frac{a}{2} \le |x| \le \ell + a \end{aligned} \tag{29}$$

with $\ell = 8$ and $a = 4$. This gives a near-top-hat variation in $S(x)$ centred at $x = 0$. From the definition of compliance we can write

$$\eta = -\mathcal{R}\{\sigma S(x)\tilde{p}_w \mathrm{e}^{-i\beta t}\} \tag{30}$$

The relationship between the unknown wall oscillation and the pressure can be obtained by analogy with the initial wave-cancellation problem in the form:

$$\tilde{p}_w = \mathrm{e}^{i\alpha_1 x} + \frac{1}{2\pi}\int \tilde{p}^*(0)dk \int w(\xi)\mathrm{e}^{i(x-\xi)}d\xi \tag{31}$$

where the first term on the right-hand side is the normalized, incident, rigid-wall eigenfunction and $\tilde{p}^*$ is a normalized Fourier Transform of the compliant-wall eigenfunction. Equating eqs. (27) and (31) and using eq. (30) leads to the following integral equation for the unknown wall oscillation w:

$$w(x) = -\sigma S(x)\Big\{\mathrm{e}^{i\alpha_1 x} + \frac{1}{2\pi}\int \tilde{p}^*(0)dk \int w(\xi)\mathrm{e}^{i(x-\xi)}d\xi\Big\}. \tag{32}$$

6. The ribbon-analogy theory of Sen *et al.*

In the first place we will consider the case below cut-off where the modified TS wave is the only propagating eigenfunction over the compliant wall. Furthermore, initially we will consider the case where the junction is between a rigid wall extending from $x = -\infty$ to $x = 0$ and a compliant wall extending from $x = 0$ to $x = \infty$. Let the perturbation stream-function take the form:

$$\psi(x, y, t) = f(x, y)\mathrm{e}^{-i\beta t}, \tag{33a}$$

where

$$f(x, y) = H(-x)f_1(x, y) + H(x)f_2(x, y) \tag{33b}$$;

$H(x)$ is the Heaviside step function, so that $f_1(x)$ and $f_2(x)$ are the solutions for the rigid and compliant walls respectively, and take the forms of half-range Fourier transforms (Noble 1958), such that

$$f_{1,2} = \int_{-\infty}^{+\infty} F_{1,2}(y; \alpha)\mathrm{e}^{i\alpha x}d\alpha, \tag{34a, b}$$

where

$$F_1(y; \alpha) = \frac{1}{2\pi}\int_{-\infty}^{0} f_1(x, y)\mathrm{e}^{-i\alpha x}dx, \qquad F_2(y; \alpha) = \frac{1}{2\pi}\int_{0}^{\infty} f_2(x, y)\mathrm{e}^{-i\alpha x}dx. \tag{35a, b}$$

The form of these solutions must be such that away from the immediate vicinity of the junction they reduce to TS waves. Thus we expect them to take the following forms:

$$f_1(x,y) = \mathcal{R}\{\phi_1(y)e^{i\alpha_1 x} + f_{n1}(\xi,y)\}, \qquad f_2(x,y) = \mathcal{R}\{\lambda\phi_2(y)e^{i\alpha_2 x} + f_{n2}(\xi,y)\}, \tag{36a,b}$$

where $\xi = x/\delta$ and δ is the width of the junction region; it is assumed that $\delta \to 0$ as $R \to \infty$. ϕ_1 and ϕ_2 are the known normalized eigenfunctions corresponding to TS waves over the rigid and compliant walls respectively; α_1 and α_2 are the corresponding known eigenvalues. These eigenfunctions and eigenvalues are determined by solving the coupled Orr-Sommerfeld/compliant-wall eigenproblem. The unknown quantity that has to be determined is the relative amplitude λ of the (transmitted) wave over the compliant wall. The functions $f_{n1,2}$ are the near-field solutions in the vicinity of the junction and are expected to decay to zero as $|\xi| \to \infty$.

The function f and all its derivatives are continuous at $x = 0$, so it follows from the definition of ξ and the expected solution forms (36a,b) that

$$\left.\frac{\partial^m f_1}{\partial x^m}\right|_{x=0} = (i\alpha_1)^m \phi_1(y) + \frac{1}{\delta^m}\left.\frac{\partial^m f_n}{\partial \xi^m}\right|_{x=0}, \tag{37a}$$

$$\left.\frac{\partial^m f_2}{\partial x^m}\right|_{x=0} = (i\alpha_2)^m \lambda\phi_2(y) + \frac{1}{\delta^m}\left.\frac{\partial^m f_n}{\partial \xi^m}\right|_{x=0}. \tag{37b}$$

Therefore to a good approximation

$$\left.\frac{\partial^m f_{n1}}{\partial \xi^m}\right|_{x=0} = \left.\frac{\partial^m f_{n2}}{\partial \xi^m}\right|_{x=0} \left(= \left.\frac{\partial^m f_n}{\partial \xi^m}\right|_{x=0}, \quad \text{say}\right) \quad \text{if} \quad m \geq 1. \tag{38}$$

Noting that

$$\int_{-\infty}^{0} \frac{\partial^m}{\partial x^m}(H(x)f_1)e^{-i\alpha x}dx = \left.\frac{\partial^{m-1} f_1}{\partial x^{m-1}}\right|_{x=0} + i\alpha \int_{-\infty}^{0} \frac{\partial^{m-1}}{\partial x^{m-1}}(H(x)f_1)e^{-i\alpha x}dx,$$

$$\int_{0}^{\infty} \frac{\partial^m}{\partial x^m}(H(x)f_2)e^{-i\alpha x}dx = -\left.\frac{\partial^{m-1} f_2}{\partial x^{m-1}}\right|_{x=0} + i\alpha \int_{0}^{\infty} \frac{\partial^{m-1}}{\partial x^{m-1}}(H(x)f_2)e^{-i\alpha x}dx,$$

eq. (33a,b) is substituted into eq. (18), and half-range Fourier Transforms of the resulting equation taken to obtain

$$\mathcal{L}_{os}(\alpha)F_{1,2} = \mp\frac{U(y)}{\delta^2}\left.\frac{\partial^2 f_n}{\partial \xi^2}\right|_{x=0} \simeq \mp U(y)\left.\frac{\partial^2 f_n}{\partial x^2}\right|_{x=0}, \tag{39a,b}$$

where $\mathcal{L}_{os}(\alpha)$ is the usual Orr-Sommerfeld operator. The other terms that would have been found on the right-hand side of eq. (39) are higher-order, remembering that $\delta \ll 1$, and have accordingly been dropped.

It can be seen from the form of eqs. (39a,b) that the right-hand sides ressemble the form appropriate for a wave-driver like a vibrating ribbon (e.g., see Gaster

1965 and also §3). Without solving the complete problem, which would defeat the purpose of developing a theory based on the (far-field) eigensolutions, there is no way to determine the precise functional form of the right-hand sides. Here we have taken them to be $\mp\kappa U(y)$ where κ is a function that is to be determined. We have also tried other functional forms; but, owing to the method used to estimate κ, the calculated value of λ is almost independent of the precise functional form chosen.

We now look for solutions to eqs (39a,b) where α is in the vicinity of $\alpha_{1,2}$, so that we can approximate eqs. (39a,b) as follows:

$$\mathcal{L}_{OS}(\alpha_{1,2})F_{1,2} + (\alpha - \alpha_{1,2})\mathcal{L}^*(\alpha_{1,2})F_{1,2} = \mp\kappa U(y) \qquad (40a, b)$$

where the operator $\mathcal{L}^*(\alpha) \equiv \partial\mathcal{L}_{OS}(\alpha)/\partial\alpha$. It follows from the assumed forms (36a,b) that we can write

$$F_1(y;\alpha) = A_1\phi_1(y;\alpha_1) + F_{1f}, \qquad F_2(y;\alpha) = A_2\phi_2(y;\alpha_2) + F_{2f}, \qquad (41a, b)$$

where the first terms on the right-hand side are the singular solutions, corresponding to the expected, known eigenfunctions, multiplied by contants $A_{1,2}$; and the second terms are the 'forced' terms contributing to the near field around the junction. In practice, we are only interested in obtaining the singular solutions corresponding to the eigenfunctions.

Similarly we can decompose the right-hand sides of (40a,b) into resonating and regular components, $\mathcal{R}_{1,2}$, as follows:

$$\mp\kappa U(y) = C_{1,2}\mathcal{L}^*(\alpha_{1,2})\phi_{1,2} + \mathcal{R}_{1,2} \qquad (42a, b)$$

This enables the singular solutions to be simply determined as:

$$A_1\phi_1 = \frac{C_1\phi_1}{\alpha - \alpha_1}, \qquad A_2\phi_2 = \frac{C_2\phi_2}{\alpha - \alpha_2} \qquad (43a, b)$$

In practice, we only require the resonating part of eqs. (42a,b), and these can be determined from $\mp\kappa U(y)$ by introducing suitably normalized, adjoint eigenfunctions $\theta_{1,2}$, corresponding to $\phi_{1,2}$ (that have been determined in advance) and exploiting their properties. Thus if each side of eqs. (42a,b) is multiplied by its respective adjoint eigenfunction and then integrated with respect to y, we obtain

$$\mp\kappa \int_0^1 \theta_{1,2}(y;\alpha_{1,2})U(y)dy = C_{1,2}\int_0^1 \theta_{1,2}(y;\alpha_{1,2})\mathcal{L}^*(\alpha_{1,2})\phi_{1,2}(y;\alpha_{1,2})dy. \quad (44a, b)$$

The contribution of the singular solutions (43a,b) to the streamfunction (33a,b) is obtained by taken an inverse Fourier Transform and using the Residue Theorem, thus

$$\frac{C_{1,2}}{2\pi}\int_{-\infty}^{\infty} \frac{\phi_{1,2}}{\alpha - \alpha_{1,2}} e^{i\alpha x} d\alpha = iC_{1,2}\phi_{1,2}e^{i\alpha_{1,2}x} \qquad (45a, b)$$

But, if we are to obtain the desired solution forms given in eqs. (36a,b), it follows that $C_1 = -i$ and $C_2 = -i\lambda$. This allows us to determine the strength, κ, of the pseudo-driver from eq. (44a), giving

$$\kappa = \frac{i \int_0^1 \theta_1 \mathcal{L}^*(\alpha_1)\phi_1 dy}{\int_0^1 \theta_1 U(y) dy}. \tag{46}$$

Then, with κ evaluated, the amplitude of the transmitted wave can be determined from eq. (44b), as

$$\lambda = i\kappa \frac{\int_0^1 \theta_2 U(y) dy}{\int_0^1 \theta_2 \mathcal{L}^*(\alpha_2)\phi_2 dy}. \tag{47}$$

The junction at the trailing edge of the compliant panel can be treated in much the same way. From symmetry, it can be seen that at the trailing edge, if the amplitude of the TS wave over the compliant wall is a_2, it having decayed or grown while propagating over the compliant panel, the amplitude of the TS wave emerging on the rigid wall just downstream of the trailing edge is (a_2/λ). This explains why the jumps in kinetic energy and enstrophy at the leading and trailing edges of the compliant panel, seen in the numerical simulations of Davies & Carpenter (1997) when the frequency of the TS wave is below cut-off, are always opposite in direction. Results obtained with the method outlined above give almost perfect agreement with the results of numerical simulations, as can be seen from Figures 3 and 4.

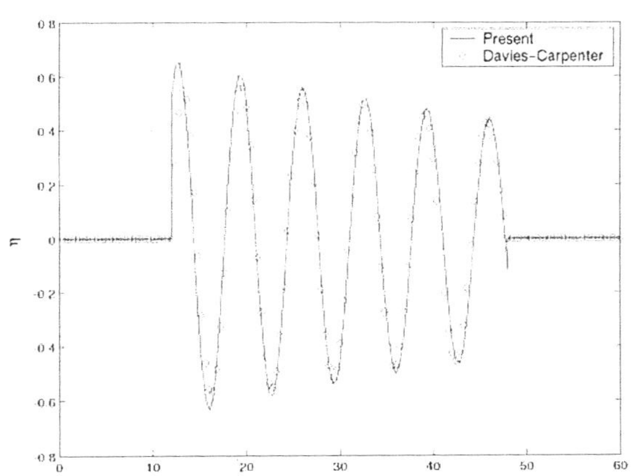

Figure 3. Wall displacement for the case below cut-off illustrated in Figs. 6 and 7 of Davies & Carpenter (1997). $R = 12,000$ and $\beta = 0.24$. Wall parameters: $m = 1/3$, $d = 0$, $B = 1.92 \times 10^7$, $K = 4B$. The compliant panel extends from $x = 12$ to $x = 48$.

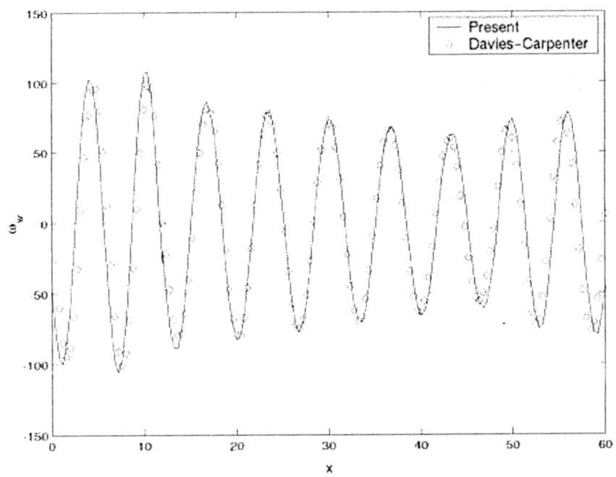

Figure 4. Vorticity at the wall for the case in Figure 2.

We now turn to the case where the TS frequency is above cut-off, so that there are now three propagating eigenfunctions over the compliant panel. In most respects it turns out to be perfectly straightforward to extend the method outlined above to this case. At the leading edge, the additional propagating eigenmode that is created, is the slowly decaying, relatively long wave. This is virtually vorticity-free and, accordingly, cannot be created directly by the TS wave, i.e., by a pseudo-driver at the leading edge. Thus the TS wave just downstream of the leading edge is determined exactly the same way as described above. The amplitudes of the remaining two propagating eigenmodes can then be determined by satisfying the boundary conditions imposed on the wall deflection at $x = \ell_1$ and $x = \ell_2$ (see eq. 22). At first sight the problem is now over-determined, as we have only two unknown amplitudes of the two remaining propagating eigenmodes but four boundary conditions. Accordingly, it is necessary to also include (rapidly decaying) evanescent eigenmodes, one at each end of the compliant panel. These are true eignemodes of the coupled Orr-Sommerfeld/compliant-wall eigensystem (see Carpenter & Morris 1990) that are analogous to the evanescent modes used in eqs. (12a,b). The appropriate evanescent modes must be chosen with care. But once that is done, the unknown amplitudes of the two other propagating eigenmodes can be readily determined. Although it is difficult to make a direct comparision, the results in Figure 5 are in close agreement with the corresponding numerical simulations of Davies & Carpenter (1997).

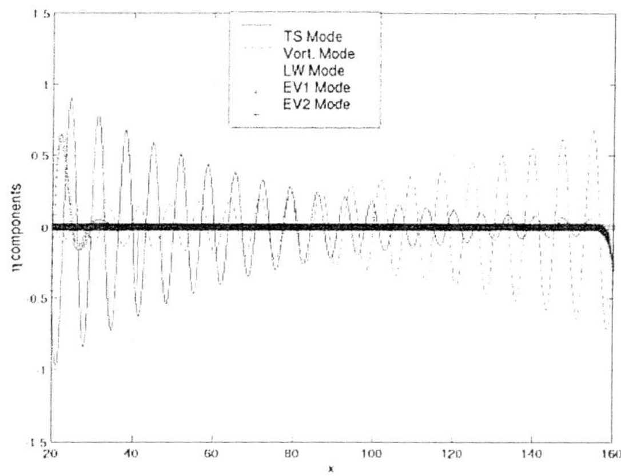

Figure 5. The various components of the wall displacement for a typical case above cut-off. The parameters are the same as for Fig. 3 except that $K = 1.92\times10^7$, $B = 4K$ and the compliant panel extends from $x = 20$ to $x = 160$. The key denotes the various modes as follows: TS, Tollmien-Schlichting wave; Vort., the damped vortical wall mode that propagates upstream from the trailing edge; LW, the near-neutral long wall wave that propagates downstream from the leading edge; EV1, EV2, the evanescent modes in the vicinity of the leading and trailing edges respectively.

There is one remaining twist to this problem. The decaying eigenmode that propagates up from the trailing edge of the compliant panel is vortical. Consequently, it can be coupled to the TS wave via a pseudo-driver. We do not need to take this into account for determining its amplitude, but it is another matter when it comes to determining the amplitude of the TS wave just downstream of the trailing edge. This TS wave results from a superposition of waves generated by pseudo-drivers corresponding to the TS wave and the vortical 'wall' mode over the compliant panel. The strength of this second pseudo-driver can be determined by following the methods described above with fairly obvious modifications. This second pseudo-driver is not merely a theoretical nicety. Under certain circumstances, it can be sole source of the TS wave downstream of the compliant panel. Such a case is illustrated in Figures 6 to 8. There the compliant panel is sufficiently long for the TS wave to have virtually disappeared completely by the time it reaches the trailing edge. Nevertheless, a TS wave of finite amplitude is found just downstream of the trailing edge. As the results shown in Figure 8 make plain, in this case the downstream TS wave is generated by the pseudo-driver created by the vortical 'wall ' mode.

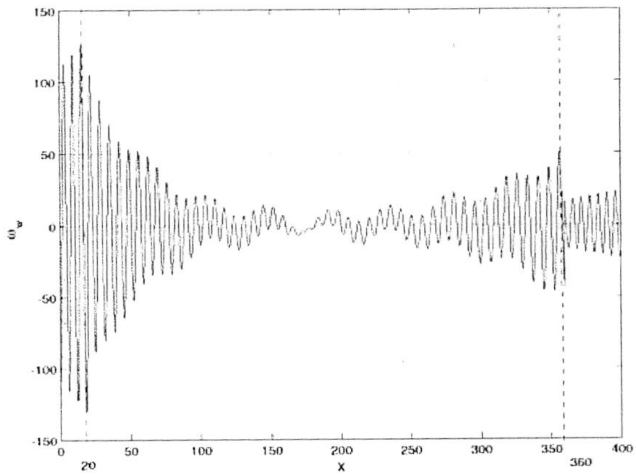

Figure 6. Streamwise variation of the vorticity at the wall for a very long compliant panel above cut-off. The parameters are the same as Fig. 5 except that the panel extends from $x = 20$ to $x = 360$.

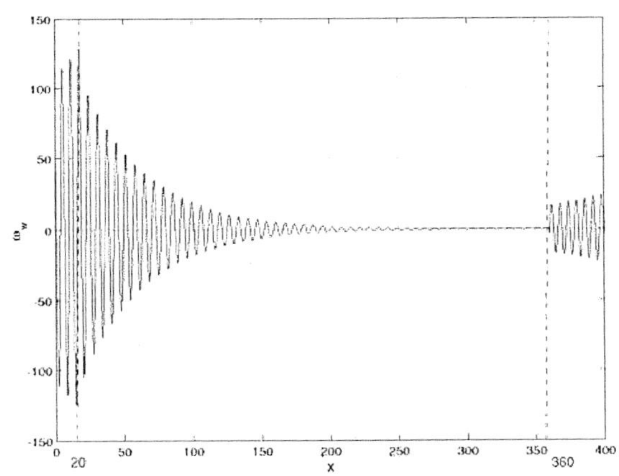

Figure 7. The Tollmien-Schlichting component alone for the case in Fig. 6

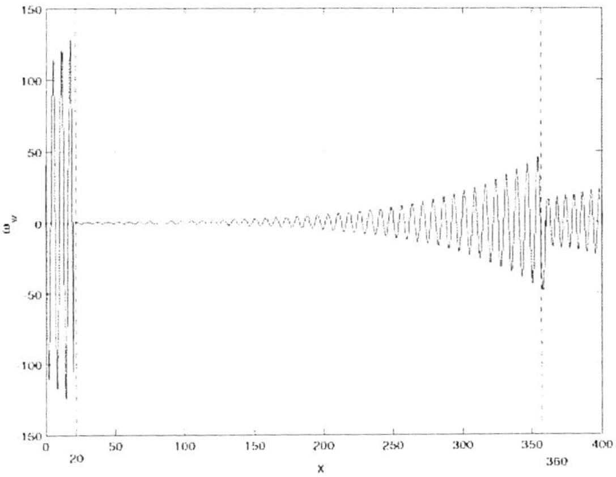

Figure 8. The vertical-mode component alone over the compliant section for the case in Fig. 6. Note that the TS wave is plotted upstream and downstream of the compliant section.

Acknowledgements

This work was supported by the U.K. Engineering and Physical Sciences Research Council through Visiting Research Fellow grants that allowed P.K. Sen to visit the University of Warwick.

References

BENJAMIN, T.B. (1960) Effects of a flexible boundary on hydrodynamic stability, *J. Fluid Mech.* **9**, 513-532.

CARPENTER, P.W., DAVIES, C. and LUCEY, A.D. (2000) Hydrodynamics and compliant walls: Does the dolphin have a secret?, *Current Science* **79**, 758-765.

CARPENTER, P.W. and GARRAD, A.D. (1985) The hydrodynamic stability of flow over Kramer-type compliant surfaces. Part 1. Tollmien-Schlichting instabilities, *J. Fluid Mech.* **155**, 465-510.

CARPENTER, P.W., LUCEY, A.D. and DAVIES, C. (2001) Progress on the use of compliant walls for laminar-flow control, *AIAA J. of Aircraft* **38**, 504-512.

CARPENTER, P.W. and MORRIS, P.J. (1990) The effect of anisotropic wall compliance on boundary-layer stability and transition, *J. Fluid Mech.* **218**, 171-223.

CARPENTER, P.W., SEN P.K., HEGDE, S and DAVIES, C. (2002) Wave propagation in flows across joins between rigid and flexible walls, *ASME Paper IMECE2002-32490.*

DAVIES, C. and CARPENTER, P.W. (1997) Numerical simulation of the evolution of Tollmien-Schlichting waves over finite compliant panels, *J. Fluid Mech.* **335**, 361-392.

DRAZIN, P.G. and REID, W.H. (1981) *Hydrodynamics Stability*, Cambridge University Press.

GASTER, M. (1965) On the generation of spatially growing waves in a bounfary layer, *J. Fluid Mech.* **22**, 433-441.

HENNINGSON, D.S. and SCHMID, P.J. (1992) Vector eigenfunction-expansions for plane channel flows, *Sudies in Applied Maths.* **87**, 15-43.

HOWE, M.S. (1998) *Acoustics of Fluid-Structure Interactions*, Cambridge University Press.

LANDAHL, M.T. (1962) On the stability of a laminar incompressible boundary layer over a flexible surface, *J. Fluid Mech.* **13**, 609-632.

LIGHTHILL, J. (1978) *Waves in fluids*, Cambridge University Press.

LUCEY, A.D., SEN, P.K. and CARPENTER, P.W. (2002) Wall excitation on a flexible surface in the presence of a uniform mean flow, *ASME Paper IMECE2002-32287.*

LUO, X.Y. and PEDLEY, T.J. (1995) A numerical simulation of steady flow in a 2-D collapsible channel, *J. Fluids Struct.* **9**, 149-174.

LUO, X.Y. and PEDLEY, T.J. (1996) A numerical simulation of unsteady flow in a 2-D collapsible channel, *J. Fluid Mech.* **314**, 191-225.

MANUILOVICH, S.V. (1992) Passage of an instability wave through a channel sectiion of variable width, (in Russian), *Izv. Ros. Akad Nauk, Mekh. Zhidk. Gaza*, No. 2, 34-41 (Translation in *Fluid Dynamics* **27**,177-182).

MANUILOVICH, S.V. (1995) Passage of an instability wave over a compliant section of a wall, Unpublished paper.

MANUILOVICH, S.V. (2001) Propagation of Tollmien-Schlichting wave in a boundary layer over flexible path of a wall, In book of abstracts for *IUTAM Symp. on Flow in Collapsible Tubes and Past Other Highly Compliant Boundaries*, Univ. of Warwick, 26-30 March, 2001.

NOBLE, B. (1958) *Methods based on the Wiener-Hopf technique*, Pergamon Press. NGUYEN, V.B., PAIDOUSSIS, M.P. and MISRA, A.K. (1994) A CFD-based model for the study of the stability of cantilevered coaxial cylindrical shells, *J. Sound and Vibration* **176**, 105-125.

PAIDOUSSIS, M.P. (1998) *Fluid-Structure Interactions. Slender structures and axial flow. Vol. 1*, Academic Press.

PEDRIZZETTI, G. (1998) Fluid flow in a tube with an elastic membrane insertion, *J. Fluid Mech.* **375**, 39-64.

SEN, P.K., CARPENTER, P.W., GAJJAR, J.S.B. and HEGDE, S. (2001) The jump conditions for instability waves at a rigid-complant joint, In book of abstracts for *IUTAM Symp. on Flow in Collapsible Tubes and Past Other Highly Compliant Boundaries*, Univ. of Warwick, 26-30 March, 2001.

WIPLIER, O. and EHRENSTEIN, U. (2000) Numerical simulation of linear and nonlinear disturbance evolution in a boundary layer with compliant walls, *J. Fluids Struct.* **14**, 157-182.

Chapter 8

ROTATING FLOWS OVER COMPLIANT WALLS

PETER W. CARPENTER & PETER J. THOMAS
School of Engineering, University of Warwick
Coventry, CV4 7AL, U.K.

MASATO NAGATA
Department of Aeronautics and Astronautics
Kyoto University, Kyoto, Japan

Abstract

This chapter reviews the experimental and theoretical research on rotating flows over compliant walls. The most widely studied such flow is the one over a rotating disc. Its popularity stems from its relative simplicity and also because the cross-flow vortices, which is the dominant route to laminar-turbulent transition, are also found in flows over swept leading edges. After an introduction, the chapter begins with Section 2 which is a review of experimental investigations of compliant rotating discs that are predominantly confined to measuring driving torques. This is followed by Section 3 which reviews the theoretical work on the effects of a compliant disc on flow stability. Section 4 reviews the corrresponding experimental studies of disturbance evolution over rotating compliant discs. Recently thin-gap rotating flows with compliant boundaries have been investigated experimentally and this work is covered in Section 5. The chapter ends with an account of Taylor-Görtler flows with compliant boundaries.

P.W. Carpenter and T.J. Pedley (eds.),
Flow past Highly Compliant Boundaries and in Collapsible Tubes, 167-187.

1. Introduction

It has been clearly established over the last twenty years, through experimental and theoretical investigations, that wall compliance can suppress the growth of Tollmien-Schlichting (TS) waves in boundary layers leading to substantial delays in the onset of laminar-turbulent transition. (See reviews by Riley *et al.* (1988), Carpenter (1990), Carpenter *et al.* (2000, 2001), and Gad-el-Hak (1996, 2002)). But this favourable evidence really only applies to low-noise flow environments that are similar to the flat-plate boundary layer, for which amplification of TS waves is the primary route to transition. In many applications other, quite different, routes to transition may be dominant. This is likely to be the case, for example, for three-dimensional boundary layers, particularly those developing over bodies with swept leading edges. In such cases inviscid instability mechanisms dominate that are much more powerful than TS waves. These are often associated with velocity profiles having inflexion points. An important example is the cross-flow vortices that develop near the leading edge of a swept wing. These also develop in the three-dimensional boundary layer over a rotating disc. And this is the main reason why flow instability and transition in the rotating-disc flow has been so widely studied.

Another attraction of the rotating disc is that it offers a relatively simple, perhaps deceptively simple, method for testing the drag-reducing capabilites of compliant coatings. All that is apparently required is a facility whereby a flat compliant disc rotates in a sufficiently large volume of water. In addition accurate instrumentation is required for measuring the rotational speed and driving torque of the disc. Several investigations of this type have been carried out, some resulting in drag reduction. These are reviewed in Section 2.

The problem with just collecting data on torque as a function of rotational speed is that, because no information is available on the instabilities or turbulence characteristics, any change in effective drag cannot really be explained. For this reason the effects of wall compliance on flow stability and transition have recently been investigated fairly intensively through stability theory, numerical simulation, and experiment. The theoretical research will be reviewed in Section 3 and the experimental studies in Section 4. Also Cros *et al.* (2001) have reported an experimental study of the effect of wall compliance on the flow structures previously observed by Schouveiler *et al.* (1999, 2001) to develop in the narrow gap between rotating and stationary rigid discs. This work will be described in Section 5.

Finally, yet other classes of instability and routes to transition are found in the flows between concentric rotating cylinders and boundary layers over concave surfaces. These are often called Taylor-Görtler Instabilities. The research on the effects of wall compliance on such instabilities will be reviewed in Section 6.

2. Torque Measurements on Compliant Rotating Discs

Several investigators have covered discs with compliant material and rotated them in a controlled fashion in a volume of water. This type of arrangement allows one to make the most straightforward drag measurement for a compliant surface by means of accurately measuring the torque required to drive the disc over a range of rotational speeds. A series of experiments of this kind were carried out by Hansen and Hunston (1974, 1976, 1983) (a good overview is available in Hansen *et al.* 1980). They used polyvinyl chloride (plastisol) with a wide range of elastic moduli as the material to make single-layer, viscoelastic, compliant walls. This material tends to have a high level of viscoelastic damping. They did not observe any torque reduction, indicative of drag reduction, for any of their compliant discs. In fact, in many cases there was a drag increase. This was probably attributable to the formation of divergence-type instability waves. The critical flow speed for the onset of these waves appeared to be consistent with what was known for flat-plate-type boundary layers over compliant walls (see Gad-el-Hak *et al.* 1984, Gad-el-Hak 2002). It is probably fair to say that the main contribution of the work of Hansen and Hunston was the light it shed on the hydroelastic instabilities.

In a later experimental study Chung (1985) designed his composite compliant walls with some care and according to certain basic principles suggested by earlier investigators. Namely, that the response of the compliant wall should be such as to produce small-amplitude, short-wavelength, high-frequency displacements. Following these principles his composite compliant walls consisted of three main layers:

(i) An inner layer of velvet fabric, 2.5 mm thick, bonded to the metal supporting plate with the velour facing outward. The idea was to produce an anisotropic compliant wall that was resistant to the sort of divergence instability observed by Hansen and Hunston. The fibres were susceptible to buckling but could not be stretched appreciably.

(ii) A layer of silicone rubber, 0.75 mm thick, was cast over the velvet into which it presumably penetrated to act as the matrix material. Lightly cross-linked silicone networks were used in combination with high-viscosity oil.

(iii) Finally a thin, stiff, polymer film was used as an outer covering. The best results were obtained with a 125 μm thick ptfe (teflon) film.

The concept of the inner anisotropic layer remained unproven, but the thin outer film was found to play an essential role in obtaining a drag reduction. This agrees with the practical findings of Gaster (1987) and his co-workers, and the theoretical conclusions of Dixon *et al.* (1994). Chung also found that high levels of damping were necessary to obtain a drag reduction. For his best compliant walls he obtained the equivalent of 20% drag reduction for a compliant disc rotating in water at an equivalent Reynolds number of 8.92×10^5.

More recently Fitzgerald *et el.* (1995a,b,1998a,b,c, 2001) in a series of papers have described research on the material properties of whale and dolphin blubber with a view to mimicking these with artificial compliant walls. They have proposed a model based on electric-circuit theory to describe the fundamental aspects of the interaction between the turbulent boundary layer and a compliant wall. They regard the boundary layer as a 'generator' of fluctuating shear stress that is coupled through the viscous near-wall layer to the 'load' in the form of the compliant wall. In order for appreciable energy, or power, to be transmitted the shear impedances of the boundary layer (generator) and compliant wall (load) must be matched. And this is thought to be an essential condition for the wall to act to delay laminar-turbulent transition. They use a complex shear compliance, $J^* = J' - iJ''$ as the material property to characterize the wall shear impedance. J^* is closely linked with the complex shear modulus, G^*, i.e. $G^* = 1/J*$, and shear wave speed is proportional to $\sqrt{G'}$. (Incidentally, the measurements of Fitzgerald and Fitzgerald 1995a indicate that the shear wave speed for pilot-whale blubber is about 9 m/s.)

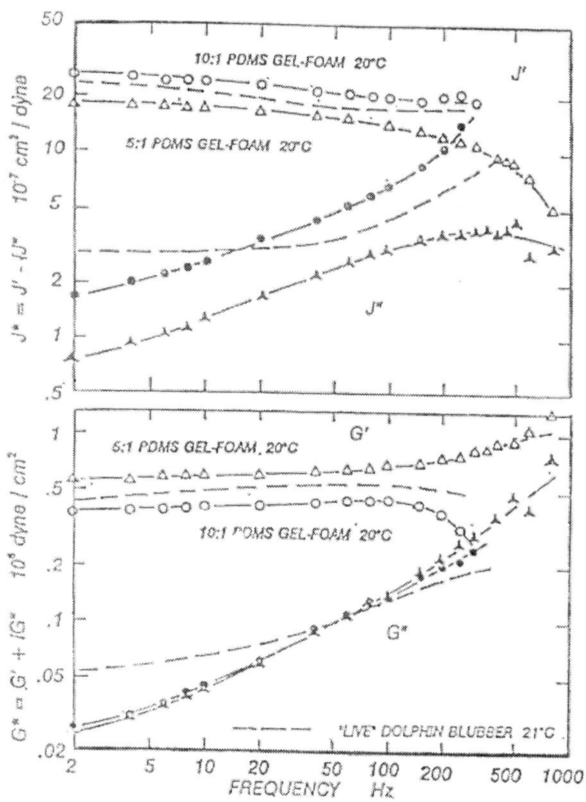

Figure 1. Frequency variation of shear compliance and modulus J*, G*, for 'live' dolphin blubber (dashed lines) compared with those for 10:1 (circles) and 5:1 (triangles) polymer:curing agent polydimethyl siloxane (PDMS) gel-foam composites.
[Based on Fig. 2 of Fitzgerald et al. (1998b)]

Figure 1 presents the frequency variation of the elastic and viscous components of the complex shear compliance $J^* = J' + iJ''$ and shear modulus $G^* = G' + iG''$ for 'live' dolphin blubber compared with two artificial compliant materials. (The shear compliance for 'live' dolphin blubber is obtained by measuring it at various times after death and extrapolating back to the time of death.) It can be seen that the gel-foam composites display characteristics that are fairly close to dolphin blubber.

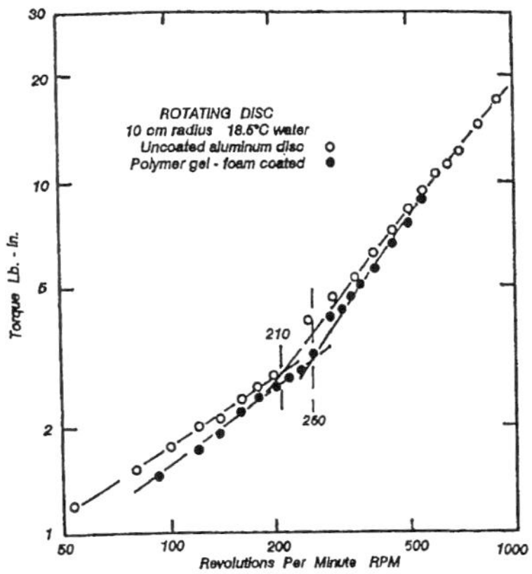

Figure 2. Drag torque vs. rotational speed for a rigid reference disc (open circles) and for a disc coated with a 4.85 mm thick polymethyl siloxane gel-polyurethane foam and a smooth, 0.15 mm thick, outer layer of natural rubber latex (filled circles).
[Based on Fig. 6 of Fitzgerald et al. (1998b)]

A 20-cm aluminium disc was coated with the PMS compliant materials in order to test their drag-reducing capabilities. Coating thicknesses ranging from 5 to 20 mm were used. The discs were then rotated in a volume of water at speeds ranging from 60 to 2250 RPM. Typical results for torque versus rotational speed are displayed in Fig. 2. There the results for the coated disc are compared with a rigid disc. It can be seen that an abrupt change of slope occurs at a particular rotational speed indicative of laminar-turbulent transition. This change occurs at 210 RPM for the rigid disc and at 260 RPM for the compliant disc. This is equivalent to raising the transitional Reynolds number from $2.2. \times 10^5$ to 2.7×10^5. So, use of the compliant disc led to a modest (23 percent) delay in transitional Reynolds number.

3. Flow Stability Theory for Compliant Rotating Disc Flows

In this section we shall review the recent theoretical research on the effects of wall compliance on the instabilities of the rotating-disc boundary layer. This three-dimensional flow develops over a rotating disc in an initially still ambient fluid. The undisturbed flow field $(\bar{u}, \bar{v}, \bar{w})$ (where these are the velocity components corresponding to the radial r, azimuthal θ, and axial z co-ordinates respectively) can be modelled by an exact similarity solution of the Navier-Stokes equations first obtained by von Kármán (1921) and given by

$$\bar{u} = r\Omega F(z), \qquad \bar{v} = r\Omega G(z), \qquad \bar{w} = \sqrt{\nu\Omega} H(z) \tag{1}$$

where Ω is the rotational speed of the disc and ν is the kinematic viscosity of the fluid. The velocity-profile functions are plotted in Fig. 3.

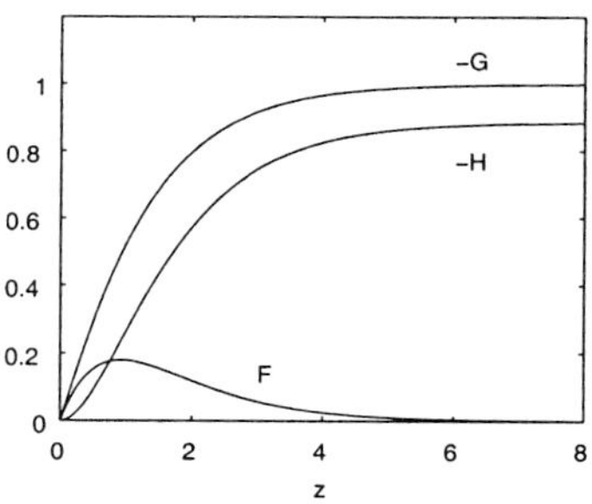

Figure 3. The Kármán (1921) solution for the radial (F), azimuthal (G), and axial (H) velocity components for boundary-flow flow over a rotating disc.

This relatively simple idealised flow exhibits many of the features found in boundary layers in practical applications. Of particular significance is the cross-flow-vortex instability that results from the inflexion point in the radial component of the velocity profile. It is this form of instability that can lead to transition for flows over swept leading edges. Although the primary instability mechanism for the cross-flow vortices is essentially inviscid, viscous effects must be considered in the stability analysis if growth rates are to be represented accurately. Including viscous terms in the usual sense results in the well-known Orr-Sommerfeld equation. In the case of the rotating disc, however, this fourth-order equation is not sufficient for a good representation of the instability characteristics because it omits effects like Coriolis acceleration. Lilly (1966) first established that the inclusion of Coriolis acceleration terms significantly affected the stability characteristics of the related Ekman boundary layer. Later Malik *et al.* (1981) demonstrated that Coriolis and streamline-curvature effects have a significant stabilizing effect on the disturbances developing in the rotating-disc boundary layer. Including both of these effects results in a sixth-order system of ordinary differential equations which reduces to the Orr-Sommerfeld equation when the Coriolis and streamline curvature terms are omitted. Cooper and Carpenter (1997a) showed how this system of instability equations may be coupled to the Navier equations of motion for a viscoelastic solid in order to model the dynamics of a boundary-layer disturbance developing interactively over a single viscoelastic layer.

The instability of the boundary layer over a rigid rotating disc has been studied extensively and the characteristic features of the eigenmode spectrum are quite complex; at least three distinct families of eigensolutions have been identified. One is the inviscid or Type I instability (i.e. the cross-flow vortices mentioned above) first studied by Gregory *et al.* (1955) who were followed by notable contributions from Malik *et al.* (1981), Wilkinson and Malik (1985) and Mack (1985). This form of instability is considered to be the commonest route to transition and has been well studied experimentally. The early work by Gregory *et al.* using a china-clay visualisation technique was able to reveal the cross-flow instability clearly as a series of closely spaced spiral streaks that are stationary relative to the disc surface. Mack successfully modelled theoretically the modulated disturbance forms observed experimentally by Wilkinson and Malik, thereby establishing that they result from a superposition of the complete, zero-frequency, azimuthal wavenumber spectrum. He also noted that, although the experimentally observed spiral streaks correspond to zero-frequency disturbances, they are not the most unstable modes in the boundary layer. Some travelling modes were found to have a higher growth rate than the stationary ones. More comprehensive information on the theoretical stability characteristics of travelling modes was supplied by Balakumar and Malik (1990) who confirmed Mack's results.

Until fairly recently the role of the travelling disturbances in experiments on transition remained fairly obscure. This was despite the fact that in the first experimental study of instability of the rotating-disc boundary layer Smith (1946) used two hot-wire probes and detected travelling disturbances. Recently, however, travelling modes have been detected and studied experimentally. Jarre *et al.* (1996a,b) (see also Jarre *et al.* 1991 and Le Gal 1992) have carried out 2-point measurements with hot-film probes enabling them to measure phase speeds. Jarre *et al.* (1996a) found that, for the natural transition process over a nominally smooth disc, travelling waves dominated in the early stages of instability. For the later stages, when the disturbances had reached a larger amplitude, stationary disturbances (i.e. zero-frequency modes) dominated. Jarre *et al.* (1996b) then studied the case when transition is forced with a roughness element placed at a position just inboard of the location corresponding to the linear threshold for Type I instability. The roughness element was of a size comparable to the boundary-layer thickness and accordingly generated finite-amplitude disturbances. Nevertheless one might have expected strong stationary disturbances to have dominated the transition process. What was found instead was that the dominant disturbances had a small, but definite, negative phase speed, i.e. they travelled with a circumferential phase speed slightly less than the disc rotation speed. The range of propagation angles at this phase speed agreed well with the theoretical predictions of Balakumar & Malik (1990). But, owing to finite-amplitude effects, good agreement with linear stability theory was not found in other respects. It is probably true, however, that the most dramatic role played by the travelling modes is to generate absolute instability. This was discovered by Lingwood (1995,1996) and will be discussed below.

A second family of eigenmodes, termed Type II, contains a weaker, viscous instability that appears only when Coriolis effects are included. A similar instability was first discovered in the Ekman boundary-layer by Faller (1963). In the rotating-disc boundary layer the stationary form of this instability typically appears at higher Reynolds numbers than the Type I mode and Coriolis acceleration is essential for its occurrence, but little else is known about the precise physical instability mechanism or the role of this mode in the transition process. However, clear evidence of the existence of Type II instabilities has been revealed by Lingwood (1996) in her experimental study of the impulsively perturbed rotating-disc flow. In Lingwood (1995) she also noted that the Types I and II coalesce at a point just inboard of the critical Reynolds number (defined as $R = r\sqrt{\Omega/\nu}$) where both eigenmodes are convectively stable. In this case algebraic growth results (as first suggested in the context of the Orr-Sommerfeld equation by Koch 1986) rather than absolute instability. This could have been physically significant and has recently been confirmed in numerical simulations reported by Davies and Carpenter (2001). But for the rigid disc, at least, their simulations show that the algebraic growth is dominated by the exponentially growing Type I instability that sets in at a slightly more outboard radial position.

A third family of eigenmodes, which we will term Type III, was first identified by Mack (1985) and briefly mentioned by Balakumar & Malik (1990). These eigenmodes are found to have large negative values of α_i (where $\alpha = \alpha_r + i\alpha_i$ is the radial wavenumber) but are considered to be strongly stable modes in view of their inwardly directed group velocity. Recently Lingwood (1995) has shown that this Type III can coalesce with the Type I instability to form an absolute instability, thereby providing another route to transition. This has been confirmed experimentally in Lingwood (1996).

The effects of wall compliance on the Types I and II instability have been studied theoretically by Cooper and Carpenter (1997a). The form of compliant wall used in their study was a single layer of viscoelastic material. Since any disturbances created by the interaction of the boundary layer and the compliant wall tend to attenuate rapidly as they propagate into the wall, Cooper and Carpenter carried out their instability calculations for an infinitely deep compliant layer. They chose a modulus of rigidity, $G = 1000$ N/m^2 and varied the effective compliance of the disc by varying the rotational speed. Typical results for neutral-stability curves of the stationary eigenmodes are displayed in Fig. 4. It is found that as the rotational speed (effective wall compliance) is increased there is an increasing stabilizing effect on the Type I instability. This is confirmed by the results for growth rate given in Cooper and Carpenter (1997a) and the results of the numerical simulations by Davies and Carpenter (2001) plotted in Fig. 5. This depicts the stabilizing effect of a compliant annulus on the Type I instability. In this case the wall dynamics are modelled theoretically by the plate-spring compliant wall (Carpenter and Garrad 1985). The wall properties are chosen so that there is marginal stability with respect to the hydroelastic instabilities at the outer radius of the annulus.

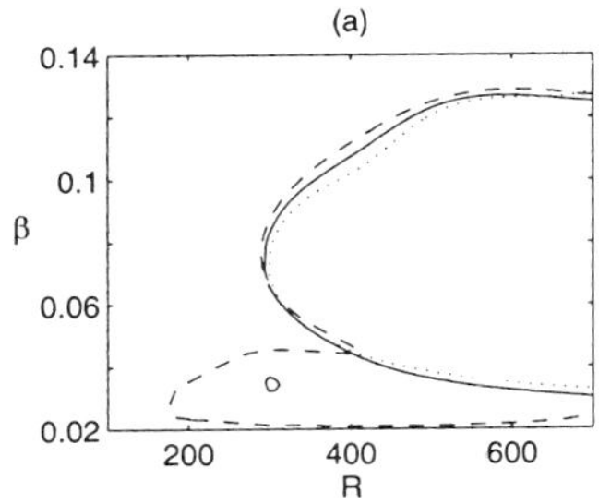

Figure 4 Neutral curves for stationary disturbances over rigid and compliant rotating discs showing the variation of the azimuthal wave-number β with Reynolds number R according to the theory of Cooper and Carpenter (1997a). Rigid wall: solid grey curve; compliant wall: - -, $\Omega = 20$ rad/s; —, $\Omega = 40$ rad/s; · · ·, 60 rad/s.
[Based on Fig. 2a of Cooper and Carpenter(1997a)]

It is clear from Fig. 4 that the effect of wall compliance on the Type II instability is more complex. For the rigid wall the critical Reynolds number for the stationary Type II instability is about 440. For relatively low levels of wall compliance (low rotational speed) the effect is strongly destabilizing. Note that for $\Omega = 20$ rad/s the enlarged area of instability is quite striking and the critical Reynolds number falls to 177. However, as the degree of effective wall compliance is raised further this unstable region shrinks and ultimately vanishes for $\Omega = 60$ rad/s.

Since the Type II instability is destabilized for moderately compliant walls it is only to be expected that wall compliance will also promote the appearance of modal coalescence between the Types I and II eigenmodes. Cooper and Carpenter (1997b) showed that for rigid walls modal coalescence first occurs at $R = 437$ (where both eigenmodes are stable), just a little ahead of the value of $R = 439$ where its occurrence coincides with neutral stability of the Type I instability. But for a disc with a relatively low level of compliance (i.e., a rotation speed of 20 rad/s) the corresponding two Reynolds numbers are $R = 373$ and 387. So it is possible that the algebraic growth associated with the modal coalescence is physically more significant for a compliant disc. But this has not been confirmed.

A more important phenomenon is the absolute instability discovered by Lingwood (1995, 1996) for which the critical Reynolds number is 507 (this corrected value is given in Lingwood 1997). This comes about through a coalescence of the Type I and Type II eigenmodes. Cooper and Carpenter (1997b) showed theoretically that even a small degree of wall compliance is enough to substantially postpone the absolute instability or even suppress it completely. This strong stabilizing effect was confirmed by the numerical simulations of Davies and Carpenter (2000) (see also Carpenter and Davies 2001). The numerical simulations are also illustrated in Fig. 9 of Davies (2002).

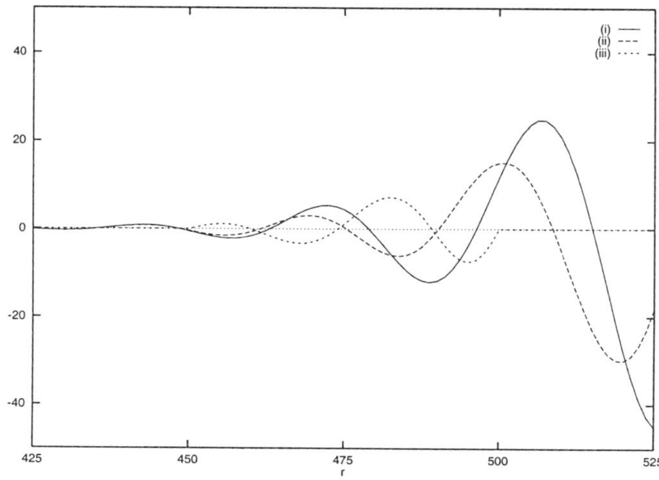

Figure 5 Instantaneous surface displacements and perturbations at the disc surface obtained by numerical simulation for a Type I disturbance with azimuthal wave-number n=43 excited impulsively at $r\sqrt{\Omega/\nu} = 400$. (i) Pressure for an entirely rigid disc, (ii) pressure at the same instant of time in the periodic forcing cycle when there is a compliant annulus extending from $r\sqrt{\Omega/\nu} = 450$ to $r\sqrt{\Omega/\nu} = 500$, (iii) corresponding compliant surface displacements. The compliant wall parameters are such that, notionally, there is marginal stability with respect to divergence and travelling-wave flutter at the outermost radius of the annulus. [Based on Fig. 8 of Davies and Carpenter (2001)]

4. Experimental Studies of Disturbance Evolution in Compliant Rotating Disc Flows

The first experiments on the evolution of boundary-layer disturbances over a compliant rotating disc were carried out by Colley *et al.* (1999). Their compliant walls were manufactured from a single layer of silicone rubber having a typical elastic modulus of 400 kPa which corresponded to a comparatively low effective compliance at the rotational speed of 7.85 rad/s used in their experiments. They followed fairly closely the methodology of Jarre *et al.* (1991, 1996a, 1996b) in terms of both the overall set-up and the use of a series of Fourier spectra (e.g., see Fig. 7) to determine the growth rates and other characteristics of the disturbances. Thus the disc rotated within an inner partial enclosure with the whole contained in a larger tank filled with water. The enclosure consisted of a lid with an annular side wall. The height of the lid above the surface was large compared with the boundary-layer displacement thickness.

Colley *et al.* showed that the compliant disc had a stabilizing effect on the Type I, cross-flow-vortex, eigenmode. Nevertheless, transition actually occurred earlier than for the rigid disc, although the turbulence level during and after transition was significantly lower for the compliant disc. These observations appear to be consistent with the theoretical study of Cooper and Carpenter (1997a) who showed that wall compliance had a stabilizing effect on the Type I instability and that low to moderate levels of wall compliance destabilized the viscous Type II mode. It could therefore be argued that it was the latter that provided the route to transition. In an important respect, however, the experimental results were not consistent with the theory. This is because the theory suggests that the effects of compliant walls as stiff as those used in the experiments should produce results that are indistinguishable from those found for a rigid disc.

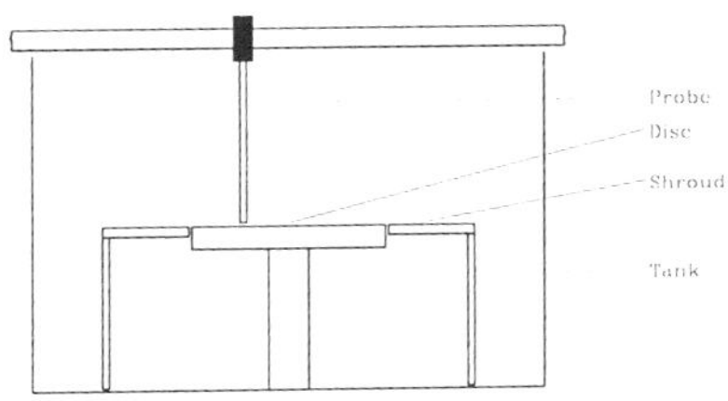

Figure 6 The new rotating-disc apparatus used by Colley et al. (2000)

In their more recent experimental study Colley *et al.* (2000) redesigned their rotating-disc apparatus as shown in Fig. 6. The new set-up has dispensed with the cover over the rotating disc and instead added an annular shroud. This shroud is designed to prevent the disturbed flow created in the region on the underside of the disc from disturbing the flow over the test surface. The new configuration appears to be much superior to the old one in terms of background noise and it is much easier to make measurements. The undisturbed velocity profiles measured by Ali (2001) are virtually identical to those reported in Fig. 2 of Colley *et al.* (1999).

The new apparatus was used to study the development of boundary-layer disturbances over a softer compliant disc. This was manufactured from MED 6340 silicone rubber (see Section 5 for further details) that had a typical modulus of elasticity, $E = 80$ kPa. Successive Fourier spectra for a range of radial positions corresponding to the rigid and compliant discs are presented in Fig. 7. It can be seen that for the rigid disc only the Type I disturbance, corresponding to the central peaks of the spectra, is present. Note that the final spectrum is broadband which is indicative of fully turbulent flow. For the compliant disc there is a Type II disturbance also present exemplified by the spectral peaks seen at lower frequencies and lower Reynolds number. (Note that the probe is stationary so that the disc with its stationary vortex-like disturbances rotates below it. Thus temporal frequency is really a measure of azimuthal wave-number.) Accordingly, it can be seen that the results in Fig. 7 are qualitatively in good agreement with the theory of Cooper and Carpenter (1997a) as illustrated in Fig. 4. In other words the theory shows that low-to-moderate levels of wall compliance greatly advance the critical Reynold number for the Type II disturbance and this is reflected in a comparison between Fig. 7a and 7b.

Unlike the earlier investigation of Colley *et al.* (1999), modest transition delays were found by Colley *et al.* (2000) and are summarized in Table 1 below.

TABLE 1. Transitional Reynolds numbers for the rigid and ($E = 80$ kPa) compliant discs

Ω rad/s	Re_t (rigid)	Re_t (compliant)
1.25π	c. 510	c. 520
2.5π	c. 510	c. 520
3.0π	c. 520	c. 540

5. Experimental Studies of Thin-gap Rotating Flows

The flow in a fluid-filled gap between a rotating disc (rotor) and a stationary enclosure (stator) is of practical and fundamental interest (Owen and Rogers 1989). The fluid motion is driven by the rotor but boundary layers are established on both the rotor and the stator. The laminar-turbulent transition process of the flow is governed by the Reynolds number ($Re = \Omega R^2/\nu$) and the ratio, $\bar{h} = h/R$, of gap width h and disc radius R.

Flow visualization experiments by Schouveiler *et al.* (1998, 1999, 2001) have revealed two different transition scenarios for gap widths larger or smaller than the combined thickness of the two boundary layers on rotor and stator. When the gap is wider than this combined thickness circular and spiral waves are observed to destabilize the stationary-disc boundary layer. Transition occurs in this case by mixing of these waves. When the gap is narrower, such that the rotor and stator boundary layers merge, finite-size turbulent structures can appear. These structures consist of turbulent spots or turbulent spirals which penetrate the laminar domains of the flow as the Reynolds number is increased (Schouveiler *et al.* 2001).

Recently the research groups at the University of Warwick and I.R.P.H.E, Marseille initiated a collaborative experimental investigation studying transition over the type of compliant discs used by Colley *et al.* (1999, 2000) but employing the narrow-gap rotating-disc facility of Schouveiler *et al.* (1998, 1999, 2001). The parameter range covered in their initial experiments was $Re \leq 4.2 \times 10^5$ with $0.005 \geq \bar{h} \geq 0.02$.

The compliant discs were manufactured from Silicon Gel (NuSil MED-6340, Polymer Systems Technology Ltd., High Wycombe, UK). The gel is supplied in two separate components. Varying the mixing ratio of the components enables compliant surfaces with different elastic moduli to be produced. Two discs were manufactured at Warwick and tested at Marseille.

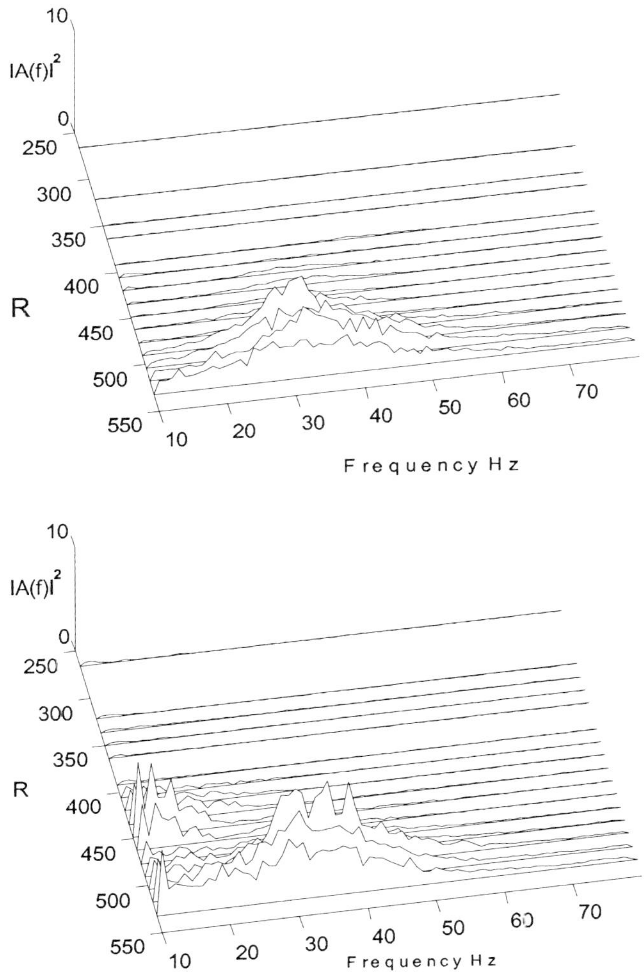

Figure 7 Spectra at various values of Reynolds number obtained by Colley et al. (2000) for a rigid disc (upper figure), and a compliant disc (lower figure), both rotating at 1.25 rev/s.

The first disc was fabricated using a mixing ratio of the gel components. The elastic modulus and the damping coefficient were determined by the techniques described in Colley *et al.* (1999). The elastic modulus was $E = 326$ kPa and the damping coefficient corresponded to $ln(h_0/h_1) = 1.33 \pm 0.05$, (where h_0 and h_1 are the initial and rebound heights of a small metal ball). The experiments revealed that this level of compliance did not measurably affect the critical Reynolds numbers for the onset and the disappearance of the wave structures observed by Schouveiler *et al.* (1998, 1999, 2001) over rigid discs.

The second disc was more compliant. It was fabricated from a 2:1 gel mixture (2 parts component A and 1 part component B). The elastic modulus was 267 kPa while its damping coefficient was $ln(h_0/h_1) = 3.11 \pm 0.05$. This level of compliance was found to affect the transition process (Cros *et al* 2001, 2002). It was, for instance, observed that the onset for the occurrence of the spiral waves was postponed to higher disc speeds. The delay depends on $\bar{h}$ and lies, typically, between 10% (for onset) 60 % (for disappearance) above the values measured for the rigid disc. The solitary waves evolve directly from defects of the primary spiral waves and, hence, delays in onset of a similar magnitude were observed for these structures. The threshold velocity for the emergence of turbulent spots over the compliant disc remained unchanged, however, compared with the values measured by Schouveiler *et al.* (1998, 1999, 2001) for rigid discs. A detailed discussion of these results is currently being prepared for publication (Cros *et al.* 2002).

6. Taylor-Görtler Instabilities

Görtler vortices are instabilities in the form of contra-rotating streamwise vortices found in boundary layers over concave walls. They are centrifugally driven, so the basic instability mechanism is inviscid. The critical flow parameter governing the growth of the streamwise vortices is the Görtler number

$$G \equiv Re\frac{\delta}{\mathcal{R}} \qquad \text{where} \qquad Re \equiv \frac{U_\infty \delta}{\nu}. \tag{2}$$

where δ is the boundary-layer thickness and $\mathcal{R}$ the radius of curvature of the concave wall. Denier *et al.* (1991) showed that the most unstable Görtler mode is confined to a thin wall layer of thickness $\mathcal{O}(G^{-1/5})$. It might therefore be expected that this most unstable mode would be likely to be fairly strongly affected by wall compliance. However, as yet the effect of wall compliance on Görtler vortices has only been subject to preliminary studies.

For example, Yurchenko and Babenko (1987) and Yurchenko *et al.* (1989) reported an experimental study of three-dimensional disturbances in the form of streamwise vortices developing of the concave floor of a water channel. This floor was covered with a compliant wall formed from a layer of porous rubber. The experimental study showed that regions of (G,k) space (where k is the wavenumber of the vortices) corresponding to the maximum amplification rates of the initial disturbances display a shift to the left over a compliant, as compared with a rigid, wall. This suggests that wall compliance has a stabilizing effect on the growth rate of the streamwise vortices.

Subsequent to this experimental work, Denier and Hall (1991) developed an asymptotic theory for studying the effect of wall compliance on the most unstable Görtler modes. Their 'compliant-wall' model consisted only of an unsupported thin plate. Nevertheless their theory agreed qualitatively with the experiments of Yurnenko *et al.* in that the growth rate of the streamwise vortices was reduced by wall compliance. But the stabilization was fairly weak.

Another commonly studied system that is subject to centrifugal instability is Taylor-Couette flow, i.e. is the flow between two concentric cylinders. Recently Koga and Nagata (2001) have investigated the linear stability of this flow theoretically when the inner *rigid* cylinder of radius r_i rotates at angular speed of Ω_i and the outer compliant (elastic) cylinder of radius r_o rotates at an angular speed of Ω_o.

They introduced a cylindrical co-ordinate system (r, θ, z) to analyse the stability of this flow which in its undisturbed state is purely azimuthal with the velocity profile given by

$$V(r) = R\Big(\frac{\tilde{\Omega} - \eta^2}{1-\eta^2}\frac{1-\eta}{2\eta}r + \frac{1-\tilde{\Omega}}{1-\eta^2}\frac{2\eta}{1-\eta^2}\frac{1}{r}\Big) \tag{3}$$

where $\eta = r_i/r_o$, $\tilde{\Omega} = \Omega_o/\Omega_i$, and the Reynolds number is defined by $R = r_i\Omega_i l/\nu$ where $2l = r_o - r_i$. The velocity perturbation is decomposed into the toroidal and poloidal parts as:

$$\mathbf{u} = (u_r, u_\theta, u_z) = \nabla \times (\psi r\hat{e}_r) + \nabla \times \nabla \times (\phi r\hat{e}_r) \tag{4}$$

(where $\hat{e}_r$ is the unit vector in the radial direction) so that the continuity equation is automatically satisfied.

By carrying out the operations $(\hat{e}_r/r)\cdot\nabla\times$ and $(\hat{e}_r/r)\cdot\nabla\times\nabla\times$ on the linearized Navier-Stokes equations in cylindrical co-ordinates the governing equations for ψ and ϕ are obtained in the forms:

$$\begin{aligned}&\frac{\partial}{\partial t}\Big(-\Delta_c\psi + \frac{2}{r^2}\phi_{\theta z}\Big) + V(r)\Big(\frac{2}{r^3} + \frac{1}{r}\Delta_c\phi_z - \frac{1}{r}\Delta_c\psi_\theta\Big) + \frac{dV}{dR}\Delta_c\phi_z\\ &= \frac{4}{r^2}\Delta\phi_{\theta z} - \bar{\Delta}\Delta_c\psi - \frac{2}{r^3}\psi_{r\theta\theta},\end{aligned} \tag{5}$$

$$\begin{aligned}&\frac{\partial}{\partial t}\Big(\bar{\Delta}\Delta_c\phi + \frac{2}{r^2}\psi_{\theta z} + \frac{2}{r^3}\phi_{r\theta\theta}\Big) + V(r)\Big(\frac{1}{r}\bar{\Delta}\phi_{\theta zz} + \frac{1}{r^3}\phi_{\theta\theta\theta zz} + \frac{2}{r}\Delta_c\psi_z + \frac{2}{r^3}\psi_{\theta\theta z}\\ &+ \frac{1}{r^3}\phi_{rr\theta\theta\theta} + \frac{1}{r^5}\phi_{\theta\theta\theta} - \frac{3}{r^3}\phi_{\theta zz} + \frac{1}{r^5}\phi_{\theta\theta\theta\theta\theta} + \frac{1}{r^4}\phi_{r\theta\theta\theta}\Big) - \frac{dV}{dr}\Big(\frac{1}{r^4}\phi_{\theta\theta\theta} + \frac{1}{r^2}\phi_{\theta zz}\Big)\\ &- \frac{d^2V}{dr^2}\frac{1}{r}\Delta_c\phi_\theta = \bar{\Delta}\Big(\bar{\Delta}\Delta_c\phi + \frac{2}{r^3}\Big) + \frac{4}{r^2}\Delta\psi_{\theta z} + \frac{2}{r^3}\Big(\frac{\partial}{\partial r}\Delta\phi_{\partial\theta} - \frac{2}{r}\phi_{\theta\theta zz}\Big),\end{aligned} \tag{6}$$

where

$$\Delta_c \equiv \frac{\partial^2}{\partial r^2} + \frac{1}{r^2}\frac{\partial^2}{\partial \theta^2}, \qquad \bar{\Delta} \equiv \frac{3}{r}\frac{\partial}{\partial r} + \frac{\partial^2}{\partial r^2} + \frac{1}{r^2}\frac{\partial^2}{\partial \theta^2} + \frac{\partial^2}{\partial z^2}$$

The plate-spring model of the Kramer compliant wall (Carpenter and Garrad 1985) was extended by replacing thin-plate theory with cylindrical-shell theory. Accordingly the equation of motion for the compliant wall is written in the form:

$$m\frac{\partial^2}{\partial t^2}\Delta_h^2\xi + d\frac{\partial}{\partial t}\Delta_h^2\xi + B\Delta_h^4\xi + \frac{Eh}{r^2}\frac{\partial^4\xi}{\partial z^4} + K\Delta_h^2\xi = \Delta_h^2 p_w, \tag{7}$$

where

$$\Delta_h \equiv \frac{1}{r^2}\frac{\partial^2}{\partial \theta^2} + \frac{\partial^2}{\partial z^2},$$

p_w is the perturbation wall pressure generated by the fluid flow; $\xi(\theta, z)$ is the wall displacement from its undisturbed position; m is the mass per unit area, d the damping rate, B the flexural rigidity, and E and h are respectively the modulus of elasticity of the shell and its thickness; and K is the spring stiffness of the compliant wall.

The no-slip condition and that on the wall-normal velocity component imposed at $r = r_0$ are implemented in a similar way to previous investigators (e.g., see Carpenter 1990, Carpenter and Morris 1990), namely

$$u_r = \frac{\partial \xi}{\partial t} + \frac{V(r)}{r}\frac{\partial \xi}{\partial \theta}, \qquad u_\theta + \frac{dV}{dr}\xi = 0, \qquad , u_z = 0 \tag{8}$$

A normal-mode analysis of these governng equations is carried out whereby the perturbations are assumed to be periodic in the axial and azimuthal directions with wave-numbers γ and n (an integer) respectively. A spectral method whereby the variations of ψ and ϕ in the radial direction are approximated by series of Chebyshev polynomials and a collocation method is used. This is coupled with numerical solution of eq. (7) taking account of the boundary conditions (8). The resulting matrix eigenvalue problem is solved numerically for growth rate after specifying the values of (n, γ) and the various parameters of the system.

Based on the methods outlined above, Koga and Nagata (2001) carried out a numerical investigation of the effects of wall compliance on the Taylor-Couette system. They found that when the compliance of the outer cylinder is small with relatively large spring stiffness, K, the $n = 0$ mode has the lowest critical Reynolds number, R_c. When the degree of compliance was increased, R_c for the $n = 0$ mode fell monotonically. They were unable to find any values of the compliant-wall parameters that led to a delay in the onset of linear instability. In fact, in comparison with rigid-wall behaviour, wall compliance tended to extend the range of values of η for which instability could be observed. See Fig. 8. The flow pattern in the extended region of instability is characterized by a strong longitudinal vortex structure concentrated near the outer compliant boundary (see Fig. 9).

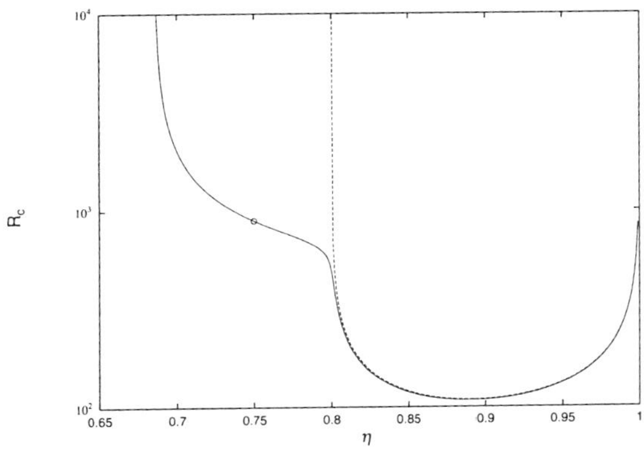

Figure 8. The variation of critical Reynolds number, R_c with η. —, compliant boundaray; $\cdots$, rigid boundaries. $m = 1$, $d = 1$, $B = 10^2$, $Eh = 10^3$, $K = 10^3$, $\tilde{\Omega} = 0.64$, $n = 0$, and $\gamma = 1.5$. For rigid boundaries the basic flow is stable for all Reynolds numbers if the Rayleigh criterion $\eta \leq 0.8$ is satisfied.

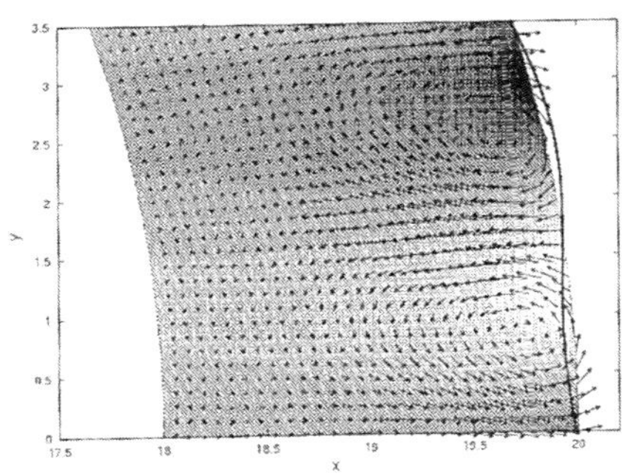

Figure 9. Taylor vortices modified by the compliant outer cylinder. Velocity vectors are denoted by arrows. The shades of grey depict the pressure distribution. The compliant wall displacement ξ is plotted as the thick wavy curve near $r = 8$. The parameters are the same as for Fig. 8 with $\eta = 0.75$.

When the spring constant K is sufficiently small the critical Reynolds number for the non-axisymmetric modes with $n \neq 0$ become less than that for $n = 0$. In such cases the most unstable mode takes the form of an axially independent, transverse vortex flow travelling in the azimuthal direction (see Fig. 10). This form of disturbance is consistent with the Taylor-Proudman theorem of rotating fluid motion.

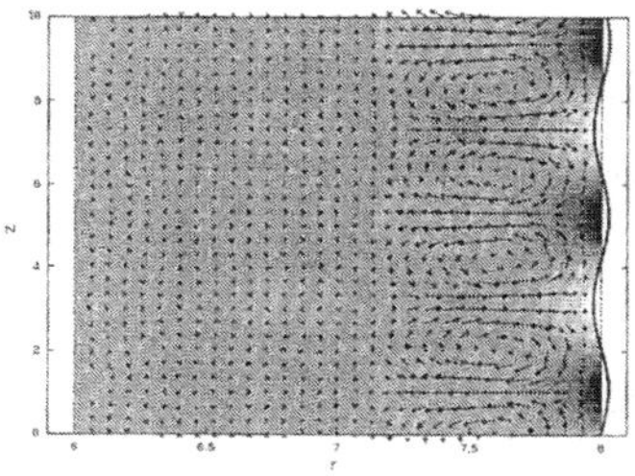

Figure 10. Disturbance in the form of a transverse vortex travelling in the azimuthal direction The notation is the same as for Fig. 9. The wall parameters are similar to those in Fig. 8, except that $\eta = 0.9, n = 35, \gamma = 0$; $(x, y) = (r\cos\theta, r\sin\theta)$.

References

ALI, R. (2001) Private communication, University of Warwick.

BALAKUMAR, P. and MALIK, M.R. (1990) Traveling disturbances in rotating-disk flow, *Theoretical and Computational Fluid Dynamics* **2**, 125-137.

CARPENTER, P.W. (1990) Status of transition delay using compliant walls, *Progress in Astronautics and Aeronautics* **123**, 79-113.

CARPENTER, P.W. and DAVIES, C. (2001) Global behaviour corresponding to the absolute instability of rotating-disk flow, *Bull. Amer. Phys. Soc.* **46**, 30.

CARPENTER, P.W., DAVIES, C. and LUCEY, A.D. (2000) Hydrodynamics and compliant walls: Does the dolphin have a secret?, *Current Science* **79**, 758-795.

CARPENTER, P.W. and GARRAD, A.D. (1985) The hydrodynamic stability of flows over Kramer-type compliant walls. Pt 1. Tollmien-Schlichting Instabilities, *J. Fluid Mech.* **155**, 465-510.

CARPENTER, P.W., LUCEY, A.D. and DAVIES, C. (2001) Progress on the use of compliant walls for laminar-flow control, *J. Aircraft* **38**, 504- 512.

CARPENTER, P.W. and MORRIS, P.J. (1990) The effect of anisotropic wall compliance on boundary-layer stability and transition, *J. Fluid Mech.* **218**, 171-223.

CHUNG, K.H. (1985) Composite compliant coatings for drag reduction utilising low modulus high damping silicone rubber, Ph.D. Thesis, MIT.

COLLEY, A.J., CARPENTER, P.W. and THOMAS, P.J. (2000) Experiments on boundary-layer transition over a rotating compliant disc – New experimental method, in abstracts of *4th EUROMECH Fluid Mech. Conf.*, Eindhoven, 19-23 November, p. 66.

COLLEY, A.J., THOMAS, P.J., CARPENTER, P.W. and COOPER, A.J. (1999) An experimental study of boundary-layer transition over a rotating, compliant disc, *Phys. Fluids* **11**, 3340-3352.

COOPER, A.J. and CARPENTER, P.W. (1997a) The stability of rotating-disc boundary-layer flow over a compliant wall. Part 1. Type I and II instabilities, *J. Fluid Mech.* **350**, 231-259.

COOPER, A.J. and CARPENTER, P.W. (1997b) The stability of rotating-disc boundary-layer flow over a compliant wall. Part 2. Absolute instability, *J. Fluid Mech.* **350**, 261-270.

CROS, A, ALI, A., SCHOUVELIER, L., THOMAS, P.J., LE GAL, P., CHAUVE, M.P., CARPENTER, P.W. and DAVIES, C. (2001) Transition of torsional Couette flow between a compliant rotating disk and a stationary rigid wall, presented at *IUTAM Symp. on Flow in Collapsible Tubes and Past Other Highly Compliant Boundaries*, 26-30 March, University of Warwick.

CROS, A., ALI, R., LE GAL, P., THOMAS, P.J., SCHOUVEILER, L., CARPENTER, P.W. and CHAUVE, M.P. (2002) Effects of wall compliance on the laminar-turbulent transition of torsional Couette flow, submitted to *J. Fluid Mech.*

DAVIES, C. (2002) Convective and absolute instabilities of flow over compliant walls, Chap. 4 of present volume.

DAVIES, C. and CARPENTER, P.W. (2000) A novel velocity-vorticity formulation of the Navier-Stokes equations. Part II: Global behaviour corresponding to the absolute instability of the rotating-disk boundary layer, Tech. Rep. 2000/1, Fluid Dynamics Research Centre, Univ. of Warwick.

DAVIES, C. and CARPENTER, P.W. (2001) A novel velocity-vorticity formulation of the Navier-Stokes equations with applications to boundary-layer disturbance evolution, *J. Comp. Phys.* **172**, 119-165.

DENIER, J.P. and HALL, P. (1991) The effect of wall compliance on the Goertler vortex instability, *Phys. Fluids A – Fluid Dynamics* **3**, 2000-2002.

DENIER, J.P., HALL, P. and SEDDOUGUI, S.O. (1991) On the receptivity problem for Gortler vortices – vortex motion induced by wall roughness, *Philos. Trans. Roy. Soc. London Ser A* **335**, 51-85.

DIXON, A.E., LUCEY, A.D. and CARPENTER, P.W. (1994) Optimization of viscoelastic compliant walls for transition delay. *AIAA J.* **32**, 256-267.

FALLER, A.J. (1963) An experimental study of the instability of the laminar Ekman boundary layer, *J. Fluid Mech.* **15**, 560-576.

FITZGERALD, E.R. and FITZGERALD, J.W. (1995a) Blubber and compliant coatings for drag reduction in water. I. Viscoelastic properties of blubber and compliant coating materials, *Material Science and Engineering C***2**, 209-214.

FITZGERALD, J.W., FITZGERALD, E.R., CAREY W.M. and WINKLE, W.A. von (1995b) Blubber and compliant coatings for drag reduction in water. II. Matched shear impedance for compliant layer drag reduction, *Material Science and Engineering C***2**, 215-220.

FITZGERALD, J.W., FITZGERALD, E.R. and J.E. MARTIN (1998a) Blubber and compliant coatings for drag reduction in water. III. Rotating disc drag measurements that mimic dolphin blubber, in G.R. Tomlinson and W.A. Bullough (eds.), *Smart Materials and Structures*, IOP Pub. Ltd., Bristol, U.K., pp. 543-550.

FITZGERALD, E.R. and FITZGERALD, J.W. (1998b) Blubber and compliant coatings for drag reduction in water. V. Driving point shear impedance measurements on compliant surfaces, in J.C.S. Meng (ed.), *Proc. Int. Symp. on Seawater Drag Reduction*, 22-23 July, 1998, Newport, RI, U.S.A., pp. 211-214.

FITZGERALD, J.W., J.E. MARTIN and E.F. MODERT (1998c) Blubber and compliant coatings for drag reduction in water. VI. Rotating disc apparatus for drag measurement on compliant layers, in J.C.S. Meng (ed.), *Proc. Int. Symp. on Seawater Drag Reduction*, 22-23 July, Newport, RI, U.S.A., pp. 215-218.

FITZGERALD, E.R. (2001) Influence of viscoelastic material properties on turbulent flow past thin compliant coatings, presented at *IUTAM Symp. on Flow in Collapsible Tubes and Past Other Highly Compliant Boundaries*, 26-30 March, University of Warwick.

GAD-EL-HAK, M. (1996) Compliant coatings: A decade of progress, *Appl. Mech. Rev.* **49**, s1-s11.

GAD-EL-HAK, M. (2002) Drag reduction using compliant walls, Chap. 9 in present volume.

GAD-EL-HAK, M., BLACKWELDER, R.F. and RILEY, J.J. (1984) On the interaction of compliant coatings with boundary layer flows, *J. Fluid Mech.* **140**, 257-280.

GASTER, M. (1987) Is the dolphin a red herring?, in H.W. Liepmann and R. Narasimha (eds.), *Proc. IUTAM Symp. on Turbulence Management and Relaminarisation, Bangalore, India*, Springer, pp. 285-304.

GREGORY, N., STUART, J.T. and WALKER, W.S. 1955 On the stability of three-dimensional boundary layers with application to the flow due to a rotating disk, *Philosophical Transactions of the Royal Society, London, Series A* **248**, 155-199.

HANSEN, R.J. and HUNSTON, D.L. (1974) An experimental study of turbulent flows over compliant surfaces, *J. Sound Vib.* **34**, No. 3, 297-308.

HANSEN, R.J. and HUNSTON, D.L. (1976) Further observations on flow-generated surface waves in compliant surfaces, *J. Sound Vib.* **46**, 593-595.

HANSEN, R.J., HUNSTON, D.L., NI, C.C., REISCHMAN, M.M. and HOYT, J.W. (1980) Hydrodynamic drag and surface deformations generated by liquid flows over flexible surfaces, *Prog. in Astronautics and Aeronautics* **72**, 439-452.

HANSEN, R.J. and HUNSTON, D.L. (1983) Fluid-property effects on flow-generated waves on a flexible surface, *J. Fluid Mech.* **133**, 161-177.

JARRE, S., LE GAL, P. and CHAUVE, M.P. (1991) Experimental analysis of the instability of the boundary layer over a rotating disk, *Europhysics Letters* **14**, 649-654.

JARRE, S., Le GAL , P. and CHAUVE, M.P. (1996a) Experimental study of rotating disk instability. I. Natural flow, *Phys. Fluids* **8**, 496-508.

JARRE, S., LE GAL , P. and CHAUVE, M.P. (1996b) Experimental study of rotating disk flow instability. II. Forced flow, *Phys. Fluids* **8**, 2985-2994.

KÁRMÁN, Th. von (1921) Über laminare und turbulente Reibung, *Zeitschrift für angewandte Mathematik und Mechanik* **1**, 233-252.

KOCH, W. (1986) Direct resonance in Orr-Sommerfeld problems, *Acta Mechanica* **58**, 11-29.

KOGA, S. and NAGATA, M. (2001) Stability of Couette flow with a compliant cylinder, presented at *IUTAM Symp. on Flow in Collapsible Tubes and Past Other Highly Compliant Boundaries*, 26-30 March, University of Warwick.

LE GAL, P. (1992) Complex demodulation applied to the transition to turbulence of the flow over a rotating disk, *Phys. Fluids A* **4**, 2523-2528.

LILLY, D.K. (1966) On the instability of Ekman boundary flow, *Journal of Atmospheric Sciences* **23**, 481-494.

LINGWOOD, R.J. (1995) Absolute instability of the boundary layer on a rotating disk, *J. Fluid Mech.* **299**, 17-33.

LINGWOOD, R.J. (1996) An experimental-study of absolute instability of the rotating-disk boundary-layer flow, *J. Fluid Mech.* **314**, 373-405.

LINGWOOD, R.J. (1997) Absolute instability of the Ekman layer and related rotating flows, *J. Fluid Mech.* **331**, 405-428.

MACK, L.M. (1985) The wave pattern produced by point source on a rotating disk, *AIAA Paper 85-0490.*

MALIK, M.R., WILKINSON, S.P. and ORSZAG, S.A. (1981) Instability and transition in rotating disk flow, *AIAA J.* **19**, 1131-1138.

OWEN, J.M. and ROGERS, R.H. (1989) *Flow and heat transfer in rotating-disc systems, Vol. 1 - Rotor-stator systems*, Research Studies Press Ltd., Taunton, Somerset, UK.

RILEY, J.J., GAD-EL-HAK, M. and METCALFE, R.W. (1988) Compliant coatings, *Annual Review Fluid Mech.* **20**, 393-420.

SCHOUVEILER, L., LE GAL, P. and CHAUVE, M.P. (1998) Stability of a traveling roll system in a rotating disk flow, *Phys. Fluids* **10**, 2695-2697.

SCHOUVEILER, L., LE GAL, P. and CHAUVE, M.P. (1999) Spiral and circular waves in the flow between a rotating and a stationary disk, *Exp. Fluids* **26**, 179-187.

SCHOUVEILER, L., LE GAL, P. and CHAUVE, M.P. (2001) Instabilities of the flow between a rotating and a stationary disk, *J. Fluid Mech.* **443**, 329-350.

SMITH, N.H. (1946) Exploratory investigation of laminar-boundary-layer oscillations on a rotating disk, *N.A.C.A. TN 1227.*

WILKINSON, S.P. and MALIK, M.R. (1985) Stability experiments in the flow over a rotating disk, *AIAA J.* **23**, 588-595.

YURCHENKO, N.F. and BABENKO, V.V. (1987) Instability criteria for three-dimensional disturbances over concave compliant surfaces, (in Russian), *Zh. Inzhenerno-Fizik.* **52**, 781-787.

YURCHENKO, N.F., BABENKO, V.V. and KOZLOV, L.F. (1989) Development of three-dimensional disturbances over compliant surface, in *Proc. Int. Conf. on Engineering Aero-Hydroelasticity*, Prague, pp. 63-70.

SECTION C

DRAG MODIFICATION AND TURBULENCE MODIFICATION

Chapter 9

DRAG REDUCTION USING COMPLIANT WALLS

MOHAMED GAD-EL-HAK
University of Notre Dame
Notre Dame, IN 46556
U.S.A.

Abstract

Compliant coating research is one of those areas which experienced its fair share of triumphs and debacles. For over forty years, the subject has fascinated, frustrated and occasionally gratified scientists and engineers searching for methods to delay laminar-to-turbulence transition, to reduce skin-friction drag in turbulent wall-bounded flows, to quell vibrations, and to suppress flow-induced noise. In its simplest form, the technique is passive, relatively easy to apply to an existing vehicle or device, and perhaps not too expensive. Through the years, however, claims for substantial drag and noise reductions were made, only to be later refuted when the results were more critically examined. There are several important issues with regard to the reliability of available analytical, numerical and experimental results. In this chapter, some of these issues, particularly as they relate to the search for drag-reducing compliant coatings, will be addressed with the objective of elucidating the potential pitfalls to newcomers to the field. Problem formulation with the proper boundary conditions, impossibility of obtaining first-principles analytical solutions when the wall-bounded flow is turbulent, and limitations of existing numerical simulations will be elaborated. The effects of background turbulence in a wind or water tunnel, accurate drag measurements, compliant wall motion, and the geometry and properties of the coatings used will be among the outstanding experimental issues discussed. Attempts will be made to explain some of the seemingly contradictory results available in the open literature.

P.W. Carpenter and T.J. Pedley (eds.),
Flow past Highly Compliant Boundaries and in Collapsible Tubes, 191-229.

Prologue

He was an old man who fished alone in a skiff in the Gulf Stream and he had gone eighty-four days now without taking a fish. In the first forty days a boy had been with him. But after forty days without a fish the boy's parents had told him that the old man was now definitely and finally salao, *which is the worst form of unlucky, and the boy had gone at their orders in another boat which caught three good fish the first week. It made the boy sad to see the old man come in each day with his skiff empty and he always went down to help him carry either the coiled lines or the gaff and harpoon and the sail that was furled around the mast. The sail was patched with flour sacks and, furled, it looked like the flag of permanent defeat.*

......

"Dolphin," the old man said aloud. "Big dolphin."

He shipped his oars and brought a small line from under the bow. It had a wire leader and a medium-sized hook and he baited it with one of the sardines. He let it go over the side and then made it fast to a ring bolt in the stern. Then he baited another line and left it coiled in the shade of the bow. He went back to rowing and to watching the long-winged black bird who was working, now, low over the water.

As he watched the bird dipped again slanting his wings for the dive and then swinging them wildly and ineffectually as he followed the flying fish. The old man could see the slight bulge in the water that the big dolphin raised as they followed the escaping fish. The dolphin were cutting through the water below the flight of the fish and would be in the water, driving at speed, when the fish dropped. It is a big school of dolphin, he thought. They are widespread and the flying fish have little chance. The bird has no chance. The flying fish are too big for him and they go too fast.

(Ernest Miller Hemingway, 1899–1961, in *The Old Man and the Sea*, 1952)

1. Introduction

A compliant wall, as opposed to a rigid one, offers the potential for favorable interference with a wall-bounded flow. Laminar-to-turbulence transition may be delayed or advanced, boundary-layer separation may be prevented or triggered, flow-induced noise may be modulated, and skin-friction drag in both laminar and turbulent flows may be altered. The challenge is of course to find a coating with the optimum physical properties to achieve a desired goal. Despite past frustrations and *salao*, Hemingway's words quoted in the Prologue deserve some reflection. Would hydrodynamists ever be lucky or tenacious enough to catch their prize fish, the right kind of compliant coating to reduce skin friction, the jackpot? Certainly existing experimental and analytical studies indicate that we can, at least in the laboratory, but field trials are yet to be conducted and many technical challenges remain. The present article commences with a résumé of past and recent attempts to search for drag-reducing compliant coatings and concludes with the future prospects of the research field.

For better or worse, hydrodynamically speaking, the epidermis of most nekton is pliable at their typical swimming speeds. For close to half a century the science and technology of compliant coatings has fascinated, frustrated and occasionally gratified *Homo sapiens* searching for methods to delay laminar-to-turbulence transition, to reduce skin-friction drag in turbulent wall-bounded flows, to quell vibrations, and to suppress flow-induced noise. Compliant coatings offer a rather simple method to delay laminar-to-turbulence transition, and therefore to benefit from the order-of-magnitude lesser skin friction in a laminar, in contrast to a turbulent, wall-bounded flow, as well as to interact favorably with a turbulent boundary layer or pipe flow. In its simplest form, the technique is passive, relatively easy to apply to an existing vehicle or device, and perhaps not too expensive. Unlike other drag-reducing techniques such as suction, injection, polymer or particle additives, passive compliant coatings do not require slots, ducts or internal equipment of any kind. Likewise, unlike the exceedingly complex reactive control strategies now in vogue (Gad-el-Hak, 2000), passive compliant coatings do not require sensors, actuators, feedforward/feedback loops or control algorithms. Aside from reducing drag, other reasons for the perennial interest in studying compliant coatings are their many other useful applications, for example as sound-absorbent materials in noisy flow-carrying ducts in aero-engines, and as flexible surfaces to coat naval vessels for the purposes of shielding their sonar arrays from the sound generated by the boundary-layer pressure fluctuations and of reducing the efficiency of their vibrating metal hulls as sound radiators.

The original interest in the field was spurred by the experiments of Kramer (1957) who demonstrated a compliant coating design based on the dolphin's epidermis and claimed substantial transition delay and drag reduction in hydrodynamic flows. Those experiments were conducted in the seemingly less-than-ideal environment of Long Beach Harbor, California. Subsequent laboratory attempts to substantiate Kramer's results failed, and the initial interest in the idea fizzled. A similar bout of excitement and frustration that dealt mostly with the reduction of skin-friction drag in turbulent flows for aeronautical applications followed. Those results were summarized in the comprehensive review by Bushnell et al. (1977). During the early 1980s, interest in the subject was rejuvenated mostly due to modest investment in resources by the Office of Naval Research in the United States and the Procurement Executive of the Ministry of Defence in Great Britain.[1] Significant advances were made during this period in numerical and analytical methods to solve the coupled fluid–structure problem. New experimental tools were developed to measure the minute yet important surface deformation caused by the unsteady fluid forces. Coherent structures in turbulent wall-bounded flows were routinely identified, and their modulation by the surface compliance could readily be quantified.

Careful analyses by Carpenter and Garrad (1985) and Willis (1986) as well as the well-controlled experiments reported by Daniel et al. (1987) and Gaster (1988)[2] have, for the first time, provided direct confirmation of the transition-delaying potential of compliant coatings, convincingly made a case for the validity

[1]Through all the ups and downs in the West, compliant coating research continued at more or less steady pace in the former Soviet Union, but open-literature publications resulting from this work, either in English or in Russian, are rather scarce. For some valuable references, see, for example, the books by Aleyev (1977), and Choi (1991), and the articles by Babenko and Kozlov (1972) and Carpenter and Garrad (1985).

[2]Originally reported in the thesis by Willis (1986) and later contrasted to theory in the paper by Lucey and Carpenter (1995).

of Kramer's original claims, and offered a plausible explanation for the failure of the subsequent laboratory experiments. There is little doubt now that compliant coatings can be *rationally* designed to delay transition and to suppress noise on marine vehicles and other practical hydrodynamic devices. Transition Reynolds numbers that exceed by an order of magnitude those on rigid-surface boundary layers can be readily achieved. Although the number of active researchers in the field continues to dwindle, new promising results are being produced. Recent theoretical work by Davies and Carpenter (1997a) and Carpenter (1998) indicates that transition to turbulence can be delayed indefinitely, at least in principle, provided that optimized multiple-panel compliant walls are used and that the freestream is a low-disturbance environment. There is also recent evidence of favorable interactions of compliant coatings even for air flows (Lee et al., 1995) and even for turbulent boundary layers (Lee et al., 1993a; Choi et al., 1997).

The present chapter emphasizes the significant compliant coating research that took place during the last 10–15 years and suggests avenues for future research. The reader is referred to prior reviews for more classical work on the subject, for example those by Bushnell et al. (1977), Gad-el-Hak (1986a; 1987; 1996a), Riley et al. (1988), Carpenter (1990), and Metcalfe (1994). The book by Gad-el-Hak (2000) places compliant coatings within the broader area of flow control, and the very recent article by Carpenter et al. (2001) focuses on the use of compliant walls for laminar-flow control. Following these introductory remarks, a somewhat sketchy history of the subject, particularly prior to 1985, is recalled. This will help place more recent developments in proper perspective.

2. Compliant Coatings Prior to 1985

Before embarking on describing the recent accomplishments in the field of compliant coatings, we first elaborate on its history prior to 1985. This seemingly arbitrary date is chosen because it demarks the time after which tools for rationally designing a compliant coating to delay transition became more readily available. The victories and defeats of the subject matter will become clear through the discussion that follows. The idea of using compliant coatings for drag reduction motivated much of the earlier work in this area and was first introduced by Kramer (1957) based on his earlier observation, while crossing the Atlantic ocean in 1946, of dolphins swimming in water. He advanced the concept that the stability and transition characteristics of a boundary layer may be influenced by coupling it hydroelastically to a compliant coating. In his pioneering paper and several subsequent publications, Kramer (1960a; 1960b; 1960c; 1961; 1962; 1965; 1969) reported substantial drag reduction for towed underwater bodies covered with compliant coating modeled after the dolphin skin. He hypothesized that by tuning the elastic wall damping to a frequency near that of the most unstable Tollmien–Schlichting wave, it would be possible to dissipate partially the instability waves, thus delaying the transition to turbulence. Kramer's tests were performed by towing a test model behind a motor boat in Long Beach Harbor. Unfortunately, many attempts by other investigators to repeat Kramer's experiments under more controlled conditions failed to yield similar conclusions (e.g., Puryear's, 1962, experiment in a towing tank). This so-called Kramer controversy will be revisited in Section 5.

Theoretical work by Benjamin (1960a; 1960b), Betchov (1960), Landahl (1962), and Kaplan (1964) indicated that drag reduction by delaying transition is possible.

However, the theoretically predicted successful coatings had specific characteristics that would be extremely difficult to match in practice. It is important to stress that almost all this early work addressed the delay of transition and ignored the potential for reducing turbulence skin friction with compliant coatings.

During the mid-1960s, Benjamin (1966) explored the possibility that a compliant coating may affect the skin-friction drag in a fully-developed turbulent boundary layer without necessarily delaying transition. Dinkelacker (1966) conducted careful tests of a compliant surface in a water pipe flow. He systematically attempted to determine the repeatability of rigid-tube data, the influence of small steps in the tube wall, and the possible occurrence of organ pipe acoustic modes. Dinkelacker's results seemed to indicate a modest reduction in drag by using a compliant wall.

Blick and his co-workers at the University of Oklahoma experimentally demonstrated significant reductions in turbulence skin friction for compliant surfaces in air (Fisher and Blick, 1966; Looney and Blick, 1966; Smith and Blick, 1966; Blick and Walters, 1968; Chu and Blick, 1969). Subsequent tests by Lissaman and Harris (1969), who attempted to substantiate Blick's conclusions, yielded only extremely modest gains. In another study, McMichael et al. (1980) demonstrated that the apparent reduction in turbulence skin friction in the University of Oklahoma's experiments could be a consequence of experimental deficiencies coupled with the improper interpretation of data. McMichael et al. concluded that drag reduction via compliant coating in gaseous flows would not be as successful as in liquids.

During the 1970s various compliant materials were tested in water at the Naval Ocean Systems Center, the Naval Research Laboratory, the Naval Undersea Systems Center, and the Advanced Technology Center of the LTV Corporation, all in the United States. In no case was a statistically significant reduction in drag measured. Fischer and Ash (1974) presented a general review of concepts for reducing skin friction, including the use of compliant coatings. Bushnell et al. (1977), in summarizing the work conducted at the NASA Langley Research Center and the general status of compliant surface drag reduction, stated that, while it was possible to increase the transition Reynolds number by perhaps a factor of two, there was no definitive reduction of drag for *turbulent* flows in air. They also stated that drag reduction in turbulent flows in water is potentially feasible and can be accomplished using surfaces that can be practically built. It is of particular interest to note that much of the research on compliant coatings has been based on materials that attempt to replicate dolphin skin. Yet, in the Russian book *Nekton* (Aleyev, 1977) it is indicated that the "wrinkling" (Figure 1) of the dolphin skin has no hydrodynamic-drag advantage. Other characteristics of the dolphin's skin may, however, be beneficial. This subject will be revisited in Section 6.4.

Bushnell et al. (1977) put forward the possibility of a feedback mechanism in turbulent wall-bounded flows through which the quasi-periodic, coherent structures termed bursts regenerate. Older bursts grow, migrate away from the wall, and interact to produce a pressure field which contains pulses of sufficient duration and amplitude to induce new bursts in the near-wall region. This model is supported by the measurements of Burton (1974), who reported a strong correlation between the occurrence of a burst and the imposition on the wall flow of a large, moving adverse pressure gradient followed by a favorable pressure-gradient. Bushnell et al. hypothesized that a successful compliant coating would modulate the preburst flow in the turbulent boundary layer by providing a pressure field that would tend to block the feedback mechanism and, thus, inhibit burst formation.

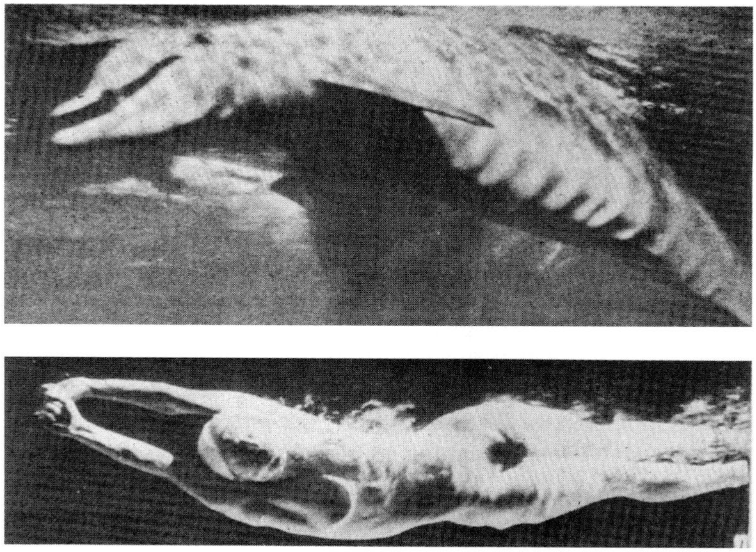

Figure 1. Wavelike folds in compliant skin appearing during rapid swimming of a *delphinidae* (top photograph) and a *Homo sapiens* (bottom photograph). From Aleyev (1977).

This would result in a reduction in the number of bursts occurring per unit time and also in the skin-friction drag. Orszag (1979) assumed this conceptual model and performed numerical calculations of wall boundary-layer instability to explore the effects of compliant surfaces. His results, although preliminary, indicated that turbulence drag reduction may be possible for certain classes of materials. He concluded that compliant walls which support only short wavelengths may have an appreciable effect in inhibiting further bursts in a turbulent boundary layer.

During the U.S. Navy-sponsored research program conducted over the period 1980–1985, the subject of boundary-layer interaction with compliant coatings has been re-examined to answer the question: can compliant coatings delay transition and/or significantly reduce turbulence skin friction on bodies at high Reynolds numbers? Several significant developments have been achieved by the many investigators participating in this research program. Although irrefutable experimental evidence of compliant-coating drag reduction was still lacking by 1985, our understanding of boundary-layer flow over a compliant surface has increased dramatically over this period. That understanding proved crucial to the subsequent successes in the field, a subject which will be emphasized throughout this chapter.

3. Free-Surface Waves

Before addressing the complex issue of stability of the coupled fluid–structure system, it is instructive to study the solid as a wave-bearing medium when no fluid (another wave-bearing medium) is present. Some understanding of how a compliant surface will respond to the flow above it can be obtained by examining the free-surface waves of the coating. As pointed out by Rayleigh (1887), the surface waves can be modeled as a linear combination of waves having displacements perpendicular and parallel to the propagation direction. These are called transverse and longitudinal displacement waves, respectively. Assuming that the coating is a single-layer, elastic solid of thickness d attached to a rigid half-space at its lower boundary and bounded by vacuum on its upper surface, both wave systems satisfy the wave equation (see, for example, Landau and Liftshitz, 1987)

$$\frac{\partial^2}{\partial t^2}\eta - c^2\left(\frac{\partial^2}{\partial x^2} + \frac{\partial^2}{\partial y^2}\right)\eta = 0 \tag{1}$$

where η is a component of the displacement vector, and x and y are coordinates parallel and normal to the undisturbed surface, respectively. The propagation velocities are $c = c_t = \sqrt{G/\rho_s}$ for the transverse waves, and $c = c_\ell = \sqrt{(\Theta + 2G)/\rho_s}$ for the longitudinal waves, where G and Θ are elastic constants, and ρ_s is the density of the solid.

The free-surface wave dispersion relationship is obtained by assuming that the wave solutions have exponential dependence on the distance measured normal to the surface y, and have harmonic dependence on the distance measured along the surface x, and on time t

$$\xi = (A \sinh \alpha y + B \cosh \alpha y)\, e^{i(k_\ell x - \omega t)} \tag{2}$$

$$\eta = (C \sinh \alpha y + D \cosh \alpha y)\, e^{i(k_\ell x - \omega t)} \tag{3}$$

where ξ and η are respectively the displacement in the x- and y-direction, and k_ℓ is the longitudinal wavenumber. For real α the waves decay exponentially with depth, whereas for imaginary α they oscillate. Substituting these solutions into the wave equation (1), and applying the boundary conditions: no normal or tangential stress at the surface $y = 0$ and no displacement at the bottom $y = -d$, respectively,

$$(c_\ell^2 - 2\,c_t^2)\,\frac{\partial \xi}{\partial x} + c_\ell^2\,\frac{\partial \eta}{\partial y} = 0 \qquad \text{at} \quad y = 0 \tag{4}$$

$$\frac{\partial \xi}{\partial y} + \frac{\partial \eta}{\partial x} = 0 \qquad \text{at} \quad y = 0 \tag{5}$$

$$\xi = \eta = 0 \qquad \text{at} \quad y = -d \tag{6}$$

Gad-el-Hak et al. (1984) obtained the following dispersion relationship

$$\begin{aligned} M(\zeta) \equiv\ & 4\,\frac{\alpha_t\,\alpha_\ell}{k_\ell^2}\,(2 - \zeta^2) \\ & - \frac{\alpha_t\,\alpha_\ell}{k_\ell^2}\left[4 + (2 - \zeta^2)^2\right] \cosh \alpha_t d \cosh \alpha_\ell d \\ & + \left[4\,\frac{\alpha_t^2\,\alpha_\ell^2}{k_\ell^4} + (2 - \zeta^2)^2\right] \sinh \alpha_t d \sinh \alpha_\ell d = 0 \end{aligned} \tag{7}$$

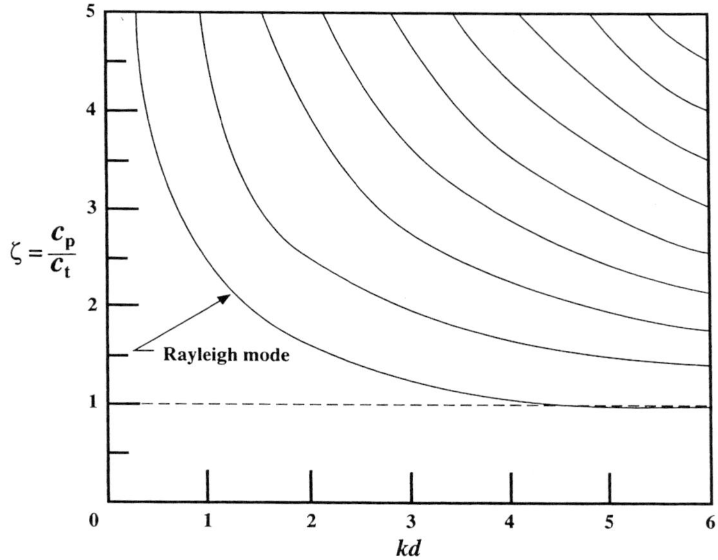

Figure 2. Solutions of the dispersion equation for free-surface waves on a single-layer, homogenous coating. From Gad-el-Hak et al. (1984).

where $\alpha_t = k_\ell \sqrt{1 - \zeta^2}$, $\alpha_\ell = k_\ell \sqrt{1 - \zeta^2 R^2}$, R is the ratio of transverse to longitudinal wave speed, and $\zeta = c_p/c_t$ is the ratio of the surface wave speed to transverse wave speed.

Plots of the dispersion curves resulting from Equation (7) are given in Figure 2 for the case $R = 0$, which is close to typical experimental conditions. Only one solution has a surface wave speed below c_t ($\zeta < 1$). This solution approaches its asymptote $\zeta = 0.956$ for large $k_\ell\, d$, which is the value of Rayleigh's infinite half-layer solution. Also note that as ζ increases for a given $k_\ell\, d$, the possibility exists for a richer range of interactions between the coating and a flow. Typical values for $k_\ell\, d$ in experiments, with the wavelength $2\,\pi/k_\ell$ taken as the boundary-layer thickness, are in the range of 1–5. Figure 2 indicates that interaction between a fluid flow and a compliant surface may not be expected for flow speeds much below c_t, and that the best opportunity for interactions will be for flow speeds well above c_t. Unfortunately, hydroelastic instability waves appear for freestream speeds somewhat above c_t, thus causing large surface deformations and limiting the opportunities for favorable interactions. The waveform can be obtained by determining the constants A, B, C and D in Equations (2) and (3) using the four boundary conditions (4)–(6).

Duncan and Hsu (1984) calculated the dispersion relations for a two-layer coating by finding the zeroes of the determinant of the boundary condition coefficients. Their results for a thin, stiff coating placed on a much thicker, soft coating are shown in Figure 3. The upper coating has about 100 times the shear modulus of the lower surface. In this figure, the wavenumber is normalized by the total coating thickness, while the phase speed of the free-surface waves is normalized by the transverse wave speed of the lower layer. This normalization allows an easy

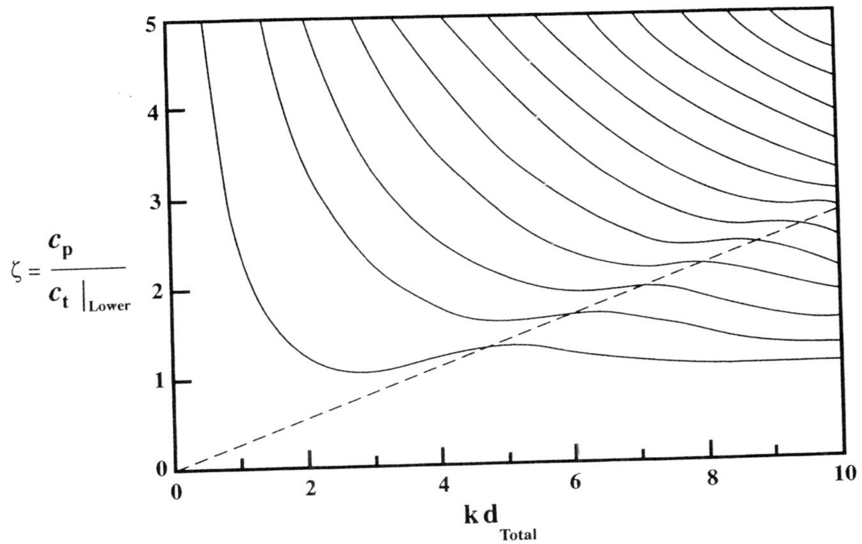

Figure 3. Solutions of the dispersion equation for free-surface waves on a two-layer coating. From Duncan and Hsu (1984).

comparison with the single-layer case when the upper portion of the coating is replaced by a layer of stiffer material. The gross shapes of the curves in Figure 3 are similar to those in Figure 2, although a wrinkle is clearly seen in the two-layer case. The dashed line in Figure 3 connects the maxima of the wrinkle in each modal curve and is a solution of the equation for long bending waves in a free plate. Duncan and Hsu (1984) have hypothesized that the upper coating acts like a plate that is influenced by the lower layer. They have attributed the cutoff of the response of the bending waves in the upper layer at small wavenumbers to the lack of slow-moving waves in the lower layer.

4. System Instabilities

From a fundamental viewpoint, a rich variety of fluid–structure interactions exists when a fluid flows over a surface that is able to interact with the flow. Not surprisingly, instability modes proliferate when two wave-bearing media are coupled. Some waves are flow-based, some are wall-based, and some are a result of the coalescence of both kind of waves. What is most appealing about compliant coatings is their potential to inhibit, or to foster, the dynamic instabilities that characterize both transitional and turbulent boundary-layer flows, and in turn to modify the mass, heat and momentum fluxes and change the drag and the acoustic properties. While it is relatively easy to suppress a particular instability mode, the challenge is of course to prevent other modes from growing if the aim is, say, to delay laminar-to-turbulence transition. From a practical point of view, it is obvious that an in-depth understanding of the coupled system instabilities is a prerequisite to rationally designing a coating that meets a given objective.

There are at least three classification schemes for the fluid–structure waves, each with its own merits. The original scheme is attributable to Landahl (1962) and Benjamin (1963). It divides the waves into three classes according to their response to irreversible energy transfer to and from the compliant wall. Both class **A** and class **B** disturbances are essentially oscillations involving conservative energy exchanges between the fluid and solid, but their stability is determined by the net effect of irreversible processes such as dissipation in the coating or energy transfer to the solid by non-conservative hydrodynamic forces. Class **A** oscillations are Tollmien–Schlichting-type waves in the boundary layer modified by the wall compliance, in other words by the motion of the solid in response to the pressure and shear-stress fluctuations in the flow. The disturbance eignenfunction for class **A** waves has its maximum amplitude within the fluid region. Such waves are stabilized by the irreversible energy transfer from the fluid to the coating, but destabilized by dissipation in the wall.

Class **B** waves are to be found in both the fluid and the wall. However, the disturbance eignenfunction has its maximum amplitude at the fluid–solid interface and thus those waves are principally wall-based modes of instability. Such instability would not exist had the wall been rigid. The instability is due to the downstream-running free wave in the solid being modified by the fluid loading. The destabilization of class **B** waves is effected by the phase-difference between the pressure perturbation and the wall deformation, which allows a flow of energy from the fluid into the wall. The behavior of Class **B** waves is the reverse of that for class **A** waves, stabilized by wall damping but destabilized by the non-conservative hydrodynamic forces. Essentially class **B** waves are amplified when the flow supplies sufficient energy to counterbalance the coating's internal dissipation.

Finally, class **C** waves are akin to the inviscid Kelvin–Helmholtz instability and occur when conservative hydrodynamic forces cause a unidirectional transfer of energy to the solid. The pressure distribution in an inviscid flow over a wavy wall is in exact antiphase with the elevation. In that case, class **C** waves can grow on the solid surface only if the pressure amplitude is so large as to outweigh the coating stiffness. Class **C** waves are the result of a modal-coalescence instability where the flow speed is sufficiently high that the originally upstream-running wall free waves are turned to travel downstream and merge with the modified downstream-running wall waves. Irreversible processes in both the fluid and solid have negligible effect on class **C** instabilities.

If one considers the total disturbance energy of the coupled fluid–solid system, a decrease in that energy leads to an increase in the amplitude of class **A** instabilities, class **B** is associated with an energy increase, and virtually no change in total energy accompanies class **C** waves. In other words, any non-conservative flow of activation energy from/to the system must be accompanied by disturbance growth of class **A**/**B** waves, while the irreversible energy transfer for class **C** instability is nearly zero.

The second classification scheme is due to Carpenter and Garrad (1985; 1986). It simply divides the waves into fluid-based and solid-based. Tollmien–Schlichting instability (TSI) is an example of fluid-based waves. The solid-based, flow-induced surface instabilities (FISI) are closely analogous to the instabilities studied in hydro- and aeroelasticity, and include both the traveling-wave flutter that moves at speeds close to the solid free-wavespeed (class **B**) and the essentially static—and more dangerous—divergence waves (class **C**). The main drawback of this classification scheme is that under certain circumstances the fluid-based T–S waves and the solid-based flutter can coalesce to form a powerful new instability termed the transitional mode by Sen and Arora (1988). According to the energy criterion advanced by Landahl (1962), this latest instability is a second kind of class **C** wave. In a physical experiment, however, it is rather difficult to distinguish between static divergence and the transitional mode.

The third scheme to classify the instability waves considers whether they are convective or absolute (Huerre and Monkewitz, 1990). An instability mode is considered to be absolute if there is a pinch point in the Fourier contour which prevents the temporal amplification rate from being reduced down to zero. In this case, the unstable mode propagates upstream as well as downstream and often has a very small (or even zero) group velocity in comparison with the velocity of the mean flow. On the other hand, the unstable development of a disturbance is said to be convective when none of its constituent modes possesses zero group velocity. Both classes **A** and **B** are convective, while class **C** divergence and the transitional modes are absolute. As Carpenter (1990) points out, the occurrence of absolute instabilities would lead to profound changes in the laminar-to-turbulence transition process. 'It is therefore pointless to consider reducing their growth rate or postponing their appearance to higher Reynolds number; nothing short of complete suppression would work.' Figure 4 combines and summarizes all three classification schemes.

5. The Kramer's Controversy

It may be worth recalling in more details the pioneering work of Max O. Kramer and the controversy surrounding it. The entire field of compliant coatings became the *Rodney Dangerfield* of fluid mechanics research, getting no respect from a skeptical community, largely because of the loss of credibility of Kramer's original experiments. However as will be seen below the most recent evidence resurrects the good name of this ingenious German-American engineer, and with it renewed confidence in this waning and waxing field.

As already mentioned, Kramer (1957; 1960a; 1960b; 1960c; 1961; 1962; 1965; 1969) conducted his original experiments by towing a model behind an outboard motor boat in Long Beach Harbor, California. His early tests showed a drag reduction of more than 50% when a dolphin-like skin was used. A typical successful coating used by Kramer consisted of a flexible inner skin, an outer diaphragm, and stubs, all made of soft natural rubber. The cavity between the outer diaphragm and the inner skin was usually filled with a highly viscous damping fluid, such as silicone oil, which in Kramer's view damped out the Tollmien–Schlichting waves.

Subsequent experiments to confirm Kramer's findings were conducted in a towing tank, a lake, or a water tunnel (Puryear, 1962; Nisewanger, 1964; Ritter and Messum, 1964; Ritter and Porteous, 1964). No significant drag reduction was observed in any of these investigations. Since then, many researchers have assumed that Kramer's results were in error and that his observed drag reduction could have come about as a result of favorable changes to the form drag or the accidental excretion of the silicone oil used as the damping fluid during the tests. Surface discontinuities could have favorably altered the pressure drag, and the released oil could have acted as a drag-reducing polymer when released into the boundary layer and the ambient fluid.

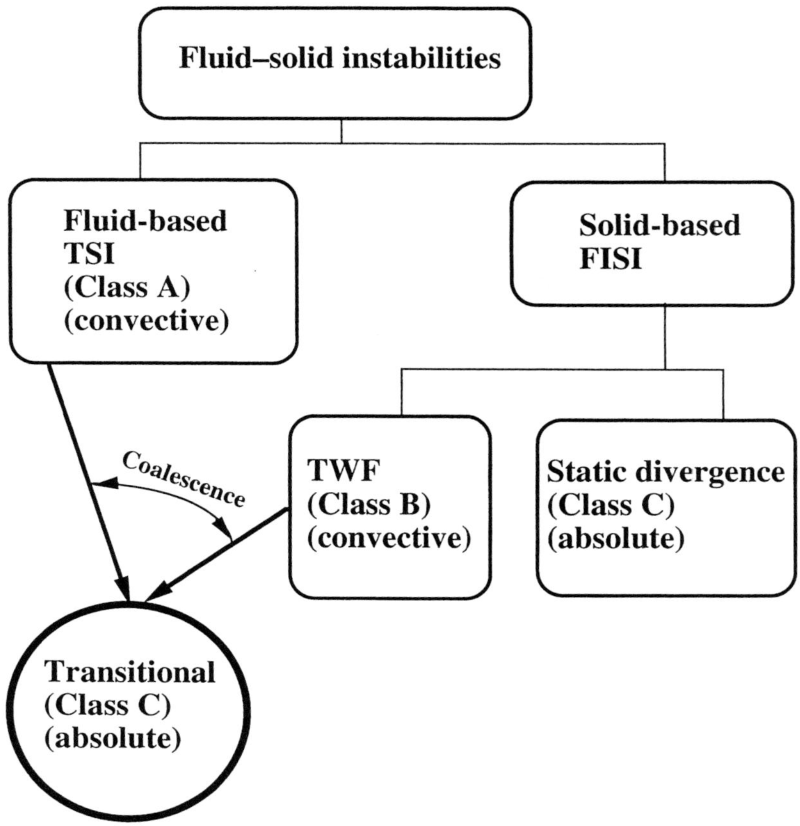

Figure 4. Summary of all three classification schemes.

Carpenter and Garrad (1985) stated that: "It is probably no exaggeration to suggest that the credibility of Kramer's coatings is now rather low." Acceptance of his results was not granted by the scientific community because the rigorous standards of scientific investigation were not met and the gradual improvements by Kramer to meet these standards were not adequate (Johnson, 1980). It did not help his cause any that Kramer's explanations of his own empirical results, though intuitively appealing, were proven physically incorrect. For example, we now know that damping in the solid *destabilizes* TSI.

Almost 30 years after Kramer's original investigation, Carpenter and Garrad (1985) presented a very careful analysis of his experiments (e.g., Kramer, 1957) and the subsequent tests (Puryear, 1962; Nisewanger, 1964; Ritter and Messum, 1964; Ritter and Porteous, 1964) that attempted to provide independent evidence of the drag-reducing capabilities of Kramer's coatings. Based on their own rigorous analysis of the hydrodynamic stability of flows over Kramer-type compliant surfaces, Carpenter and Garrad argue that Kramer's coatings were only marginally capable of delaying transition. Any unfavorable factor, such as an adverse pressure gradient, a step where the compliant surface is joined to a rigid surface or an unusually high freestream turbulence level, could badly affect the performance of the coating. Also, a particular coating was designed for a restricted range of Reynolds number and was therefore unlikely to delay transition outside that range.

Carpenter and Garrad (1985) contend that one or more of the above adverse factors may have existed in the experiments conducted by Puryear (1962), Nisewanger (1964), Ritter and Messum (1964), and Ritter and Porteous (1964). Puryear's (1962) experiments were conducted using a prolate spheroid in a towing tank. He did not use Kramer's coating with the best performance, and serious problems were encountered in making a smooth joint between the rigid and compliant surfaces. Nisewanger's (1964) tests were conducted by releasing a lighter-than-water body of revolution from the bottom of a lake. His Kramer-like coating contained a fluid with a viscosity below that of the optimum fluid as determined from Kramer's results. Ritter and Messum (1964), and Ritter and Porteous (1964) conducted their experiments in a water tunnel using either a flat plate or a cylindrical model with an elliptic nose. The conventional flume had a relatively high freestream turbulence level, which may render the facility unsuited for transition experiments.

In addition to these adverse effects, some evidence existed in the tests conducted to confirm Kramer's results for the onset of a hydroelastic instability in the coating. Such large-amplitude waves would certainly lead to drag increase, and their presence may indicate that the boundary layer was already turbulent. Based on these experiments, Carpenter and Garrad (1985) concluded that the results presented in these tests should not be taken as conclusive evidence that the Kramer coatings are not capable of delaying transition and that "the case against Kramer's coating may not be so strong as popularly supposed."

Carpenter's (1988) optimization procedure described in Section 6.2, results in a compliant coating that is capable of delaying transition by a factor of 4–6 in Reynolds number. It is therefore quite conceivable to design a Kramer-type coating that may lead to a drag reduction of the order reported by the original inventor himself. The above analysis of Kramer's tests illustrates the importance of carefully selecting the flow facility to conduct compliant coating experiments. The background turbulence in the facility should particularly be monitored if transition delay is sought. This is precisely what was done in the seminal experiments conducted by Gaster (1988) to confirm the theoretical prediction of Carpenter and Garrad (1985), both described in more details in the following section.

6. Transitional Flows

6.1. LINEAR STABILITY THEORY

Both the hydrodynamic and the hydroelastic stability theories have reached an impressive level of maturity during the last two decades. The linear theories can be handled, for the most part, analytically, while the nonlinear stability theories are more computer intensive. Perhaps no one has contributed more to the recent application of the stability theory to compliant coatings than Peter W. Carpenter, originally with the University of Exeter and presently with the University of Warwick. His list of relevant publications includes 80 papers and growing; obviously only a selected few will be cited in the present short chapter.

Within the framework of the linear stability theory, two-dimensional small disturbances are assumed to be superimposed upon a steady, unidirectional mean flow. The nonlinear, partial Navier–Stokes equations are then reduced to the well-known Orr–Sommerfeld equation which is a fourth-order, linear, ordinary differential equation. The order of this equation increases when additional complexities are included in the problem. For rotating-disk flows, for example, Coriolis and streamline-curvature terms are incorporated leading to a sixth-order stability equation. The major difficulty in integrating the Orr–Sommerfeld equation is that it is highly stiff and unstable, which makes it virtually impossible to apply conventional numerical schemes. Explicit codes with step size that is commensurate with the global behavior of the solution lead to numerical instabilities, and alternative routines have been developed to handle this stiff eigenvalue problem.

An added difficulty when the walls are compliant is the interfacial conditions which require continuity of velocity and stress. Those boundary conditions can also be linearized, but special care should still be exercised in handling them. Appropriate equations must be used for the compliant walls to be able to fully couple the fluid and solid dynamics. Many types of compliant surfaces exist, so that there are numerous models for the solid. Those models can be either surface-based or volume-based (Figure 5). The former model reduces the spatial dimensions by one, and is therefore less computationally demanding. An example is the thin plate-spring model used by Garrad and Carpenter (1982), McMurray et al. (1983), Carpenter and Garrad (1985; 1986), Domaradzki and Metcalfe (1987), Metcalfe et al. (1991), and Davies and Carpenter (1997a; 1997b), among others, to simulate Kramer-type coatings. This model is relatively simple yet contains characteristics representative of a broad range of surfaces. If a coordinate system is chosen with the x-axis lying along the undisturbed free surface and the y-axis normal to this surface, then the equation for the y-component[3] of the momentum of the compliant coating reads

$$\frac{\partial^2 \eta}{\partial t^2} = \frac{T_\ell}{m} \frac{\partial^2 \eta}{\partial x^2} - D \frac{\partial \eta}{\partial t} - \frac{\mathcal{F}}{m} \frac{\partial^4 \eta}{\partial x^4} - \frac{k}{m} \eta + F \tag{8}$$

where $\eta(x,t)$ is the y-displacement of the surface from its equilibrium state at time t and position x, T_ℓ is the longitudinal tension and $\mathcal{F}$ the flexural rigidity of the thin plate, m is the mass per unit area, D is the damping coefficient, k is the spring constant, and F is an external forcing term.

The volume-based models are based on the Navier equation and include single and multi-layer coatings (Duncan et al., 1985; Fraser and Carpenter, 1985; Buckingham et al., 1985; Yeo, 1988) as well as isotropic and anisotropic materials (Yeo and Dowling, 1987; Yeo, 1990; 1992; Duncan, 1988). The equations describing the stability of the coupled system form a numerical eigenvalue problem for the complex wavenumber of the disturbance. Duncan (1987) offers a useful comparison between the results obtained from a surface-based model and a corresponding volume-based one.

Compliant walls do suppress the Tollmien–Schlichting waves due to the irreversible energy transfer from the fluid to the solid, but solid-based instabilities proliferate if the coating becomes too soft. For the class **A** T–S waves, the wall compliance reduces the rate of production (via Reynolds stress) of the disturbance

[3]The x-component of momentum is usually neglected in such a model.

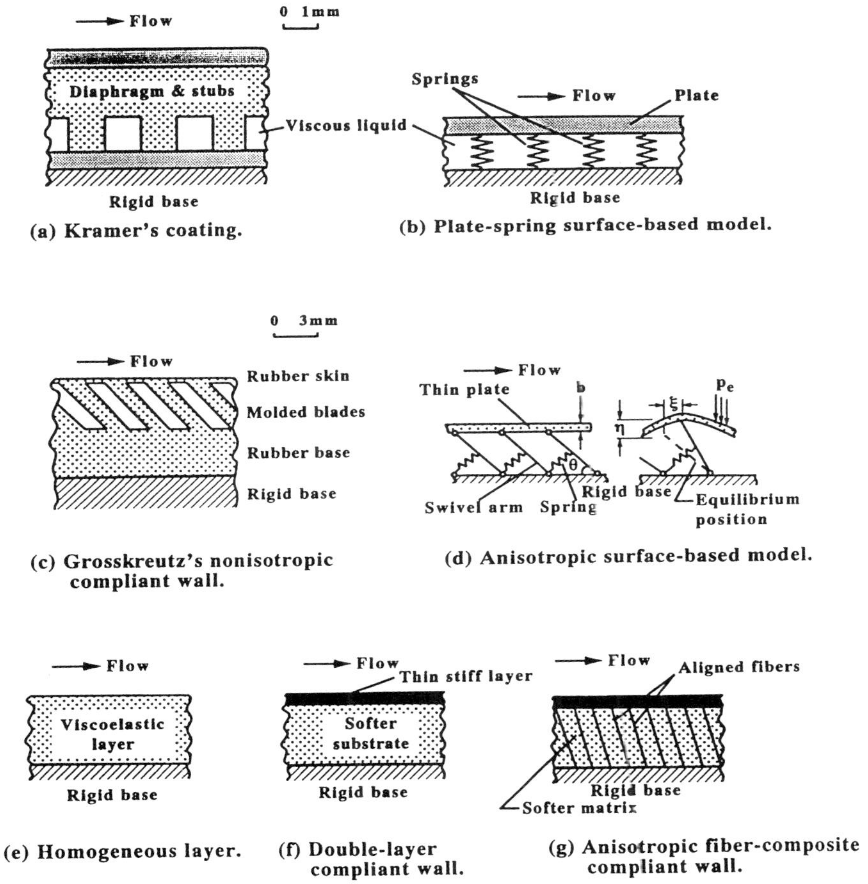

Figure 5. Volume-based and surface-based models of compliant coatings. Adapted from Carpenter (1990).

kinetic energy. Simultaneously, the viscous dissipation is increased and thus the balance between the energy production and removal mechanisms is altered in favor of wave suppression.

Experimental validation of the stability calculations is rather difficult and requires well-controlled tests in a quiet water or wind tunnel. Several careful experiments to test the flow stability to two-dimensional as well as three-dimensional controlled disturbances have been reported in the past few years (Daniel et al., 1987; Gaster, 1988; Lee et al., 1995; 1997). Figures 6 and 7 show the remarkable agreement between theory and towing tank experiments for both rigid-wall and

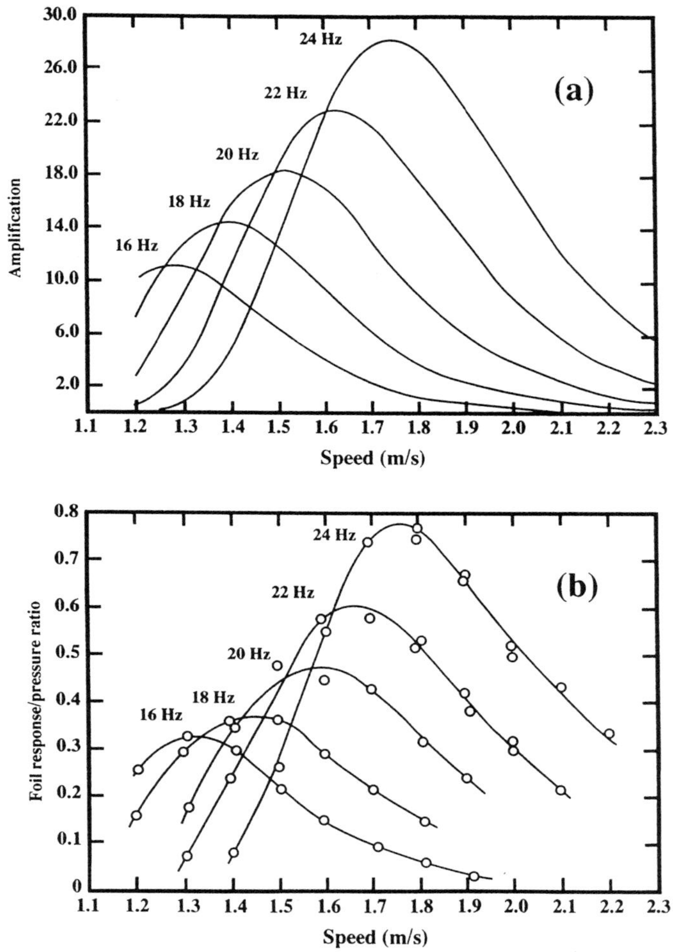

Figure 6. Growth curves for the rigid-wall case. (a) Theoretical prediction; (b) Experimental results. From Willis (1986).

compliant-wall cases. The top part of each figure is the predicted amplification factors as function of flow speed for a range of modal frequencies, while the bottom part is the measured growth-decay cycle of artificially induced T–S waves. A simple compliant model predicts a dramatic decrease in the instability of the flow, and this prediction agrees well with the experimental observations when a thick, soft coating is covered with a thin, stiff layer.

The papers by Lee et al. (1995; 1997) report the results of wind tunnel experiments and actually demonstrate the stabilizing potential of compliant coatings in aerodynamic flows; a remarkable achievement that has been deemed imprac-

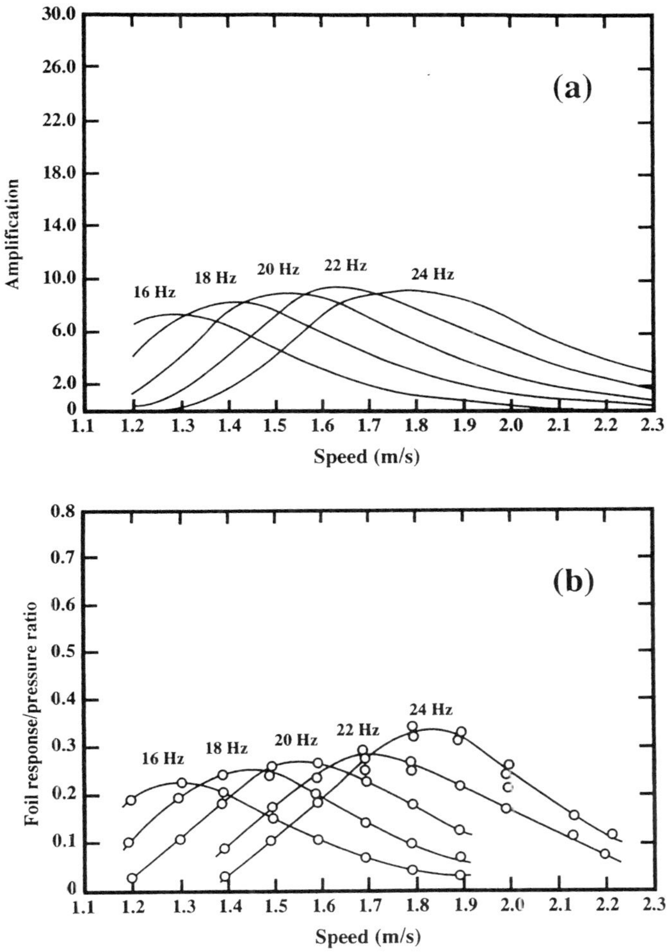

Figure 7. Growth curves for the compliant-wall case. A two-layer coating with an elastic modulus of the lower layer of $E = 5000\,\mathrm{N\,m^{-2}}$. (a) Theoretical prediction; (b) Experimental results. From Willis (1986).

tical in the past (Bushnell et al., 1977; Carpenter, 1990). Excellent agreements are reported between the results of the stability theory and the hydrodynamic experiments (Willis, 1986; Daniel et al., 1987; Gaster, 1988; Riley et al., 1988; Carpenter, 1990; Lucey and Carpenter, 1995). The paper by Lucey and Carpenter, in particular, applies the linear stability theory to predict the experimentally observed evolution of both Tollmien–Schlichting waves and traveling-wave flutter in water flows. For the wind tunnel experiments, Carpenter (1998) has conducted the corresponding calculations, but his preliminary results thus far are negative:

the density of an effective coating must be comparable to the fluid density otherwise no transition delaying benefits are observed. This theoretical result leaves open the question of explaining the positive experimental findings of Lee et al. (1995). In here, we show one example of the suppression of T–S waves in an air boundary layer developing on top of a silicone elastomer-silicone oil compliant surface. Figure 8 depicts the wind tunnel results of Lee et al. (1995). The coating in Figure 8a was made by mixing 91% by weight of 100 mm^2/s silicone oil with 9% of silicone elastomer. In Figures 8b, 8c and 8d, the corresponding mix was 90% and 10%, yielding about 35% higher shear modulus. As compared to the rigid wall, the single-layer, isotropic, viscoelastic compliant coating significantly suppresses the rms-amplitude of the artificially-generated Tollmien–Schlichting waves across the entire boundary layer, for a range of displacement-thickness Reynolds numbers. Reductions in the maximum rms-amplitude of as much as 40% are observed for the softer coating (Figure 8a), which may lead to delayed transition.

6.2. COATING OPTIMIZATION

If a compliant coating is to be designed for use on an actual vehicle, a relevant question may be: what are the optimum wall properties to give the greatest transition delay? The large number of available parameters makes it imperative that a rational (i.e., one derived from first principles) selection process be conducted. For obvious reasons, the trial-and-error empirical approach used in the past *(if it is soft, let us try it!)* should not even be contemplated. This should be particularly true now that rational optimization procedures are becoming readily available as described below. A wall that is too compliant (i.e., too soft) can substantially delay transition via TSI by shrinking its unstable region in the frequency-Reynolds number plane, but rapid breakdown can occur through the amplification of wall-based instabilities (Lucey and Carpenter, 1995). Both kinds of FISI are potentially harmful. The divergence instabilities are absolute, nearly static, and yield to wholesale deformations of the surface which are likely to trigger premature transition due to a roughness-like effect (Figure 9). Flutter instabilities, though convective, are also dangerous. As shown in the stability diagram in Figure 10, their narrow band of unstable frequencies extends indefinitely as Reynolds number increases downstream. Thus, once these instabilities are encountered at some downstream location, sustained growth follows. This is unlike the broad-band Tollmien–Schlichting instabilities which grow then decay as the different waves travel downstream and pass through the lower and upper branches of their neutral-stability curve (Figure 10).

A workable strategy for coating optimization suggested by Carpenter (1988) is to choose a restricted set of wall properties such that the coating is marginally stable with respect to FISI (both flutter and divergence). The remaining disposable wall parameters can then be varied to obtain the greatest possible transition (via TSI) delay. For the plate-spring, surface-based model, for example, there are two disposable parameters: the wall damping and the critical wavenumber for divergence. The downstream location of the transition region is estimated from an e^n-criterion, where n is typically chosen in the range of 7–10. The lower exponent represents the approximate limit of validity of the linear stability theory for a low-disturbance environment, and provides a rather conservative calculation.

Although wall dissipation destabilizes Tollmien–Schlichting waves, a viscoelas-

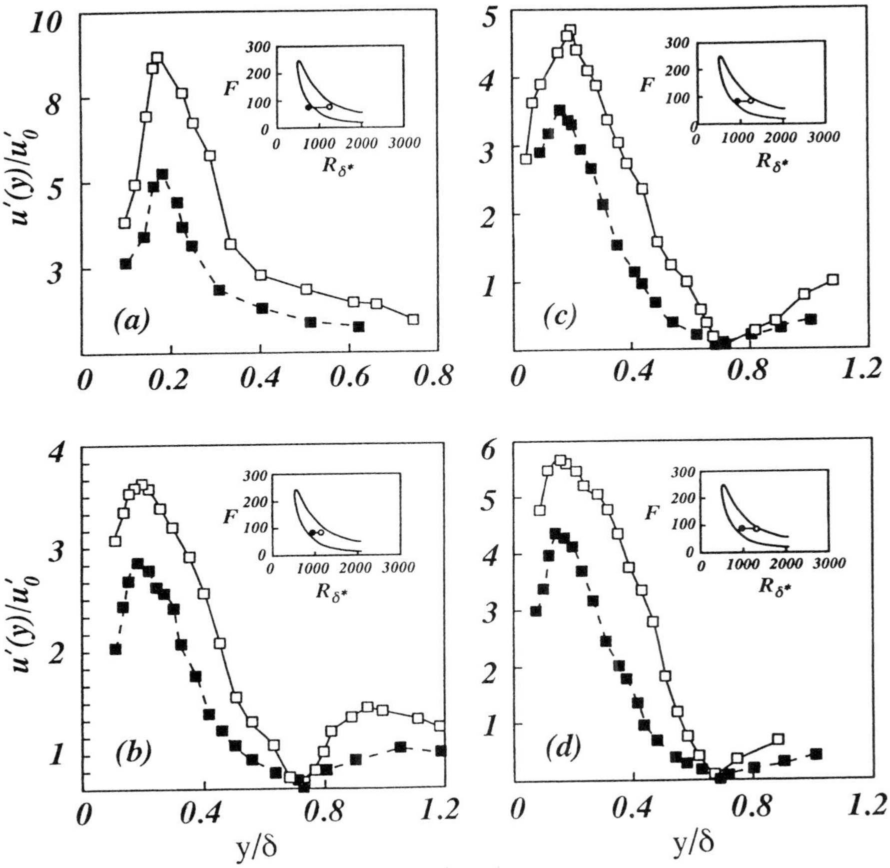

Figure 8. Comparison of distribution of rms-amplitude of the TSI of rigid surface and compliant surface across boundary layer. Wind tunnel experiments of Lee et al. (1995). □, rigid surface; filled □, compliant surface. (a) $Re_{\delta^*} = 1274$; (b) $Re_{\delta^*} = 1105$; (c) $Re_{\delta^*} = 1225$; (d) $Re_{\delta^*} = 1350$. Inset shows locations of ribbon (•) and probe (○) relative to neutral-stability curve.

tic coating with moderate level of damping leads to greater delay in transition as compared with purely elastic surfaces. Apparently the stabilizing effects of wall damping on traveling-wave flutter allow a softer wall to be used, and thus more than offset the adverse effects of coating dissipation on TSI.

Coating optimization with respect to TSI growth rate is performed at a rather narrow range of Reynolds numbers. On a growing boundary layer, the Reynolds number increases monotonically, and a compliant coating will not be optimum over the whole length of a vehicle. Carpenter (1993) suggests that a multiple-panel wall, placed in series, with each panel optimized for a particular range of

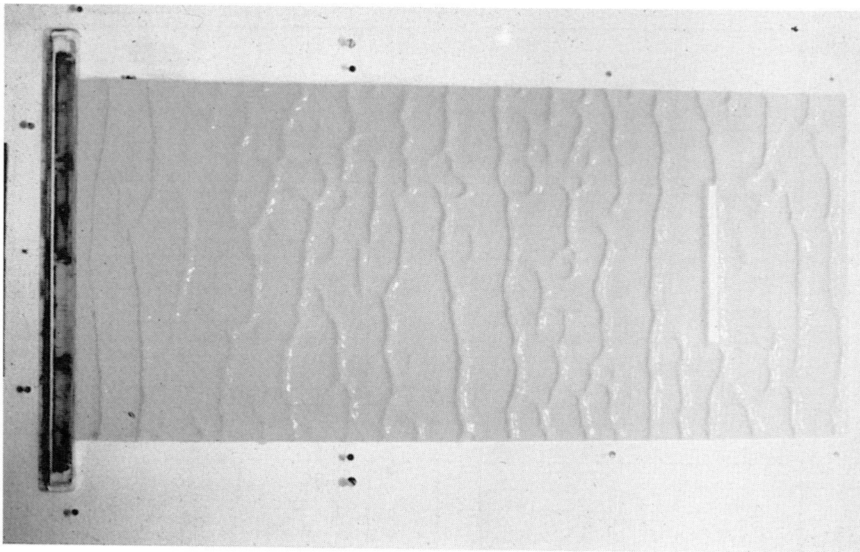

Figure 9. Static-divergence waves on a single-layer viscoelastic coating subjected to the pressure fluctuations of a turbulent boundary layer. Freestream velocity is 6.3 times the transverse-wave speed in the solid. From Gad-el-Hak et al. (1984).

Reynolds numbers, is likely to produce larger transition delays than a single-panel wall. His calculations for a two-panel, plate-spring-type compliant wall indicate an additional performance improvement of over 30% over an optimized single-panel wall.

It seems reasonable that a large number of panels, say 10, in series would lead to superior performance, but of course the calculations involved become prohibitive very quickly. An additional benefit from using multi-panels is that shorter panels are more resistant to both static-divergence waves and traveling-wave flutter (Carpenter, 1993; Lucey and Carpenter, 1993; Dixon et al., 1994), thus allowing softer panels to be used which further suppress TSI and improve the coating performance.

In the flat-plate and similar boundary layers in low-disturbance environments, the quasi-two-dimensional Tollmien–Schlichting waves dominate the laminar-to-turbulence transition. Various receptivity processes are responsible for generating 2-D instabilities which are probably initially three-dimensional and randomly distributed. Low-disturbance environments could be realized, for example, in free flight and marine vehicles. Very recently, Davies and Carpenter (1997a) and Carpenter (1998) have shown that in such environment, complete suppression of the Tollmien–Schlichting waves is possible provided that optimized multiple-panel compliant walls are used with each panel tailored to suit its local surrounding. Assured by the experimental observations that static-divergence waves are only observed when the wall-bounded flow is turbulent (see Figure 11, reproduced from

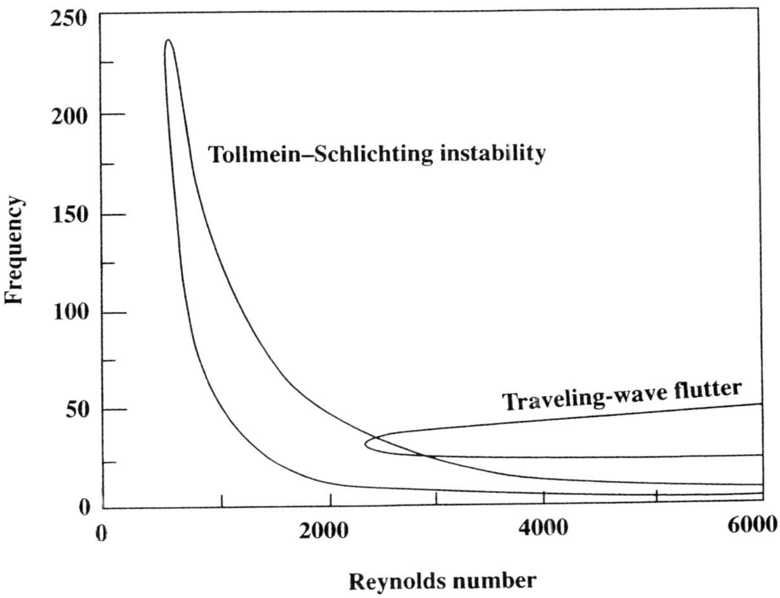

Figure 10. Typical neutral-stability curves for Tollmien–Schlichting waves and for traveling-wave flutter. The solid is a double-layer, Gaster-type compliant coating. From Lucey and Carpenter (1995).

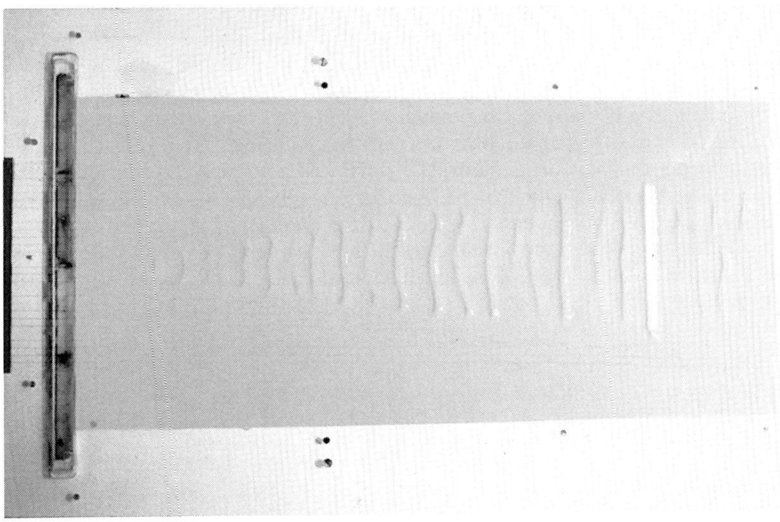

Figure 11. Static-divergence waves are clearly seen under the turbulent wedge generated by a single roughness element embedded into an otherwise laminar boundary layer. From Gad-el-Hak et al. (1984).

Gad-el-Hak et al., 1984),[4] Carpenter's (1998) new assumptions are somewhat less conservative than those used in earlier theories (Carpenter, 1993; Dixon et al., 1994). The new finding raises the possibility of maintaining laminar flow, in situations where the T–S instabilities are the primary cause of transition,[5] to indefinitely high Reynolds numbers, a very profound prospect indeed.

Work on nonlinear stability theory has recently been in the forefront and confirms that transition-delaying coatings, optimized using the linear theory, maintain their beneficial effects into the latter stages of transition to turbulence (Metcalfe et al., 1991; Joslin and Morris, 1992; Thomas; 1992a; 1992b). Lee et al. (1997) studied experimentally the effects of a compliant surface on the growth rates of both the subharmonic and three-dimensional fluid-based instabilities of a laminar boundary layer in air. Their results suggest that a delay of the excitement of the secondary instability can be achieved by suppressing the growth of the primary waves using surface compliance.

6.3. PRACTICAL EXAMPLES

Most of the theoretical as well as experimental compliant coating research has been concerned with canonical boundary layers. Nevertheless, an attempt is made in here to estimate the potential benefit of applying the technique for field applications where strong three-dimensional and pressure-gradient effects and, for aeronautical applications, compressibility effects may be present. The typical Reynolds numbers, based on vehicle speed and overall length, for a hydrofoil, a torpedo and a nuclear submarine are, respectively, of the order of 10 million, 50 million and 1 billion. Applying an e^n-type calculations (with the exponent chosen conservatively to be $n = 7$) to an optimum two-panel, plate-spring-type compliant wall, Carpenter (1993) computes a transition Reynolds number of 13.62×10^6, as compared with 2.25×10^6 for a rigid wall.[6] This means that the laminar region that would normally extend over 23%, 5% and 0.2% of the respective vehicle lengths would, with the use of an optimum coating, extend over a larger length of 100%, 27% and 1%. Computing the corresponding overall drag coefficients using standard methods for a mixed laminar-turbulent boundary layer over a flat plate, the potential reduction in skin-friction drag using the optimum two-panel compliant wall can be as much as 83%, 19% and 0% for the three respective vehicles. Obviously the large submarine does not benefit, as far as drag reduction is concerned, from the use of transition-delaying compliant coating but the smaller vehicles do. However, extending the laminar region on a submarine even by 1 m can be significant for sonar applications requiring longer quiet regions of the boundary layer.

[4]In other words, one less instability mode to worry about when the flow is laminar and the objective is to keep it so.

[5]Excludes, for example, flows where the walls and/or the streamlines are concave (Görtler instabilities), where the boundary layer is three-dimensional and the cross-flow velocity profile is inflectional (cross-flow instabilities), or where the wall is rough or the freestream turbulence levels are unusually high (by-pass transition).

[6]Contrast this six-fold increase to the 30% higher transition Reynolds number reported in the experiments of Gaster (1988), who did not attempt to optimize his single-panel, two-layer, silicone rubber/latex rubber coating. Theoretical calculations by Dixon et al. (1994) indicate that an optimum Gaster-type coating would provide a 500% higher transition Reynolds number.

For aeronautical applications, a cruising commercial jet aircraft has a fuselage Reynolds number of the order of 0.5 billion and a wing Reynolds number of the order of 50 million. Again, increasing the transition Reynolds number by a factor of 5 or so is significant for the wing but not for the fuselage. Skin-friction reduction of the order of 20% is achievable for the wings (whose skin-friction drag accounts for about 50% of the skin friction of the entire aircraft and 25% of the total drag).[7] Finding a compliant coating that would reduce the turbulent skin-friction drag would of course be very beneficial for both the typical fuselage and long submarine.

The estimates above were made for a simple plate-spring model. Using more than two panels can provide further transition delay. More complex compliant surfaces, particularly anisotropic ones designed specifically to suppress the Reynolds stress fluctuations, can conceivably offer more spectacular savings. Such custom-designed coatings can also favorably interact with fully-turbulent flows. Even for laminar flows, the calculations involved when complex, wall-based models are used, though straightforward in principle, are quite demanding in practice.

6.4. THE DOLPHIN'S SECRET

The ability to swim or to fly with minimum skin-friction and pressure drag is of extreme importance to the Darwinian survival of certain nektonic and avian species. *Homo sapiens* interested in building the fastest submarine or the most fuel-efficient aircraft have much to learn about alternative drag-reducing approaches from their humble earthlings. As Max O. Kramer has remarked close to half a century ago, a school of porpoises, including the young and the old, the weak and the strong, showing off its seemingly effortless glide along a fast ocean-liner is a sight to behold.

Cetaceans appear to possess unusually low overall drag coefficients. This is the basis for the so-called Gray's (1936) paradox, in which a steady-state energy balance based on the anticipated muscle power of various nekton, including the dolphin, failed to explain their unusually fast swimming speeds. Gray clocked bottlenose dolphins, *Tursiops trucatus*, swimming at speeds exceeding 10 m/s for a period of 7 s. If one assumes that cetaceans power output is equal to that of other mammals ($\sim$ 35 W/kg of body weight), then such speeds are reached under turbulent flow conditions only if dolphins can expend several times more power than their muscles can generate. Lang (1963) concluded that based on energy considerations, dolphins could not exceed a speed of 6 m/s for periods greater than 2 hours.

Transition delay is of course an obvious albeit arduous technique for achieving about an order of magnitude lower skin-friction drag, but does the dolphin posses an exotic means by which such difficult flow control goal can be accomplished? Obviously the dolphin is not sharing its secrets with other fellow mammals. Kramer's (1957; 1961) invention of a compliant coating tried to mimic the dolphin's epidermis and claimed drag reduction of as much as 60%. His explanation for the dolphin's secret is that their skin, like his successful compliant coating, is capable

[7]Note that the dominant mechanism leading to transition on a swept wing is cross-flow instability and not T–S waves. However, recent evidence (Cooper and Carpenter, 1995; 1997a; 1997b; 1997c) indicates beneficial effects of compliant surfaces even for the former kind of instability.

of substantially delaying laminar-to-turbulence transition. Kramer's work was discredited for a while but now seems to be back in vogue as remarked in Section 5. The calculations presented in Section 6.3 indicate that it is quite conceivable to design a Kramer-type coating that delays transition by a factor of 4–6 in Reynolds number and that drag reduction of the order reported by Kramer is also quite possible. Does the dolphin or other similar fast swimmers posses such a coating?

In a recent article, Bushnell and Moore (1991) quote the relevant energetic and controlled swimming studies, but conclude by supporting the explanation offered by Au and Weihs (1980) that dolphins, which must periodically breath air, achieve high-speed swimming by simply *porpoising*, i.e., momentarily leaping out of the water thereby reducing their drag force by a factor of 800 (density ratio of air and water). This more than pays for the additional interfacial or wave drag and accounts for the abnormally low apparent drag-coefficients inferred from the assumption of fully-submerged travel.

The present author, however, does not concur with the above *final solution* to the Gray's paradox. Dolphins have been clocked at sustained and burst speeds of close to 10 and 20 m/s, respectively. *Delphinus delphis* has a typical length of 2 m. This leads to sustained and burst Reynolds numbers based on the overall length of the order of 20 million and 40 million, respectively. Carpenter (1993) reports the results of optimizing a rather simple plate-spring coating. Using a single panel, as compared with a rigid surface a 4.6-fold increase in transition Reynolds number is estimated, which leads to a drag reduction of 36% at the typical dolphin's sustained speed and 20% at burst speed. Using a mere two-panel coating, the transition Reynolds number becomes 6.1 times the value for a rigid surface, and the potential drag reductions for the sustained and burst speeds are now 52% and 30%, respectively.

These lower levels of skin friction are compatible with the available muscle power for a dolphin of the size used above. Admittedly, the above estimates were made for a flat-plate boundary layer and may not hold when pressure-gradient and other shape effects are taken into account. Additionally, the dolphin has also pressure drag on top of the (much larger) skin friction. On the other hand, cetaceans have had millions of years of evolutionary adaptations to hone their coatings for maximum speed and efficiency, and it is quite conceivable that their epidermis is considerably more complex, and hydrodynamically beneficial, than the simple ones computed in the examples above. Moreover, each portion of the skin could have been optimized for the appropriate range of local Reynolds numbers. Therefore, the dolphin's apparent success is not incompatible with having optimum compliant coatings to substantially delay laminar-to-turbulence transition, and therefore to attain inordinately low coefficients of drag. A very recent theoretical investigation by Carpenter et al. (2000) agrees with our basic conclusion. Finally, Chapter 13 of the present book provides much more details on the subject of dolphin hydrodynamics.

Other fascinating questions related to the amazing swimming abilities of the dolphin include the possibility that its excreted mucin is a drag-reducing additive. Is there a hydrodynamic advantage to the warm-blooded cetaceans because their epidermis temperature is higher than the ambient one (in which case the near-wall water viscosity is lowered and the turbulent boundary layer may be relaminarized)? Does the dolphin's particular body shape during coasting (with no attendant overall body deformation) or actual swimming (accompanied by appropriate body oscillations) offer additional drag-reducing advantages? Also, what are the potential benefits to the porpoise when it uses ship-generated bow waves for body surfing? These subjects, though related to the above discussion, are outside the scope of the present brief and are therefore left for another circumstance.

7. Turbulent Wall-Bounded Flows

Unlike the laminar and transitional flows investigated in Section 6, compliant coating effects on turbulent boundary layers are rather difficult to study theoretically. In fact *any* turbulent flow is largely unapproachable analytically. For a turbulent flow, the dependent variables are random functions of space and time, and no straightforward method exists for analytically obtaining stochastic solutions to the governing nonlinear, partial differential equations. The statistical approach to solving the Navier–Stokes equations always leads to more unknowns than equations (the closure problem), and solutions based on first principles are again not possible. Direct numerical simulations (DNS) of the canonical turbulent boundary layer have thus far been carried out up to a very modest momentum-thickness Reynolds number of 1410 (Spalart, 1988).

How would one go about rationally choosing a coating to achieve a particular control goal for a turbulent boundary layer? Analytical optimization procedures such as those used to delay transition (Section 6.2) would not work for fully-turbulent flows. In order to analyze the full problem, direct numerical simulations of the turbulent boundary layer should be coupled to a finite-element model of the compliant coating, a task that is extremely time consuming, expensive and taxes the fastest supercomputer around. Modeling the turbulence by an eddy-viscosity or even a more sophisticated closure scheme is less computationally demanding, but there is no guarantee that turbulence models developed primarily for rigid surfaces would work for a compliant surface. In fact, it is not difficult to argue that closure models based on mean quantities miss completely the all-important spectral contents of a fluid–solid interaction, and will therefore never work.

A turbulent boundary layer is characterized by a hierarchy of coherent structures. Near the wall, the dynamics are dominated by the quasi-periodic bursting events (Robinson, 1991). A crude albeit resourceful attempt to model a turbulent boundary-layer interaction with a single-layer, isotropic, viscoelastic coating has been advanced by Duncan (1986). He approximates the turbulent flow over the coating by a potential flow with a superimposed pressure pulse, convecting downstream, that mimics the pressure footprint of a single bursting event. In order to relate the problem to a real turbulent flow, the pressure pulse characteristics are taken from actual boundary-layer measurements and the potential flow is modified to incorporate the reduced magnitudes and phase shifts found experimentally in boundary-layer flows over moving wavy walls. At low flow speeds (relative to the transverse-wave speed in the solid), the coating response to the pressure pulse is stable and primarily localized under it. At intermediate speeds, the response is still stable but includes a discernible wave pattern tagging along behind the pressure pulse. At the highest speed studied, large-amplitude, unstable waves develop on the compliant surface, much the same as the FISI observed experimentally. Duncan and Sirkis (1992) have recently extended the above model to anisotropic compliant coatings. They report that certain anisotropic surfaces provide more effective control over the amplitude and angular extent of the generated stable response pattern. Larger amplitudes are generated as compared with isotropic surfaces, thus providing for greater potential for modifying the turbulence.

Whenever the flow speed in a turbulent boundary layer becomes sufficiently large compared with the transverse free-wave speed in the solid, flow-induced surface instabilities proliferate. The pressure fluctuations within the flow are an order of magnitude larger than the normal and tangential viscous stresses, and drive the coating response. In laminar wall-bounded flows it is difficult to observe the hydroelastic waves in their unstable state. As soon as flutter or divergence waves grow, rapid breakdown to turbulence takes place in the boundary layer and the flow is no longer laminar.

Most of the experimental studies concerning compliant coating effects on turbulent boundary layers focused on documenting the unstable flow-induced surface instabilities. When divergence waves or flutter are unstable, the effects, though adverse, are pronounced and are somewhat easier to document. Only recently few hardy souls have attempted to investigate the wall-bounded flows when these FISI are stable or neutrally stable. Obviously the latter kind of studies have to await the development of refined techniques to measure the minuscule surface deformation and the associated coherent structure modulation when the FISI are neutrally stable.

Both Gad-el-Hak (1986a; 1986b) and Hess et al. (1993) introduce nonintrusive methods for the point measurement of the instantaneous vertical surface-displacement of a compliant coating, while Lee et al. (1993b) offer an optical holographic interferometer, in connection with an interactive fringe-processing system, to capture whole-field random topographic features. The latter technique is more expensive to set up but offers higher spatial resolution, of the order of 1 micron, and yield simultaneous surface displacement information on a large section of the compliant coating. Both the local and global methods were initially employed to document the unstable surface response to the pressure fluctuations in turbulent boundary layers. The holographic interferometer was recently used to record the surface topography in the presence of *stable* flow-induced deformations (Lee et al., 1993a).

The onset speed and wave characteristics of the solid-based class **B** and **C** instabilities were systematically documented in a series of towing-tank experiments (Gad-el-Hak, 1986b). Divergence waves were observed on a single-layer viscoelastic coating made from a PVC plastisol. The flutter appeared on an elastic coating made from common household gelatin, but, in the absence of damping, its threshold speed was consistently lower than that for divergence (Figure 12). The damping in the PVC coating stabilized the traveling-wave flutter and hence only divergence was observed there.[8] For the elastic coating, flutter appeared first and dominated the observed surface deformation. For both kind of waves, the threshold speed decreases with coating thickness, in other words thin surfaces (relative to the displacement thickness of the boundary layer) are less susceptible to hydroelastic instabilities than thick ones.

Typical profiles of unstable class **B** and **C** waves were also recorded in the same hydrodynamic experiments using a laser displacement gauge (Figure 13). The vertical displacement at a point associated with the slow moving, asymmet-

[8]Parenthetically, this and similar earlier observations led Gad-el-Hak (1986a) and others to the wrong conclusion, as stated in the footnote in Section 4, regarding the classification of the divergence waves. To reiterate, the class **B** flutter is stabilized by damping, while the class **C** divergence is largely unaffected.

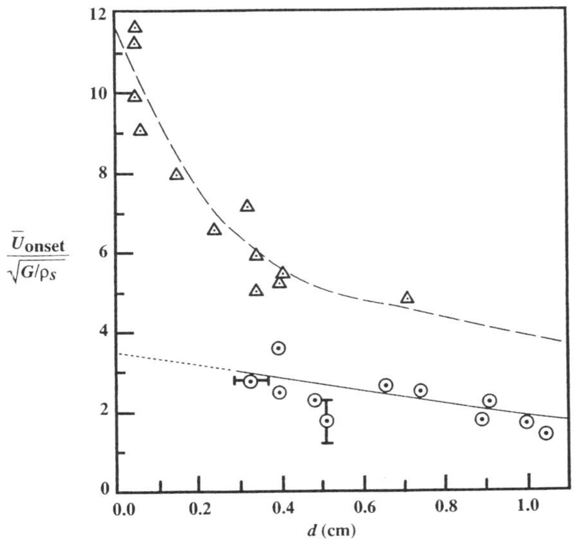

Figure 12. Onset speed dependence on thickness. ○, traveling-wave flutter on an elastic coating; △, static-divergence waves on a coating with damping. From Gad-el-Hak (1986b).

ric, large-amplitude divergence waves contrasts the faster, more-or-less symmetric, smaller-amplitude flutter. Both waves cause roughness-like effect, but the static divergence is the more dangerous instability. The phase speed of the static-divergence waves is of the order of 1% of the freestream speed, and their wavelength is about 5–10 times the coating thickness. The corresponding quantities for the flutter are 40% and 1.5–3, respectively.

Hess (1990) and Lee et al. (1993a) also investigated compliant coating effects on turbulent boundary layers. Both experiments were conducted in the same water tunnel, but the second paper focused on the stable interaction between the fluid and a single-layer, homogeneous, viscoelastic coating made of a mixture of silicone rubber and silicone oil. Lee et al.'s coating was chosen based on the criterion established by Duncan (1986). In the presence of a stable wave pattern on the compliant surface, the flow visualization experiments indicated low-speed streaks with increased spanwise spacing (by as much as 80%; see Figure 14) and elongated spatial coherence compared with those obtained on a rigid surface. More significantly, for the particular compliant coating investigated an intermittent relaminarization-like phenomenon was observed at low Reynolds numbers. Lee et al. (1993a) also report a slight thickening of the buffer region and viscous sublayer and an upward vertical shift in the compliant law-of-the-wall. The streamwise turbulence intensity, the local skin-friction coefficient and the Reynolds stress across the boundary layer were all reduced, indicating a possible interruption of the feedback loop which allows the turbulence to be self-sustaining. Thus, potentially favorable interaction between a compliant coating and a turbulent boundary layer has been demonstrated for the first time. The more recent hydrodynamic experiments by Choi et

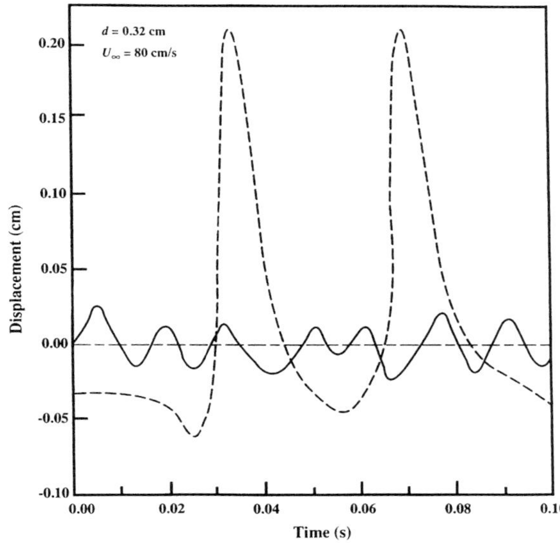

Figure 13. Typical unstable displacements of an elastic coating (———) and viscoelastic coating (− − − −). From Gad-el-Hak (1986b).

al. (1997) provide additional evidence for favorable interaction and indicate a total drag reduction for a long, slender body of revolution of the order of 7%.

8. What Works and What Doesn't?

For over forty years, compliant coating research has fascinated, frustrated and occasionally gratified scientists and engineers searching for methods to delay laminar-to-turbulence transition or to reduce skin-friction drag in turbulent wall-bounded flows. Through the years, claims for substantial drag reduction were made, only to be later refuted when the results were more critically examined. There are several important issues with regard to the reliability of available analytical, numerical and experimental results. Problem formulation with proper boundary conditions particularly at the fluid–solid interface, analytical solutions when the fluid side is turbulent, and limitations of existing numerical simulations are all problematic. The effects of background turbulence in a wind or water tunnel, accurate drag measurements, compliant wall motion, and the geometry and properties of the coating used are among the outstanding experimental issues.

In here we provide a brief listing of what works and what does not in compliant coating research. The articles by Gad-el-Hak (1987) and Carpenter (1990) provide more details of the pitfalls commonly encountered in, respectively, experiments and stability calculations related to compliant coating research aiming particularly for drag reduction. Many of the items listed in this section can be equally said about *any* research topic, but are particularly acute for compliant coating research because of its complexity as well as its history of false starts.

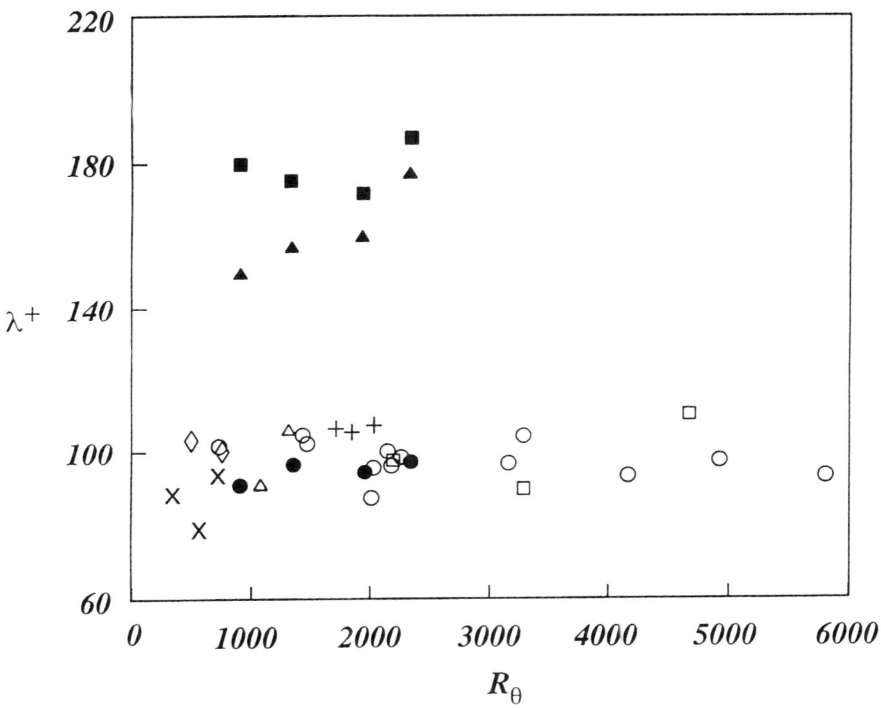

Figure 14. Variations of mean dimensionless spanwise spacing of the low-speed streaks in a turbulent boundary layer. Filled ○, rigid surface; filled △, compliant surface, friction velocity u_τ obtained from rigid-surface mean-velocity measurements; filled □, compliant surface, u_τ obtained from compliant-surface mean-velocity measurements. These data are from the water tunnel experiments of Lee et al. (1993a). Other symbols in the figure are from classical rigid-surface measurements.

ANALYTICAL RESEARCH

- The unsteady, three-dimensional, nonlinear problem coupling the Navier–Stokes equations in the fluid side and the Navier equation in the solid side is intractable.

- There is no known methodology by which stochastic solutions can be gleaned from the closed set of nonlinear, partial differential equations describing a turbulent flow. Therefore, there is no possibility of being able to obtain first-principles (i.e. without reverting to heuristic models to close the non-stochastic Reynolds-averaged equations of fluid motion) analytical solutions when the wall-bounded flow is turbulent (i.e. instantaneously random and three dimensional).
- Matching velocity and stress components at the unknown fluid–solid interface is a highly complex, nonlinear problem.
- Linearizing the interface conditions has to be done cautiously.
- Provided that the steady, laminar base flow is first determined analytically or numerically, stability calculations are doable if done with lots of TLC.
- The stability problem is extremely stiff particularly at high *Re* and high *G*, thus requiring specialized algorithms.
- Both linear and nonlinear stability calculations can, in principle, be carried out. The nonlinear problem is extremely computer intensive, however.

NUMERICAL RESEARCH

- Without sufficient validation, any numerical result of the fluid–structure interaction problem is suspect at best and wrong at worst.
- Numerical schemes which couple direct numerical simulations (DNS) in the turbulent flow side and finite-element (FE) algorithms in the compliant wall side are extremely expensive, and, for the foreseeable future, perhaps even impractical at field Reynolds numbers and realistic geometries. DNS–FE is possible, albeit expensive, at laboratory-scale Reynolds numbers and simple geometries. Strategies to extrapolate properly the low-Reynolds-number DNS to high *Re* are yet to be developed.
- Periodic boundary conditions, while convenient, cannot be used since the compliant coating is capable of propagating waves upstream as well as downstream. One can argue that this restriction can be relaxed if the propagation speeds of the solid waves are much higher than the convective speeds in the fluid side. However, by its very definition, a solid is compliant when its characteristic wave speed is of the same order of magnitude as the flow speed.
- Similar comments apply when attempting to parabolize the momentum equation in the fluid side.
- Large-eddy simulations (LES) is unable to resolve the all important fluid–solid interface region. Hybrid DNS (at the interface region) and LES (elsewhere in the flow) may be possible, but such strategies are yet to be developed properly. The hybrid DNS–LES in the fluid side must of course still be coupled with a finite-element algorithm in the solid side.
- Averaging the instantaneous equations of fluid motion and modeling the Reynolds stresses to close the resulting Reynolds-averaged Navier–Stokes

equations (RANS) will not work in principle—though is about the only analytical/numerical tool available in practice to handle high-Reynolds-number turbulent flows over compliant coatings—because the averaging process eliminates the all important frequency contents of the instantaneous flow.

EXPERIMENTAL RESEARCH

- Trial-and-error approach to finding a drag-reducing compliant coating does not work.
- Optimization strategies for rationally selecting a compliant coating capable of substantially delaying laminar-to-turbulence transition do work. In principle at least, similar strategies can be developed to design the optimum compliant coating to achieve skin-friction reduction in a turbulent flow.
- Sloppy experiments do not work.
- Meticulously executed experiments do work if guided by analysis and keen understanding of physics.
- Choose test rig with care. Natural bodies of water offer relatively quiet background turbulence, but are more difficult to control. Water tunnels are perhaps too noisy to enable well-controlled transition experiments. Towing tank may be a better alternative when studying transitional flows.
- The test rig's background turbulence is less of an issue when studying the interaction of a compliant coating with a turbulent boundary layer.
- Flat-plate geometry provides a well-documented flow in the rigid case and does not add unnecessary three-dimensional effects.
- Rotating disks can also be used as they combine the test rig and the flow geometry into one relatively inexpensive, well-documented configuration. The resulting flow is three-dimensional, however.
- The coating's physical properties such as shear modulus and loss tangent have to be carefully measured, and can change as the coating ages.
- Particularly for field applications, wearing, aging and fouling are important issues that must be considered, monitored and controlled.
- Pressure gradient has to be controlled and monitored as it affects the perceived drag reduction. Minute changes in the pressure distribution over the compliant body can lead to false reading of drag reduction.
- Edges, gaps and steps as the compliant slap joins the rigid surface have to be avoided. Texture and surface roughness of the coating have to be documented and taken into account when comparing to the drag of a smooth rigid surface.
- Surface-displacement measurements have to be done with non-invasive probes, such as the laser–photodiode strategy advanced by Gad-el-Hak et al. (1984).
- Local skin-friction measurements cannot be conducted with surface-mounted probes since these inevitably disturb the surface compliance.

- Integrating mean velocity profiles at different downstream locations to compute the drag does not offer the required accuracy.
- Computing the skin friction by measuring the velocity profile extremely close to the wall is problematic for rigid walls and is not any less so for moving, compliant surfaces.
- Drag balances can be used over an area encompassing the entire compliant slap, but have to be very precise and unaffected by surface deformation.

9. The Future

The diminishing pool of researchers remaining active in the field of compliant coatings includes teams from the University of Warwick, University of Nottingham, Johns Hopkins University, University of Houston, University of Maryland, and the Institute of Thermophysics in Novosibirsk. A larger pool was involved during the early 1980s, but the realities of research funding combined with the checkered past of the field led to the present decline.

Few suggestions for future research are given in here. The optimization procedures discussed in Section 6.2 have not been validated experimentally. Gaster-type experiments should be repeated using optimized coatings, including multi-panel ones. The recent claims by Davies and Carpenter (1997a) and Carpenter (1998) regarding the possibility of maintaining laminar flow to indefinitely high Reynolds numbers are very profound. Experiments, particularly field ones in low-disturbance environments, specifically designed to test those claims would be extremely useful.

The results of the transitional-boundary-layer, wind-tunnel experiments reported by Lee et al. (1995; 1997) are intriguing and fly in the face of the conventional wisdom. They indicate that compliant coatings are capable of delaying transition even for air flows. Past calculations using a plate-spring model and considering the extremely large density of typical walls compared with the density of air indicated that very flimsy coatings would be required to achieve transition delay and that the situation gets worse as the air speed increases. This led Carpenter (1990) among others to conclude that the use of wall compliance is impractical for aeronautical applications.[9] But the plate-spring results do not apply in any straightforward way to the homogeneous, single-layer walls studied by Lee et al. (1995). Validating the recent favorable results using both independent experiments and numerical simulations would open the door for aerodynamic applications, something that was seriously considered but later abandoned by NASA and the aerospace industry. The optimization procedures developed by Carpenter (1988) for transitional hydrodynamic flows should be extended to air flows. Experiments should be conducted using the resulting optimized coatings, if it is feasible to construct such material.

More complex coatings could potentially yield superior performance as compared with the relatively simple walls studied thus far. Multi-panels, multi-layers, anisotropic coatings and combinations thereof should be investigated. In any such research program, experiment has to be guided by theoretical results. As already

[9]It should be noted that Carpenter's calculations were based on the Navier equation modified for the viscoelasticity of the coating.

mentioned, trying to pick a compliant coating by trial and error is a very inefficient use of limited resources and will perhaps never work.

Favorably modulating a fully-turbulent flow, in contrast to merely delaying transition, is also of great practical importance. The experimental results reported by Lee et al. (1993a) are very encouraging, but the coating used was chosen based on a rather simplistic model of the turbulence pressure fluctuations. In order to custom-design compliant coatings to achieve particular control goals for turbulent wall-bounded flows, direct numerical simulations of the coupled fluid–structure system have to be performed. Turbulence modeling via classical closure schemes, while sufficient for some simple flows over rigid surfaces, will perhaps not yield reliable results for compliant walls. DNS, on the other hand, requires extensive computer resources and is quite expensive to carry out. The bottom line is that relatively large investment in resources are required for this task, but the enormous potential payoffs could easily justify the expenditure.

Most of the research thus far has considered incompressible, zero-pressure-gradient, flat-plate boundary layers. Effects of compressibility, pressure gradient and three-dimensionality on the performance of compliant coatings are largely unknown. Such studies will yield invaluable information for field application of the control technique for both air and water flows. Most practical aerodynamic flows are in the moderate-to-high Mach number regime, and compressibility effects must therefore be investigated before compliant coatings are used on actual aircraft. Related to the pressure-gradient effects is the question of separated flows: does compliant coating affect separation favorably or adversely? Other stability modifiers, such as favorable pressure-gradient, suction or heating/cooling, do delay transition as well as prevent separation. It is not known whether compliant coatings also have this dual benefit, and it may be beneficial to research the possibility.

Finally, real flows are three-dimensional and involve complex geometries. A model problem for three-dimensional flows is the rotating disk. Few experiments have been conducted using a rotating disk with a compliant face (Hansen and Hunston, 1974). More recently, Cooper and Carpenter (1995; 1997a; 1997b; 1997c) analyzed the cross-flow (type **I**; inviscid) as well as the viscous (type **II**) fluid-based instabilities which develop in the same three-dimensional flow. The preliminary results are encouraging and indicate that compliant coatings can suppress the more dangerous type **I** instabilities. Very recently, Colley et al. (1999) conducted an experiment to investigate transition over a rotating, compliant disk. Their results qualitatively agree with existing theories and show that the particular elastic coating they used has an overall destabilizing effect on the flow and results in substantially lower transitional Reynolds number compared to the case of a rigid disk. The lesson here is important: compliant coatings do not always yield beneficial results.

Active compliant coatings, or more properly internally driven flexible coatings, though bringing us back to the extreme complexity of reactive control systems, is an emerging area deserving of further research. Energy expenditure is required to drive the wall, but the potential for significant net drag reduction is higher than that for passive coatings. The feasibility of the concept for stabilizing laminar boundary layers has been shown through numerical experiments (Metcalfe et al., 1986). Internally driven flexible coatings could also be used to suppress the Reynolds stress and reduce the skin-friction drag in turbulent wall-bounded flows, but any realistic field application of the technique has to await further development of reasonably-priced and rugged microfabricated sensors and actuators (Gad-el-Hak, 1994; 1996b; 2000; 2001).

10. Parting Remarks

Passive compliant coatings present a much simpler alternative to reactive flow control strategies aimed at favorably interfering with wall-bounded flows. The last 10–15 years witnessed renewed interest in compliant coatings as a means to achieve beneficial flow control goals. Significant advances were made in numerical and analytical techniques to solve the coupled fluid–structure problem. Novel experimental tools were developed to measure the stable as well as the unstable surface deformations caused by the pressure fluctuations in the boundary layer. In turbulent wall-bounded flows, coherent structures were routinely identified and their modulation by wall compliance could be quantified.

Most significant results in the field thus far were obtained when a strong cooperation existed between theory and experiment. Recent theoretical work indicates that complete suppression of the Tollmien–Schlichting waves may be possible, provided that optimized multiple-panel compliant walls are used. The new finding raises the possibility of maintaining laminar flow to indefinitely high Reynolds numbers, a very profound prospect indeed. Recent experiments indicate favorable compliant coating interactions even for aerodynamic flows and even for turbulent boundary layers. More research is needed, however, to confirm these latest results.

The coupled system instabilities are now well understood, and compliant coatings can therefore be *rationally* designed to achieve substantial, perhaps even indefinite, transition delay in hydrodynamic flows. That recent shift from random to rational search for the right kind of coating, particularly if it can be extended to tubulent flows, is not unlike the great paradigm change in synthetic chemistry that took place near the beginning of the twentieth century. Increased understanding of the molecular geometry of organic compounds changed the scene from a hapless alchemist muddling around hoping to chance the right combination of ingredients, heat, pressure and catalysts to produce something useful, to a professional chemist figuring out what she wants and working backward from the shape of a desired molecule for, say, a synthetic hormone. In fact, the present analogy is apt: the fluid dynamist working with the Navier–Stokes equations can tell the chemist the exact properties of the compliant coating to be synthesized to achieve a given goal. If the needed molecular structure is too complicated, futuristic nanoscale machines can assemble the required molecules directly, atom by atom.

Portions of the present chapter were presented as the keynote address at the *IUTAM Symposium on Flow in Collapsible Tubes and Past Other Highly Compliant Boundaries*, Coventry, England, 26–30 March 2001, and will appear as a journal aritcle in *Progress in Aeropsace Sciences.*

References

Aleyev, Yu.G. (1977) *Nekton*, Dr. W. Junk b.v. Publishers, the Hague, the Netherlands.

Au, D., and Weihs, D. (1980) "At High Speeds Dolphins Save Energy by Leaping," *Nature* **284**, no. 5756, 10 April, pp. 548–550.

Babenko, V.V., and Kozlov, L.F. (1972) "Experimental Investigation of Hydrodynamic Stability on Rigid and Elastic Damping Surfaces," *J. Hydraulic Res.* **10**, pp. 383–408.

Benjamin, T.B. (1960a) "Effects of a Flexible Boundary on Hydrodynamic Stability," *J. Fluid Mech.* **9**, pp. 513–532.

Benjamin, T.B. (1960b) "Fluid Flow with Flexible Boundaries, *Proc. Eleventh International Congress of Applied Mechanics*, ed. H. Görtler, pp. 109–128, Springer-Verlag, Berlin.

Benjamin, T.B. (1963) "The Threefold Classification of Unstable Disturbances in Flexible Surfaces Bounding Inviscid Flows," *J. Fluid Mech.* **16**, pp. 436–450.

Benjamin, T.B. (1966) "Fluid Flow with Flexible Boundaries," in *Proc. Eleventh International Congress of Applied Mechanics*, ed. H. Görtler, pp. 109–128, Springer-Verlag, Berlin.

Betchov, K. (1960) "Simplified Analysis of Boundary-Layer Oscillations," *J. Ship Res.* **4**, pp. 37–54.

Blick, E.F., and Walters, R.R. (1968) "Turbulent Boundary-Layer Characteristics of Compliant Surfaces," *J. Aircraft* **5**, pp. 11–16.

Buckingham, A.C., Hall, M.S., and Chun, R.C. (1985) "Numerical Simulations of Compliant Material Response to Turbulent Flow," *AIAA J.* **23**, pp. 1046–1052.

Burton, T.E. (1974) "The Connection Between Intermittent Turbulent Activity Near the Wall of a Turbulent Boundary Layer with Pressure Fluctuations at the Wall," Massachusetts Institute of Technology Report No. 70208-10, Cambridge, Massachusetts.

Bushnell, D.M., and Moore, K.J. (1991) "Drag Reduction in Nature," *Annu. Rev. Fluid Mech.* **23**, pp. 65–79.

Bushnell, D.M., Hefner, J.N., and Ash, R.L. (1977) "Effect of Compliant Wall Motion on Turbulent Boundary Layers," *Phys. Fluids* **20**, no. 10, part II, pp. S31–S48.

Carpenter, P.W. (1988) "The Optimization of Compliant Surfaces for Transition Delay, in *Turbulence Management and Relaminarisation*, eds. H.W. Liepmann and R. Narasimha, pp. 305–313, Springer-Verlag, Berlin.

Carpenter, P.W. (1990) "Status of Transition Delay Using Compliant Walls," in *Viscous Drag Reduction in Boundary Layers*, eds. D.M. Bushnell and J.N. Hefner, pp. 79–113, AIAA, Washington, D.C.

Carpenter, P.W. (1993) "Optimization of Multiple-Panel Compliant Walls for Delay of Laminar-Turbulent Transition," *AIAA J.* **31**, pp. 1187–1188.

Carpenter, P.W. (1998) "Current Status of the Use of Wall Compliance for Laminar-Flow Control," *Exp. Thermal & Fluid Sci.* **16**, pp. 133–140.

Carpenter, P.W., and Garrad, A.D. (1985) "The Hydrodynamic Stability of Flow over Kramer-Type Compliant Surfaces. Part 1. Tollmien–Schlichting Instabilities," *J. Fluid Mech.* **155**, pp. 465–510.

Carpenter, P.W., and Garrad, A.D. (1986) "The Hydrodynamic Stability of Flow over Kramer-Type Compliant Surfaces. Part 2. Flow-Induced Surface Instabilities," *J. Fluid Mech.* **170**, pp. 199–232.

Carpenter, P.W., Davies, C., and Lucey, A.D. (2000) "Hydrodynamics and Compliant Walls: Does the Dolphin have a Secret?," *Current Science* **79**, no. 6, pp. 758–765.

Carpenter, P.W., Lucey, A.D., and Davies, C. (2001) "Progress on the Use of Compliant Walls for Laminar-Flow Control," *J. Aircraft* **38**, pp. 504–512.

Choi, K.-S., editor (1991) *Recent Developments in Turbulence Management*, Kluwer, Dordrecht, the Netherlands.

Choi, K.-S., Yang X., Clayton B.R., Glover E.J., Atlar M., Semenov B.N., and Kulik V.M. (1997) "Turbulent Drag Reduction Using Compliant Surfaces," *Proc. R. Soc. Lond. A* **453**, 2229–2240.

Chu, H.H., and Blick, E.F. (1969) "Compliant Surface Drag as a Function of Speed," *J. Spacecraft & Rockets* **6**, pp. 763–764.

Colley, A.J., Thomas, P.J., Carpenter, P.W., and Cooper, A.J. (1999) "An Experimental Study of Boundary-Layer Transition over a Rotating, Compliant Disk," *Phys. Fluids* **11**, pp. 3340–3352.

Cooper, A.J., Carpenter, P.W. (1995) "The Effects of Wall Compliance on Instability in Rotating Disc Flow," AIAA Paper No. 95-2257, Washington, D.C.

Cooper, A.J., Carpenter, P.W. (1997a) "The Stability of Rotating-Disc Boundary-Layer Flow over a Compliant Wall. Part 1. Type I and II Instabilities," *J. Fluid Mech.* **350**, pp. 231–259.

Cooper, A.J., Carpenter, P.W. (1997b) "The Stability of Rotating-Disc Boundary-Layer Flow over a Compliant Wall. Part 2. Absolute Instability," *J. Fluid Mech.* **350**, pp. 261–270.

Cooper, A.J., Carpenter, P.W. (1997c) "The Effect of Wall Compliance on Inflexion Point Instability in Boundary Layers," *Phys. Fluids* **9**, pp. 468–470.

Daniel, A.P., Gaster, M., and Willis, G.J.K. (1987) "Boundary Layer Stability on Compliant Surfaces," Final Report No. 35020, British Maritime Technology Ltd., Teddington, Great Britain.

Davies, C., and Carpenter, P.W. (1997a) "Numerical Simulation of the Evolution of Tollmien–Schlichting Waves over Finite Compliant Panels," *J. Fluid Mech.* **335**, pp. 361–392.

Davies, C., and Carpenter, P.W. (1997b) "Instabilities in a Plane Channel Flow between Compliant Walls," *J. Fluid Mech.* **352**, pp. 205–243.

Dinkelacker, A. (1966) "Preliminary Experiments on the Influence of Flexible Walls on Boundary Layer Turbulence," *J. Sound Vibr.* **4**, pp. 187–214.

Dixon, A.E., Lucey, A.D., and Carpenter, P.W. (1994) "The Optimization of Viscoelastic Compliant Walls for Transition Delay," *AIAA J.* **32**, pp. 256–267.

Domaradzki, J.A., and Metcalfe, R.W. (1987) "Stabilization of Laminar Boundary Layers by Compliant Membranes," *Phys. Fluids* **30**, pp. 695–705.

Duncan, J.H. (1986) "The Response of an Incompressible, Viscoelastic Coating to Pressure Fluctuations in a Turbulent Boundary Layer," *J. Fluid Mech.* **171**, pp. 339–363.

Duncan, J.H. (1987) "A Comparison of Wave Propagation on the Surfaces of Simple Membrane Walls and Elastic Coatings Bounded by a Fluid Flow," *J. Sound Vibr.* **119**, pp. 565–573.

Duncan, J.H. (1988) "The Dynamics of Waves at the Interface between a Two-Layer Viscoelastic Coating and a Fluid Flow," *J. Fluids Structures* **2**, pp. 35–51.

Duncan, J.H., and Sirkis, J.S. (1992) "The Generation of Wave Patterns on Isotropic Coatings by Pressure Fluctuations in a Turbulent Boundary Layer," *J. Sound Vibr.* **157**, pp. 243–264.

Duncan, J.H., and Hsu, C.C. (1984) "The Response of a Two-Layer Viscoelastic Coating to Pressure Disturbances from a Turbulent Boundary Layer," AIAA Paper No. 84-0535, New York.

Duncan, J.H., Waxman, A.M., and Tulin, M.P. (1985) "The Dynamics of Waves at the Interface between a Viscoelastic Coating and a Fluid Flow," *J. Fluid Mech.* **158**, pp. 177–197.

Fischer, M.C., and Ash, R.L. (1974) "A General Review of Concepts for Reducing Skin Friction, Including Recommendations for Future Studies," NASA Technical Memorandum No. X-2894, Washington, D.C.

Fisher, D.H., and Blick, E.F. (1966) "Turbulent Damping by Flabby Skins," *J. Aircraft* **3**, pp. 163–164.

Fraser, L.A., and Carpenter, P.W. (1985) "A Numerical Investigation of Hydroelastic and Hydrodynamic Instabilities in Laminar Flows over Compliant Surfaces Comprising One or Two Layers of Viscoelastic Material," in *Numerical Methods in Laminar and Turbulent Flow*, eds. C. Taylor et al., pp. 1171–1181, Pineridge Press, Swansea, Great Britain.

Gad-el-Hak, M. (1986a) "Boundary Layer Interactions with Compliant Coatings: An Overview," *Appl. Mech. Rev.* **39**, pp. 511–524.

Gad-el-Hak, M. (1986b) "The Response of Elastic and Viscoelastic Surfaces to a Turbulent Boundary Layer," *J. Appl. Mech.* **53**, pp. 206–212.

Gad-el-Hak, M. (1987) "Compliant Coatings Research: A Guide to the Experimentalist," *J. Fluids Structures* **1**, pp. 55–70.

Gad-el-Hak, M. (1994) "Interactive Control of Turbulent Boundary Layers: A Futuristic Overview," *AIAA J.* **32**, pp. 1753–1765.

Gad-el-Hak, M. (1996a) "Compliant Coatings: A Decade of Progress," *Appl. Mech. Rev.* **49**, no. 10, part 2, pp. S1–S11.

Gad-el-Hak, M. (1996b) "Modern Developments in Flow Control," *Appl. Mech. Rev.* **49**, pp. 365–379.

Gad-el-Hak, M. (2000) *Flow Control: Passive, Active, and Reactive Flow Management*, Cambridge University Press, London, United Kingdom.

Gad-el-Hak, M. (editor) (2001) *The MEMS Handbook*, CRC Press, Boca Raton, Florida.

Gad-el-Hak, M., Blackwelder, R.F., and Riley, J.J. (1984) "On the Interaction of Compliant Coatings with Boundary Layer Flows," *J. Fluid Mech.* **140**, pp. 257–280.

Garrad, A.D., and Carpenter, P.W. (1982) "A Theoretical Investigation of Flow-Induced Instabilities in Compliant Coatings," *J. Sound Vibr.* **85**, pp. 483–500.

Gaster, M. (1988) "Is the Dolphin a Red Herring?," in *Turbulence Management and Relaminarisation*, eds. H.W. Liepmann and R. Narasimha, pp. 285–304, Springer-Verlag, Berlin.

Gray, J. (1936) "Studies in Animal Locomotion. VI. The Propulsive Powers of the Dolphin," *J. Exp. Biol.* **13**, pp. 192–199.

Hansen, R.J., and Hunston, D.L. (1974) "An Experimental Study of Turbulent Flows over Compliant Surfaces," *J. Sound Vibr.* **34**, pp. 297–308.

Hess, D.E. (1990) "An Experimental Investigation of a Compliant Surface beneath a Turbulent Boundary Layer," Ph.D. Dissertation, the Johns Hopkins University, Baltimore, Maryland.

Hess, D.E., Peattie, R.A., and Schwarz, W.H. (1993) "A Nonintrusive Method for the Measurement of Flow-Induced Surface Displacement of a Compliant Surface," *Exp. Fluids* **14**, pp. 78–84.

Huerre, P., and Monkewitz, P.A. (1990) "Local and Global Instabilities in Spatially Developing Flows," *Annu. Rev. Fluid Mech.* **22**, pp. 473–537.

Johnson, R.P. (1980) "Review of Compliant Coating Research of M. O. Kramer, RPJ Associates Technical Report No. RPJA-TR-0955-001, Palos Verdes Peninsula, California.

Joslin, R.D., and Morris, P.J. (1992) "Effect of Compliant Walls on Secondary Instabilities in Boundary-Layer Transition," *AIAA J.* **30**, pp. 332–339.

Kaplan, R. E. (1964) "The Stability of Laminar Incompressible Boundary Layers in the Presence of Compliant Boundaries," Sc.D. Thesis, Massachusetts Institute of Technology, Cambridge, Massachusetts.

Kramer, M.O. (1957) "Boundary-Layer Stabilization by Distributed Damping," *J. Aeronaut. Sci.* **24**, pp. 459–460.

Kramer, M.O. (1960a) "Boundary-Layer Stabilization by Distributed Damping," *J. Aero/Space Sci.* **27**, p. 69.

Kramer, M.O. (1960b) "Boundary Layer Stabilization by Distributing Damping," *J. Am. Soc. Naval Engrs.* **72**, pp. 25–33.

Kramer, M.O. (1960c) "The Dolphin's Secret," *New Scientist* **7**, 5 May, pp. 1118–1120.

Kramer, M.O. (1961) "The Dolphin's Secret," *J. Am. Soc. Nav. Engrs.* **73**, pp. 103–107.

Kramer, M.O. (1962) "Boundary Layer Stabilization by Distributed Damping," *J. Am. Soc. Naval Engrs.* **74**, pp. 341–348.

Kramer, M.O. (1965) "Hydrodynamics of the Dolphin," in *Advances in Hydroscience*, ed. V.T. Chow, vol. 2, pp. 111–130, Academic Press, New York.

Kramer, M.O. (1969) "Die Widerstandsverminderung Schneller Unterwasserkörper Mittels Künstlicher Delphinhaut," in *Jahrbuch 1969 der Deutschen Gesellschaft für Luft- und Raumfahrt*, eds. H. Blenk and W. Schulz, pp. 1–9, Köln, Federal Republic of Germany.

Landahl, M.T. (1962) "On the Stability of a Laminar Incompressible Boundary Layer over a Flexible Surface," *J. Fluid Mech.* **13**, pp. 609–632.

Landau, L.D., and Lifshitz, E.M. (1987) *Fluid Mechanics*, second edition, Pergamon Press, Oxford, Great Britain.

Lang, T.G. (1963) "Porpoise, Whales, and Fish: Comparison of Predicted and Observed Speeds," *Naval Eng. J.* **75**, pp. 437–441.

Lee, T., Fisher, M., and Schwarz, W.H. (1993a) "Investigation of the Stable Interaction of a Passive Compliant Surface with a Turbulent Boundary Layer," *J. Fluid Mech.* **257**, pp. 373–401.

Lee, T., Fisher, M., and Schwarz, W.H. (1993b) "The Measurement of Flow-Induced Surface Displacement on a Compliant Surface by Optical Holographic Interferometry," *Exp. Fluids* **14**, pp. 159–168.

Lee, T., Fisher, M., and Schwarz, W.H. (1995) "Investigation of the Effects of a Compliant Surface on Boundary-Layer Stability," *J. Fluid Mech.* **288**, pp. 37–58.

Lee, T., Fisher, M., and Schwarz, W.H. (1997) "An Experimental Study of the Boundary-Layer Nonlinear Instability over Compliant Walls," unpublished manuscript originally submitted to *J. Fluid Mech.*.

Lissaman, P.B.S., and Harris, G.L. (1969) "Turbulent Skin Friction on Compliant Surfaces," AIAA Paper No. 69-164, New York.

Looney, R.W., and Blick, E.F. (1966) "Skin Friction Coefficients of Compliant Surfaces in Turbulent Flow," *J. Spacecraft & Rockets* **3**, pp. 1562–1564.

Lucey, A.D., and Carpenter, P.W. (1993) "The Hydroelastic Stability of Three-Dimensional Disturbances of a Finite Compliant Panel," *J. Sound Vibr.* **165**, pp. 527–552.

Lucey, A.D., and Carpenter, P.W. (1995) "Boundary Layer Instability over Compliant Walls: Comparison between Theory and Experiment," *Phys. Fluids* **7**, pp. 2355–2363.

McMichael, J.M., Klebanoff, P.S., and Meese, N.E. (1980) "Experimental Investigation of Drag on a Compliant Surface," in *Viscous Flow Drag Reduction*, ed. G.R. Hough, pp. 410–438, AIAA, New York.

McMurray, J.T., Metcalfe, R.W., and Riley, J.J. (1983) "Direct Numerical Simulations of Active Stabilization of Boundary Layer Flows," *Proc. Eighth Biennial Symp. on Turbulence*, eds. J.L. Zakin and G.K. Patterson, paper no. 36, University of Missouri, Rolla, Missouri.

Metcalfe, R.W. (1994) "Boundary Layer Control: a Brief Review, in *Computational Fluid Dynamics '94*, eds. J. Periaux and E. Hirschel, pp. 52–60, Wiley, New York.

Metcalfe, R.W., Battistoni, F., Ekeroot, J., and Orszag, S.A. (1991) "Evolution of Boundary Layer Flow over a Compliant Wall during Transition to Turbulence," in *Proc. Boundary Layer Transition and Control Conference*, pp. 36.1–36.14, Royal Aeronautical Society, Cambridge, Great Britain.

Metcalfe, R.W., Rutland, C.J., Duncan, J.H., and Riley, J.J. (1986) "Numerical Simulations of Active Stabilization of Laminar Boundary Layers, *AIAA J.* **24**, pp. 1494–1501.

Nisewanger, C.R. (1964) "Flow Noise and Drag Measurements of Vehicle with Compliant Coating," U.S. Naval Ordnance Test Station Report No. 8518, NOTS No. TP-3510, China Lake, California.

Orszag, S.A. (1979) "Prediction of Compliant Wall Drag Reduction," NASA Contractor Report No. 3071, Washington, D.C.

Puryear, F.W. (1962) "Boundary Layer Control: Drag Reduction by Use of Compliant Coatings," David Taylor Model Basin Report No. 1668, Bethesda, Maryland.

Rayleigh, Lord (1887) "On Waves Propagated along the Plane Surface of an Elastic Solid," *Proc. Lond. Math. Soc.* **XVII**, pp. 4–11.

Riley, J.J., Gad-el-Hak, M., and Metcalfe, R.W. (1988) "Compliant Coatings," *Annu. Rev. Fluid Mech.* **20**, pp. 393–420.

Ritter, H., and Messum, L.T. (1964) "Water Tunnel Measurements of Turbulent Skin Friction on Six Different Compliant Surfaces of 1 Ft Length," British Admiralty Research Laboratory Report No. ARL/N4/GHY/9/7, ARL/G/N9, London, Great Britain.

Ritter, H., and Porteous, J.S. (1964) "Water Tunnel Measurements of Skin Friction on a Compliant Coating," British Admiralty Research Laboratory Report No. ARL/N3/G/HY/9/7, London, Great Britain.

Robinson, S.K. (1991) "Coherent Motions in the Turbulent Boundary Layer," *Annu. Rev. Fluid Mech.* **23**, pp. 601–639.

Sen, P.K., and Arora, D.S. (1988) "On the Stability of Laminar Boundary-Layer Flow over a Flat Plate with a Compliant Surface," *J. Fluid Mech.* **197**, pp. 201–240.

Smith, R.L., and Blick, E.F. (1966) "Skin Friction of Compliant Surfaces with Foamed Material Substrate," *J. Hydronautics* **3**, pp. 100–102.

Spalart, P.R. (1988) "Direct Simulation of a Turbulent Boundary Layer up to $R_\theta = 1410$," *J. Fluid Mech.* **187**, pp. 61–98.

Thomas, M.D. (1992a) "On the Resonant Triad Interaction in Flows over Rigid and Flexible Boundaries," *J. Fluid Mech.* **234**, pp. 417–442.

Thomas, M.D. (1992b) "On the Nonlinear Stability of Flows over Compliant Walls," *J. Fluid Mech.* **239**, pp. 657–670.

Willis, G.J.K. (1986) "Hydrodynamic Stability of Boundary Layers over Compliant Surfaces," Ph.D. Thesis, University of Exeter, Exeter, Great Britain.

Yeo, K.S. (1988) "The Stability of Boundary-Layer Flow over Single- and Multi-Layer Viscoelastic Walls," *J. Fluid Mech.* **196**, pp. 359–408.

Yeo, K.S. (1990) "The Hydrodynamic Stability of Boundary-Layer Flow over a Class of Anisotropic Compliant Walls," *J. Fluid Mech.* **220**, pp. 125–160, 1990.

Yeo, K.S. (1992) "The Three-Dimensional Stability of Boundary-Layer Flow over Compliant Walls, *J. Fluid Mech.* **238**, pp. 537–577.

Yeo, K.S., and Dowling, A.P. (1987) "The Stability of Inviscid Flows over Passive Compliant Walls," *J. Fluid Mech.* **183**, pp. 265–292.

Chapter 10

Theoretical Approaches to the Effect of Wall Compliance on Turbulent Flow

DIETMAR REMPFER
Illinois Institute of Technology
Chicago, IL
U.S.A

LOUISE PARSONS
SHENG XU
JOHN LUMLEY
Cornell University
Ithaca, NY
U.S.A

Abstract

In this paper we are reporting preliminary results of an ongoing research project that aims at getting an improved understanding of the interaction between wall-generated turbulence and compliant surface coatings. We are pursuing an integrated approach using analysis, direct numerical simulation and low-dimensional modeling, with a view towards the reduction of turbulent sound production and turbulent drag. Numerical simulations are performed in order to get some detailed information on the behavior of near-wall turbulence in the vicinity of a compliant wall, and in order to be able to determine optimized function bases that can be used to construct low-dimensional representations of the dynamics. By Galerkin projection of the Navier-Stokes equations onto such sets of basis functions that describe the structure of the near-wall flow, we construct low-dimensional models for the dynamics of the flow. The resulting flow equations are coupled to sets of linear oscillators describing damped mass-spring systems that represent the dynamics of the wall in Fourier space.

P.W. Carpenter and T.J. Pedley (eds.),
Flow past Highly Compliant Boundaries and in Collapsible Tubes, 231-252.

1. Introduction

Since the realization by Reynolds (1883) more than a century ago, that flow turbulence brings with it a very significant increase in the drag forces that a flow exerts on bounding surfaces, reducing this effect has been a desirable goal for many practical applications. Thus, significant research efforts (Lumley & Blossey 1998) have been invested in trying to determine methods to control turbulence in order to reduce the turbulent drag.

The approaches that were taken can be broadly classified into two groups, the first one attempting to manipulate the flow using active methods, by using appropriate control algorithms to govern the behavior of certain actuators in the flow, and the second one trying to modify the boundary conditions of the flow in some passive manner. While the first of these, active control of turbulence, has lead to some impressive results (Endo & Kasagi 2001; Choi, Moin & Kim 1994), the drag reduction in that case comes at a significant price in terms of the energy that is needed to implement such control schemes. Thus, to our knowledge, there are no experiments that have demonstrated an active control scheme which reduces the total of energy lost through turbulent drag plus energy spent to drive drag-reducing actuators.

Thus the idea of reducing turbulent drag through some passive means is clearly attractive. There are three possible paths that one could take in this group:

- One can introduce some special substance in the near-wall region of the turbulent flow, which would somehow act to reduce turbulent drag. This leads into the study of polymer drag reduction, which is associated with a number of proven, successful applications. However, the disadvantage in that case is that the method requires a constant supply of whatever substance is used to achieve the desired effect.
- One can try to shape the surface bounding the turbulent flow in such a way that turbulent drag is reduced. An inspiration for this idea comes from the grooved shape of the scales of fast-swimming sharks. Attempts to exploit the associated effect have indeed been successful, and the efficiency of the so-called "riblets" has been demonstrated both experimentally (Bechert *et al.* 1997) and in numerical simulations (Choi, Moin & Kim 1993). However, the amount of drag reduction that can be obtained through such devices is relatively modest, in the range of only a few percent, and the use of such "ribbed" surfaces is not always practical.
- Finally, one can try to modify the interaction of flow turbulence with a boundary using a compliant wall with a tailored dynamic response to pressure disturbances from the turbulent wall layer. That idea in fact became very popular for some time, following observations by Kramer (1961) on exceptional swimming capabilities of dolphins. While this approach looked very promising at first, unfortunately, despite extensive research efforts for a number of years as documented by Bushnell *et al.* (1977), it achieved only very limited, if any, success for the case of turbulent flows†.

Thus, from the above list, compliant surfaces seem to look like the least promising of the potential techniques for passive turbulent drag reduction. However, with respect to this assessment, there are a number of important observations that can be made:

- The exact reasons for the failure of experiments to verify drag reduction via wall compliance remain unclear, and, in particular, to this day very little can be said definitively about whether or not compliant surfaces may in fact be able to significantly reduce turbulent drag or sound production. We believe that the potential benefits that compliant

† There have been successful demonstrations of compliant-wall technology for the manipulation of transitional flows, however, see, e. g. Gad-el-Hak (1986).

surfaces may give do warrant taking another, more detailed look at this problem, in our case from a more theoretical perspective and using modern approaches based on low-dimensional modeling of turbulent flows.

- It should also be noted that in a more recent experimental paper, Choi *et al.* (1997) seem to have been able to demonstrate a drag-reducing effect of their compliant surface coating (which deviated considerably in its mechanical properties from the materials used in earlier experiments). Substantial drag reductions using the same type of compliant walls were also observed in the earlier experimental investigations by Semenov and his collaborators that are referenced by Choi *et al.*
- Finally, and most importantly, despite the relative wealth of experimental data, there are almost no theoretical results on the effect of compliant walls on the structure and properties of *turbulent* wall layers available.

Thus, in the project we are reporting on below, we want to take a fresh look at the problem of the interaction of near-wall turbulence with a compliant boundary. Since such considerations provided some guidance for our direct numerical simulations, in the following section we start by describing one of the few attempts at a theoretical understanding using an approximate linearized analysis as carried out by Semenov and co-workers, which we will complement with a section on the basic statistical analysis of the situation. We will then discuss our own approach in more detail, starting with our direct numerical simulations and some of the preliminary results obtained from them, and following up with an introduction into our efforts on low-dimensional modeling of the turbulence-compliant wall interaction.

2. Harmonic Analysis

Semenov (1991) and Kireiko (1991) analyzed the interaction of wall turbulence with a compliant coating using harmonic approximation. The idea is to describe the near-wall turbulence field by simplified linearized equations of motion for the turbulent fluctuations and then to find the harmonic solution to both the near-wall turbulence and the compliant surface motion.

Kireiko (1991) considered the problem of the turbulent flow of a Newtonian fluid along an unbounded compliant surface in a region of constant shear stress σ_w. In the Cartesian coordinate system $x_1 = x$, $x_2 = y$ and $x_3 = z$, for $y > 0$ the equations of flow motion and continuity, nondimensionalized using the friction velocity $v_* = \sqrt{\sigma_w/\rho}$, have the form

$$\frac{\partial u_i}{\partial x_i} = 0, \tag{1}$$

$$\frac{\partial u_i}{\partial t} + u_j \frac{\partial u_i}{\partial x_j} = -\frac{\partial p}{\partial x_i} + \frac{\partial^2 u_i}{\partial x_j \partial x_j}. \tag{2}$$

The compliant surface is modeled as an elastic film in the plane $y = 0$ stretched over a layer of viscoelastic fluid of thickness h. The displacement of the film η in the direction of the y axis satisfies

$$m_0 \frac{\partial^2 \eta}{\partial t^2} - T_0 \frac{\partial^2 \eta}{\partial x^2} - T_1 \frac{\partial^2 \eta}{\partial z^2} = -p + p_1. \tag{3}$$

m_0, T_0 and T_1 are defined as

$$m_0 = \frac{mv_*}{\mu}, \quad T_0 = \frac{T_x}{v_*\mu}, \quad T_1 = \frac{T_z}{v_*\mu}, \tag{4}$$

where m is the mass per unit area of the film and T_x and T_z are the tensions on the film in the x and z directions, respectively.

The motion of the underlying viscoelastic fluid is described by the equations

$$\frac{\partial v_i}{\partial x_i} = 0, \tag{5}$$

$$\frac{\partial v_i}{\partial t} + v_j \frac{\partial v_i}{\partial x_j} = -\frac{1}{\rho_1}\frac{\partial p}{\partial x_i} + \frac{\partial \sigma_{ij}}{\rho_1 \partial x_j}, \tag{6}$$

$$\sigma_{ij} + \theta_1 \frac{\partial \sigma_{ij}}{\partial t} = 2\mu_1 \left(e_{ij} + \theta_2 \frac{\partial e_{ij}}{\partial t} \right), \tag{7}$$

$$e_{ij} = \frac{1}{2}\left(\frac{\partial v_i}{\partial x_j} + \frac{\partial v_j}{\partial x_i} \right), \quad \theta_1 = \frac{\gamma_1 v_*^2}{\nu}, \quad \theta_2 = \frac{\gamma_2 v_*^2}{\nu}, \tag{8}$$

where γ_1 and γ_2 are the relaxation and retardation times.

The boundary conditions for (5)–(8) are

$$y = -H_0 \left(= \frac{hv_*}{\nu} \right): \quad v_i = 0,$$

$$y = \eta: \quad u_2 = v_2 = \frac{\partial \eta}{\partial t}, \quad u_1 = u_3 = v_1 = v_3 = 0. \tag{9}$$

The equations for the fluctuations are linearized and the solution is found in the harmonic form

$$f'(x, y, z, t) = \mathrm{Re}\left[f(y) exp(i\alpha x + i\beta z - i\alpha c t) \right], \tag{10}$$

where f' is the fluctuation of some quantity and f its complex amplitude. The following system of equations and boundary conditions can be obtained after some substitutions:

$$i\alpha(U - c)u_1 + u_2 \frac{dU}{dy} = -i\alpha p + \frac{d^2 u_1}{dy^2} - (\alpha^2 + \beta^2)u_1, \tag{11}$$

$$i\alpha(U - c)u_2 = -\frac{dp}{dy} i + \frac{d^2 u_2}{dy^2} - (\alpha^2 + \beta^2)u_2, \tag{12}$$

$$i\alpha(U - c)u_3 = -i\beta p + \frac{d^2 u_3}{dy^2} - (\alpha^2 + \beta^2)u_3, \tag{13}$$

$$i\alpha u_1 + \frac{du_2}{dy} + i\beta u_3 = 0, \tag{14}$$

$$\eta(-\alpha^2 c^2 m_0 + \alpha^2 T_0 + \beta^2 T_1 - i\alpha c Q) = -p, \tag{15}$$

$$y = 0: \quad u_2 = -i\alpha c\eta, \quad u_1 + \eta\frac{dU}{dy} = u_3 = 0. \tag{16}$$

Here, $U(y)$ is the mean flow velocity, $Q = Q(\alpha, \beta, c, H_0, \mu_0, \rho_1)$ characterizes the damping effect of the underlying fluid at $y < 0$, and $\mu_0 = \mu_1(1 - i\alpha c\theta_2)/(1 - i\alpha c\theta_1)$ is the effective viscosity. Assuming small film displacement, the boundary conditions at the moving boundary $y = \eta$ have been transferred to the plane $y = 0$ by Taylor expansion.

The solution obtained for the fluctuations at $y > 0$ is used to determine the average flow velocity profile $U(y)$ from the equation

$$\frac{dU}{dy} = 1 + < u_1 u_2 >, \tag{17}$$

which, together with the energy of the fluctuating motion,

$$E(y) = \frac{1}{2}(< u_1^2 > + < u_2^2 > + < u_3^2 >), \tag{18}$$

can be smoothly matched with the asymptotic relations outside the viscosity influence zone,

$$U = \frac{1}{\kappa}\ln(y) + B, \tag{19}$$

$$E = E_0\left(1 + \frac{1}{\kappa_0 y}\right). \tag{20}$$

The calculations show that the interaction is resonant in character and that for certain values of the parameters a considerable reduction in turbulent skin friction is possible.

In the interference analysis in (Semenov 1991), the near-wall turbulence model developed by Sternberg (1962) was used. The field of fluctuation quantities is formed as a result of the superposition of two waves, one going out of the turbulent core and one reflected from a wall. The boundary compliance is regarded as modifying the amplitude and phase of the induced waves, and leading to a change of the turbulence-generating mechanisms. The solution obtained by Semenov indicates that the phase delay of the compliant surface displacement relative to the fluctuation pressure acting on the surface plays an important role for the drag-reducing effect of a compliant coating, which is connected with a decrease of the turbulence production ΔP,

$$\Delta P = \int_0^l \left[< uv > \left(\frac{dU}{dy}\right) - < uv >_c \left(\frac{dU}{dy}\right)_c\right] dy \leqslant 0, \tag{21}$$

where the index c corresponds to the case with a compliant coating, and $< \cdot >$ signifies ensemble averages. Combined with experimental data on the change of turbulent friction and pressure fluctuations on compliant coatings, the interference analysis can give conditions for choosing compliant coatings that are suitable for skin friction reduction.

Kulik (1986) and Semenov (1991) analyzed the effect of the spectral-energy and phase-frequency oscillation characteristics of damping coatings on the friction coefficient and turbulent pressure fluctuations in this way. Figure 1 presents the qualitative phase-frequency characteristics of a coating for typical regimes. It indicates that there is a frequency range in which $\theta_p > \theta_0$ and friction reduction may be expected. The different frequencies in this range influence the friction reduction in different ways. This range should be as wide as possible and it should be within the region of energy-containing frequencies.

Since the magnitude of the Reynolds shear stress $< uv >$ is proportional to v, the effect of the coating on the characteristics of turbulent flow is approximately proportional to the

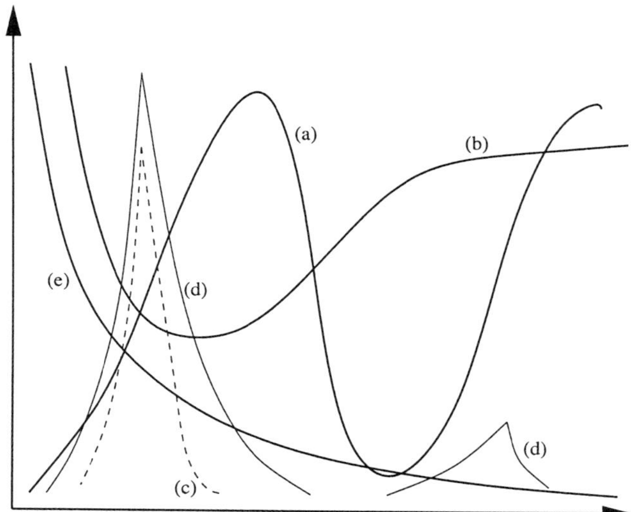

FIGURE 1. Dependence of the characteristics (a) θ_p, (b) θ_0, (c) S_ω, (d) $|v_0|$, (e) P_ω on the frequency f, where θ_p is the phase angle between the fluctuating pressure on the coating and the displacement of its surface; θ_0 is the phase angle at which the coating has a neutral effect on turbulent friction defined as the integral in 21 being equal to zero; S_ω is the spectrum of the energy dissipation rate; v_0 is the normal component of the fluctuating velocity of the flow on the surface; and P_ω is the spectrum of the pressure fluctuations.

magnitude of the deformation rate of the coating surface. The weight factor, taking the contribution of individual frequencies into account, is therefore proportional to $|v_0|$. If the fluctuating pressure is represented in the form of a set of harmonic components with a spectral density distribution of P_ω, then the magnitude of $|v_0|$ is determined as

$$|v_0| = \omega H K_d \sqrt{P_\omega}/E, \tag{22}$$

where ω is the angular frequency of a harmonic component, H and E are the thickness and elasticity modulus of the coating, and K_d is the dynamic coefficient, defined as the ratio of the forced vibration amplitude of the coating surface to the displacement under the action of an equal but static pressure. The peaks of $|v_0|$ in Figure 1 indicate the resonant nature of the coating. At the resonance frequency, the sharpness of the peak is determined by the loss coefficient of the coating and the magnitude of the peak depends on the rigidity of the coating, the resonance frequency and the fluctuating pressure at this frequency. The bandwidth of the resonant interaction region is determined at the half height of the resonance peak. It is one of the important factors influencing the effectiveness of a coating. The resonance interaction band should also be as wide as possible and lie within the region of energy-containing frequencies. In the second region of a possible positive compliance effect at the second resonance frequency, which contains the second maximum of the phase-frequency oscillation of the compliant surface, the deformation rate of the surface and consequently the effect of the wall on the characteristics of the turbulent flow are about one order of magnitude smaller than at the first frequency of resonance interaction. In the high-frequency part of the pressure fluctuations their spectral density is low and the motion of the compliant wall is therefore weak. Moreover, it is possible that the dynamics of the compliant-wall motion in the region of dissipation frequencies differs from the one in the region of energy-containing frequencies.

The specific energy dissipated in the flow adjacent to the coating is

$$E_d = \tau_w, \tag{23}$$

where τ_w is the friction stress on the coating surface. The specific energy absorbed within the coating itself is calculated by

$$E_s = \frac{5}{4\Omega}\int_0^\infty S_\omega d\omega, \tag{24}$$

where Ω is the frequency based on the characteristic flow velocity and the length of the compliant coating. Semenov (1991) has found that within a range of parameters that is promising for effective interaction of a compliant coating with the near-wall turbulence, E_s is several orders of magnitudes lower than E_d. Such small ratios of turbulent energy absorbed by the compliant coating to the energy dissipated in the flow near it demonstrate that there is no direct correlation between turbulent drag reduction (through reduction of turbulent fluctuations) and energy absorption of a compliant wall. In other words, a potential mechanism of a compliant wall reducing turbulence by turbulent energy absorption in the wall seems unlikely to have any relevance. On the contrary, one would expect to find that coatings that successfully reduce turbulent friction will have less internal damping, and absorb a smaller fraction of turbulent energy than coatings that are less effective in that respect.

The magnitude of the dynamic roughness is the r.m.s. displacement σ_w of a coating surface. It takes the form

$$\sigma_w = \frac{\sqrt{2}H}{E}\int_0^\infty K_d^2 P_\omega d\omega. \tag{25}$$

How such roughness affects the turbulent characteristics may be different from the case of static roughness distributions, and so far has not been investigated. The dynamic roughness has to be smaller than the permissible static roughness of a surface in order not to increase skin friction.

The following requirements for a compliant coating that can potentially be effective for turbulent friction reduction can be deduced from the above considerations:

(*a*) Compared to the materials that have been tried so far, more rigid materials with lower density have to be chosen and the coating thickness has to be reduced in order to increase the bandwidth of the resonance region and reduce the dynamic roughness since $\Delta f_0 H/\sqrt{E/\rho_w} = const.$, where Δf_0 is the bandwidth and ρ_w is the density.

(*b*) On the other hand, in order to have a significant interaction between the compliant wall and the turbulent flow, the deformation velocity of the wall needs to be comparable with the characteristic turbulent fluctuation velocities close to the wall. Note that this requirement may conflict with the above.

(*c*) To ensure that the resonance region lies in the region of the energy-carrying frequencies, the inequality $\sqrt{E/\rho_w}/H < 2\Omega/\pi\delta_*$ has to be satisfied (with δ_* as the viscous length scale), which is in agreement with the second requirement.

(*d*) The condition of optimal phase-frequency characteristics requires that $0.02 < \pi\nu f_0/v^{*2} < 0.06$, where ν is the kinematic viscosity, f_0 is the natural frequency of a

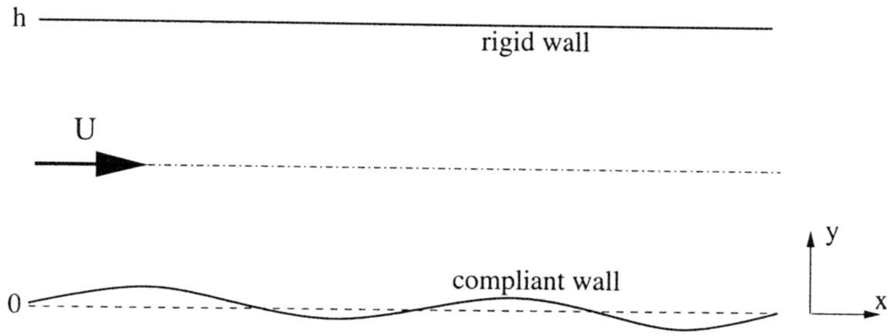

FIGURE 2. Turbulent flow in a half compliant channel

coating and v^* is the friction velocity. This imposes additional constraints on coating properties.

It should be noted that nonlinear effects can be significant even very close to a compliant boundary. The linearization that is used above is permissible only for coatings with only slight deformations, so that the surface remains hydrodynamically smooth.

3. Statistical Analysis

Let us consider the incompressible turbulent flow of a constant-property fluid in a channel composed of one rigid wall and one compliant wall as shown in Figure 2. When the fluid-solid system is in a statistically steady state, the mean displacement of the compliant wall and the mean wall-normal velocity along the compliant wall are equal to zero and so by the mean continuity equation the mean wall-normal velocity $< v >$ of the flow is also equal to zero. If $< v^2 >$ is assumed to be only a function of y, the normal mean momentum equation can be reduced to

$$-\frac{d < v^2 >}{dy} + \frac{1}{\rho}\frac{\partial p}{\partial y} = 0, \tag{26}$$

which together with the boundary condition $< v^2 >_{y=h} = 0$ can be integrated to

$$< v^2 > + \frac{< p >}{\rho} = \frac{p_{rw}}{\rho}, \tag{27}$$

where $p_{rw} =< p(x, h, z, t >$ is the mean pressure on the upper rigid wall. Differentiated with respect to x, (27) gives

$$\frac{\partial < p >}{\partial x} = \frac{dp_{rw}}{dx}. \tag{28}$$

By substitution of (28) the axial mean momentum equation can thus be written as

$$\nu\frac{d^2U}{dy^2} - \frac{d < uv >}{dy} = \frac{1}{\rho}\frac{dp_{rw}}{dx}. \tag{29}$$

where U is the mean streamwise velocity. Therefore, the total shear stress $\tau(y)$ satisfies

$$\frac{d\tau}{dy} = \frac{dp_{rw}}{dx}, \tag{30}$$

where $\tau(y)$ is defined as

$$\tau(y) = \mu \frac{dU}{dy} - \rho < uv > . \tag{31}$$

Since $\tau(y)$ is a function of y and p_{rw} is a function of x and z, it is evident that both sides in (30) should be constants, which gives

$$\tau(y) = \frac{2y - h}{h}(\tau_{rw} - \tau_0) + \tau_0, \tag{32}$$

where τ_{rw} and τ_0 are defined as

$$\tau_{rw} = \tau(h), \quad \tau_0 = \tau(\frac{h}{2}). \tag{33}$$

Nondimensionalized by $u_\tau(= \sqrt{\tau_{rw}/\rho})$ and $\delta_v(= v/u_\tau)$, (32) becomes

$$\frac{dU^+}{dy^+} - < uv >^+ = \left(\frac{\tau_0}{\tau_{rw}} - 1\right)\frac{y^+}{Re_\tau} + \left(1 - 2\frac{\tau_0}{\tau_{rw}}\right), \tag{34}$$

where $Re_\tau = u_\tau h/(2v)$. If the displacement and the velocity of the compliant wall are very small, it can be assumed that

$$\frac{\tau_0}{\tau_{rw}} \ll 1. \tag{35}$$

Thus (34) can be approximated as

$$\frac{dU^+}{dy^+} - < uv >^+ = 1 - \frac{y^+}{Re_\tau} = 1 - \frac{2y}{h}. \tag{36}$$

This relation is the same as in a canonical channel and it can be verified in DNS or experiments.

If u_{cw} is used to denote the characteristic velocity scale of the compliant wall motion, the following dimensional analysis holds

$$\frac{dU}{dy} = f\left(\frac{u_\tau^*}{y}, \frac{u_{cw}}{y}, \frac{y}{\delta_\tau^*}, \frac{y}{h}\right), \tag{37}$$

where $*$ is used to denote quantities associated with the compliant wall side. For large Reynolds number, considering the region

$$y^{*+} \gg 1, \quad \frac{y}{h} \ll 1, \tag{38}$$

if the effect of viscosity is neglected and $u_{rw} \ll u_\tau^*$, the log-law can also be derived as

$$U^{*+} = \frac{1}{\kappa}\ln(y^{*+}) + C. \tag{39}$$

The constant κ still assumes a value of 0.4 as found in experiments or DNS, but the constant C may be different, which reflects the change in skin friction. Thus, in the case of turbulence adjacent to a compliant wall, we can expect to see the same general velocity profile as for a rigid wall, but shifted by a certain amount corresponding to the effect of the wall-compliance on the turbulent drag.

4. Research Approach

The research reported in this paper rests on two main approaches, which are direct numerical simulation and low-dimensional modeling. In the following, we start by describing our simulations of a turbulent flow in a channel delimited by a compliant and a rigid wall, while our low-dimensional modeling effort will be discussed below. The time-varying computational domain we have to deal with can be handled through a time-varying coordinate transformation to eliminate the deformation of the compliant wall as described in the next section. Our choice of the values for the compliant wall parameters is guided by the type of approximate analysis presented above, but it is clear that the arguments given there are not of such a nature as to allow us to directly determine material properties that would guarantee successful turbulent drag reduction. Thus, our DNS must be thought of as an investigative tool which can be used to evaluate the effect of various combinations of materials properties, and which also allows us to obtain some detailed insight into the mechanisms that are at work in the interaction of turbulent flow with a compliant surface. However, DNS by itself will not directly allow us to find material properties that are promising for drag reduction. That role is destined to fall on our low-dimensional models which are described in § 6: Once we are able to construct low-dimensional, i. e. computationally inexpensive models of turbulence in the vicinity of a compliant wall, and once we have established that such models can describe the dynamics of the turbulence-compliant wall system with some appropriate reliability — which is a fact that will have to be established by comparing model predictions with direct numerical simulations — it is straightforward to examine the influence of characteristic parameters of the compliant coating — like stiffness and internal damping — on the dynamical behavior of the wall layer, and it will be feasible to optimize these parameters in order to maximize a desired effect, like reducing the bursting frequency of the near-wall layer which may result in a proportional reduction of the turbulent drag.

5. Direct Numerical Simulation

The problem we are considering is basically a fluid-solid problem. On the fluid side, we have the incompressible NS equation as the model for fluid motion. On the solid side, we use a spring-supported elastic plate to model the compliant wall. The equation governing the wall-normal motion nondimensionalized by the average half channel height h and the laminar centerline velocity u_0 reads

$$C_0\frac{\partial^2\eta}{\partial t^2}+C_1\frac{\partial\eta}{\partial t}+C_2\left(\frac{\partial^4\eta}{\partial x_1^4}+\frac{\partial^4\eta}{\partial x_3^4}+2\frac{\partial^4\eta}{\partial x_1^2x_3^2}\right)$$
$$+C_3\eta-C_x\frac{\partial^2\eta}{\partial x_1^2}-C_z\frac{\partial^2\eta}{\partial x_3^2}=-p_w, \tag{40}$$

where η is the wall-normal displacement, p_w is the pressure disturbance at the wall and C_0, C_1, C_2, C_3, and $C_x(C_z)$ are the nondimensional wall properties defined as

$$C_0=\frac{\rho_m b}{\rho h},\quad C_1=\frac{1}{Re}\frac{dh}{\rho\nu},\quad C_2=\frac{1}{Re^2}\frac{B}{\rho h\nu^2},$$
$$C_3=\frac{1}{Re^2}\frac{K_E h^3}{\rho\nu^2},\quad C_x=\frac{1}{Re^2}\frac{T_x h}{\rho\nu^2},\quad C_z=\frac{1}{Re^2}\frac{T_z h}{\rho\nu^2}. \tag{41}$$

The wall parameters are: the plate density ρ_m and thickness b, the wall damping coefficient d, the flexural rigidity of the plate B, the streamwise (spanwise) tension per unit width $T_x(T_z)$ and the spring stiffness K_E; ρ is the density of the fluid and ν is its kinematic viscosity. The turbulent channel flow we simulate is assumed periodic in the streamwise(x_1 or x) and spanwise(x_3 or z) directions and has a rigid upper and a compliant lower wall that are separated by an average distance of $2h$ in the wall-normal (x_2 or y) direction. Re is its Reynolds number based on h and u_0.

The main difficulty of the simulation is to handle the boundary conditions at the moving compliant wall. We use a time-varying coordinate transformation to eliminate the deformation of the compliant boundary in the computational domain. The continuity equation in the computational coordinate system becomes

$$u^i_{;i} = 0, \tag{42}$$

and the momentum equation in the rotational form becomes

$$\frac{\partial u^i}{\partial t} = -\frac{\partial T^{-1}}{\partial t}\frac{\partial u^i}{\partial x_2} + u^j g^{ik}(u_{j;k} - u_{k;j}) - g^{ij}\frac{\partial p}{\partial x_j} + \frac{1}{Re}(g^{jk}u^i_{;k})_{;j}, \tag{43}$$

where u^i is a contravariant velocity component, the semicolon denotes covariant differentiation, T^{-1} is the inverse coordinate transformation function in the wall-normal direction, and g^{ij} is the contravariant metric tensor of the transformation. The equation for pressure p is obtained by taking a covariant derivative with respect to x_i of the momentum equation and can be written as

$$\left(g^{ij}\frac{\partial p}{\partial x_j}\right)_{;i} = \left[-\frac{\partial T^{-1}}{\partial t}\frac{\partial u^i}{\partial x_2} + u^j g^{ik}(u_{j;k} - u_{k;j})\right]_{;i}. \tag{44}$$

The boundary conditions for velocity must also be transformed into the new coordinate system. The boundary conditions for pressure can then be determined indirectly from the incompressibility condition via an influence matrix technique. A Fourier-Galerkin and Chebyshev-Tau pseudospectral method for spatial discretization and a three-sub-step Runge-Kutta method for time advancement are used for solving the equations.

5.1. NUMERICAL RESULTS

We have carried out a direct numerical simulation of a turbulent flow with Reynolds number equal to 3000 in a "half-compliant" channel (lower wall compliant, upper wall rigid) with the values of the compliant wall properties given in Table 1. The computational domain sizes in the streamwise and spanwise directions are chosen to be $\frac{4\pi}{5}h$ and $32\times96\times65$ grid points are used in these two directions and the wall-normal direction.

Figure 3 shows the history of the skin friction coefficients on the compliant wall and the rigid wall. Based on the samples we have so far, the averaged skin friction coefficient on the compliant wall is 12.45, which is a little higher than the value of 12.27 on the rigid wall, but we need more samples before we can make conclusive statements as to the compliant wall effect on skin friction in this simulation. Instantaneous compliant wall shapes and the wall-normal velocity at the wall at one instant are shown in Figure 4.

TABLE 1. Values of the nondimensionalized compliant wall properties. The wall parameters are: the density ρ_m and thinkness b, the wall damping coefficient d, the flexural rigidity B, the streamwise(spanwise) tension per unit width $T_x(T_z)$ and the spring stiffness K_E. ρ is the density of the fluid and ν is its kinematic viscosity.

C_0	$\left(= \frac{\rho_m b}{\rho h}\right)$	2
C_1	$\left(= \frac{dh}{\rho \nu}\right)$	3000
C_2	$\left(= \frac{B}{\rho h \nu^2}\right)$	360
C_3	$\left(= \frac{K_E h^3}{\rho \nu^2}\right)$	1.8×10^6
C_x	$\left(= \frac{T_x h}{\rho \nu^2}\right)$	1.8×10^6
C_z	$\left(= \frac{T_z h}{\rho \nu^2}\right)$	1.8×10^6

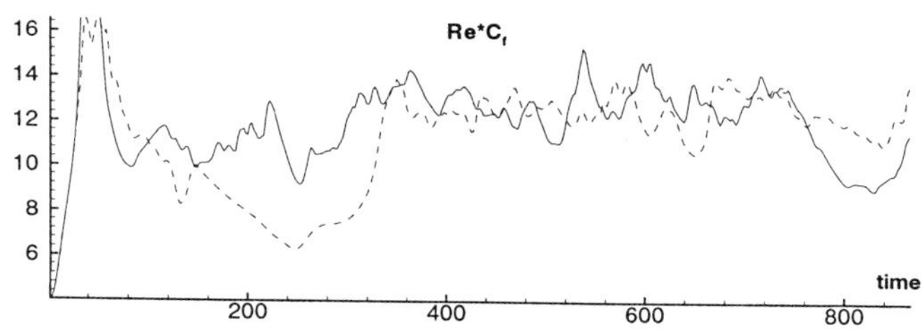

FIGURE 3. Time traces of skin friction coefficients on the compliant wall and the rigid wall. solid line: on the compliant wall; dashed line: on the rigid wall.

6. Low-Dimensional Model: Dynamical Systems Approach

6.1. INTRODUCTION

In this section we will discuss the low-dimensional models we have used to describe the interaction of a compliant surface with a turbulent wall-layer. For these models, we have restricted ourselves to consider a wall which may deform in the wall-normal direction only. This was done to simplify the equations (at least for this preliminary work) but was also motivated by the assumption that this deformation mode will be the most important one.

The model equations governing the fluid motion are derived from a hybrid POD (Proper Orthogonal Decomposition) - Galerkin Projection technique in which the Navier-Stokes equations are projected onto a set of basis functions for the fluid motion. These basis functions are usually derived from a POD, although in our early work we also used additional basis functions, which we will describe in more detail in section 6.2. It is important to note that these basis functions typically capture a large percentage of the energy in the flow. Hence, by modeling the evolution with time of the coefficients of an expansion of the flow variables in terms of these functions we can draw some significant conclusions about the time-evolution of the flow. In the system we are investigating, since the flow

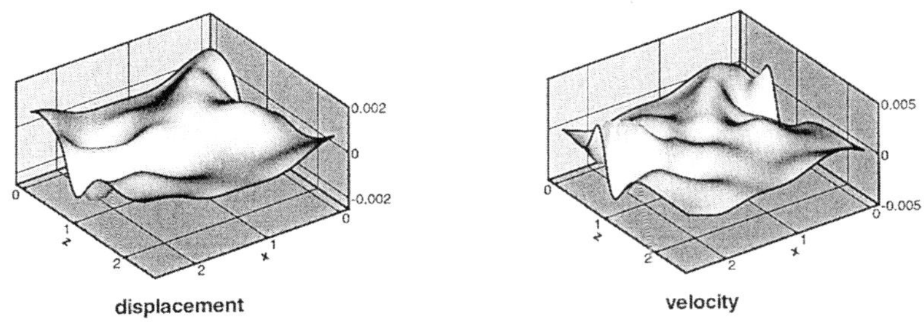

FIGURE 4. Wall-normal displacement η and velocity v_w of the compliant wall nondimensionalized by the average half channel height h and the laminar centerline velocity u_0.

is homogeneous in the spanwise and streamwise directions, the POD functions are simply Fourier modes in those directions. Further, we restrict our attention to functions in which the streamwise Fourier mode is zero since the flow structures are very long (almost unchanging) in this direction.

Thus, in order to arrive at a model that is as simple as possible, and since this assumption has proven successful in the low-dimensional models of near-wall turbulence by Aubry *et al.* (1988), we will only consider the streamwise invariant component of the flow. For this case, the model equations for the flow take the following general form:

$$\begin{aligned}\dot{a}^{(j,k)} = & \sum_n \left(b_{nk}^{j,\mathrm{meanvel}} + (1+2\pi\alpha) b_{nk}^{j,\mathrm{visc}}\right) a^{(n,k)} \\ & + \sum_{k',p,q} c^j_{(k',k-k')_{pq}} a^{(p,k')} a^{(q,k-k')} \\ & + \sum_{n,k',p,q} d^j_{nk'pq} a^{(n,k)} \mathrm{Re}\left(a^{(p,k')} a^{(q,k')*}\right) \\ & + \frac{1}{\rho}\hat{p}(k,0)\varphi_2^{j,k*}(0) - \hat{p}(k,y_m)\varphi_2^{j,k*}(y_m). \end{aligned} \tag{45}$$

The variables, $a^{(j,k)}$, represent the amplitude of the basis functions $\varphi^{(j,k)}$, where j is the index (or "quantum number") of the POD function and k represents the Fourier wavenumber in the spanwise direction.

The term involving the pressure at the wall will be coupled with the wall-pressure term arising from the equations for wall motion. The term involving the pressure at the outer boundary will be ignored at first. Its magnitude is small compared with the other terms and it has been left out of similar POD/Galerkin projection models (Aubry *et al.* 1988). Its absence from the equations will not prevent us from analyzing the fundamental system dynamics. However, in section 6.4 we will re-introduce this term and discuss its effect on the dynamics.

The wall-motion has been modeled as a damped mass-spring system in a space of spanwise Fourier modes,

$$M^k \ddot{\hat{\xi}}^k + D^k \dot{\hat{\xi}}^k + K^k \hat{\xi}^k = -\hat{p}(0,k), \tag{46}$$

where $\hat{\xi}_k$ is the wall displacement in Fourier space.

The wall-displacement is driven by the pressure of the fluid at the wall. The wall-pressure term in this equation is coupled to the wall-pressure term in the flow equations, hence we do not actually need to solve for the pressure at any time. The wall-displacement variables, ξ_k are related to the flow variables, $a_k^{(n)}$, by the boundary conditions which will be discussed in sections 6.2 and 6.5.1.

6.2. MODEL DEVELOPMENT

6.2.1. *Basis Functions*

In order to use POD basis functions in a low-dimensional model, these functions must first be extracted from a data set (either experimental or computational) for the flow of interest. When we began our early work we did not have such a data set available for a turbulent flow over a compliant wall. In order to overcome this problem we used POD basis functions from a rigid-wall flow along with *"ad-hoc"* basis functions from a Stokes flow with oscillating boundary conditions. These additional basis functions will model the flow induced by wall oscillations and allow the no-slip boundary conditions to be satisfied.

We will first approach the Stokes problem with boundary conditions corresponding to the oscillation of the wall in the streamwise direction. Since our low-dimensional model for the rigid-wall boundary layer is built upon a Fourier decomposition of the velocity field, we will use the Stokes equation to generate an eigenfunction for each wavenumber k,

$$\left[\frac{\partial^2}{\partial y^2} - \left(k^2 + \frac{1}{\nu}\frac{\partial}{\partial t}\right)\right]\hat{u}(k, y, t) = 0, \tag{47}$$

and we will then proceed to assign these solutions to our basis functions $\varphi_1^{2,k}$. The velocity field is subject to the boundary conditions

$$\hat{u}(k, 0, z) = cos(\beta t), \tag{48}$$

$$\hat{u}(k, y \to \infty, t) \to 0. \tag{49}$$

In order to satisfy the oscillating boundary conditions, the velocity field $u(y, t)$ can be represented as a superposition of left- and right-running traveling waves,

$$\begin{aligned}\hat{u}(k, y, t) &= \hat{u}^{+}(k, y)\mathrm{e}^{\mathrm{i}\beta t} + \hat{u}^{-}(k, y)\mathrm{e}^{-\mathrm{i}\beta t} \\ &= \frac{1}{2}e^{\mathrm{i}\beta t}\mathrm{e}^{-y\sqrt{k^2+\mathrm{i}\beta}} + \frac{1}{2}\mathrm{e}^{-\mathrm{i}\beta t}\mathrm{e}^{-y\sqrt{k^2-\mathrm{i}\beta}}.\end{aligned} \tag{50}$$

We choose the eigenfunction which represents streamwise compliance from the initial value of $\hat{u}(k, y, t)$, i. e. $\varphi_1^{(2,k)}(y) = \hat{u}(k, y, t = 0)$, $\varphi_2^{(2,k)}(y) = \varphi_3^{(2,k)}(y) = 0$. This choice for the form of the eigenfunction will permit the boundary conditions on the streamwise component of velocity to be satisfied.

We would ideally prefer each additional eigenfunction to have only a single component with a non-zero boundary condition. This will not always be possible when we account for the motion of the wall in all three directions, and the eigenfunctions are orthogonalized. But in the current case of strictly wall-normal motions and using "ultra-linearized" boundary conditions (see below), the additional eigenfunctions may be formulated in this fashion with $\varphi_1^{(2,k)}(y = 0) \neq 0$ and $\varphi_2^{(3,k)}(y = 0) \neq 0$. All other components are zero at

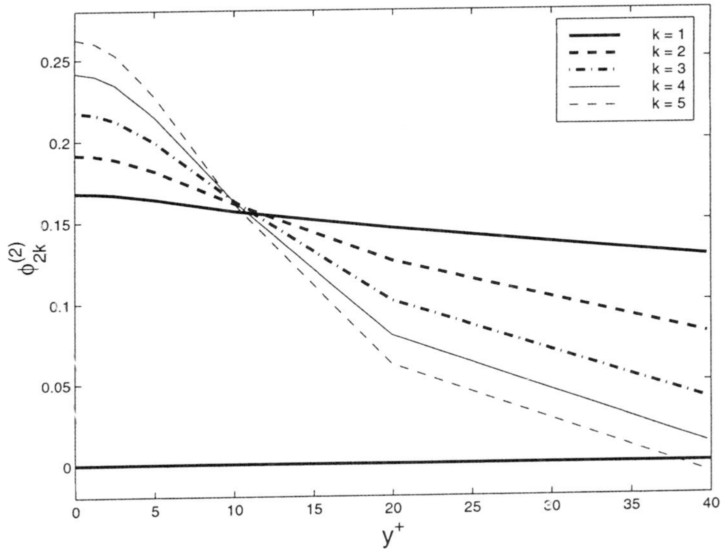

FIGURE 5. The wall-normal component of the compliant-wall eigenfunctions obtained from the Stokes equations, denoted $\varphi^{(2)}_{2_k}(y)$.

the wall. In fact, the wall-normal and spanwise components of $\varphi^{(2)}_k$ and the streamwise component of $\varphi^{(3)}_k$ are strictly zero.

We now determine the additional eigenfunctions for the wall-normal motion of the wall in a similar fashion, using the streamfunction formulation of the Stokes equation,

$$\left(\frac{\partial^2}{\partial y^2} - k^2\right)\left[\frac{\partial^2}{\partial y^2} - \left(k^2 + \frac{1}{\nu}\frac{\partial}{\partial t}\right)\right]\Psi(k, y, t) = 0. \tag{51}$$

As before, we superpose left- and right-running wave solutions to this equation, $\Psi(k, y, t) = e^{i\beta t}\Psi^+(k, y) + e^{-i\beta t}\Psi^-(y,)$, and we apply the boundary conditions at infinity to Ψ^+ and Ψ^- to eliminate the terms which grow exponentially. Next, we enforce the boundary condition on the spanwise velocity $\partial\Psi(y = 0, k)/\partial y = 0$. The two wave solutions are combined to satisfy the time-varying boundary conditions at the wall, $ik\Psi(y = 0, k) = \cos(\beta t)$. This procedure allows the formation of a divergence-free additional eigenfunction which in turn allows the low-dimensional models to satisfy the boundary conditions at the compliant surface.

After the additional eigenfunctions are obtained numerically as solutions of Stokes problems (see Figure 5), the complete set of basis functions is orthogonalized using a Gram-Schmidt procedure. The new basis functions are finally normalized so that $(\varphi^{i,k}, \varphi^{j,k}) = \delta_{i,j}$.

6.2.2. *Boundary Conditions*

Our coordinate system is such that the undisturbed compliant surface occupies the plane $y = 0$. Let $\xi(z) = [\xi_1, \xi_2, \xi_3](z)$ be the displacement vector of the compliant surface (remember that we assume streamwise invariance). The correct set of boundary conditions is given by

$$\tilde{u}_i(\mathbf{x} + \xi, t) = \dot{\xi}_i, \tag{52}$$

where $\tilde{\mathbf{u}}$ is the instantaneous fluid velocity, and $\dot{\xi}$ denotes the time derivative of the wall displacement vector. For our model, we will assume that the surface displacements are small, more specifically, we assume that $|\xi^+| = |\xi| u_\tau/\nu \leqslant 5$, so that the velocity profiles are linear†. Under this assumption, the boundary condition can be linearized, giving

$$\tilde{u}_i(0, z, t) + \tilde{u}_{i,j}(0, z, t)\xi_j(z, t) = \dot{\xi}_i(z, t). \tag{53}$$

If we express the instantaneous fluid velocity in terms of the fluctuating components $u_i(\mathbf{x}, t)$ and the mean $\mathbf{U} = [U_1(y), 0, 0]$,

$$\tilde{u}_i = U_i + u_i, \tag{54}$$

we can write (53) as

$$u_i(0, z, t) + U_{i,j}(0)\xi_j(z, t) + u_{i,j}(0, z, t)\xi_j(z, t) = \dot{\xi}_i(z, t), \tag{55}$$

where we have used $U_i(0) = 0$. Let us now assume that $u_i(0, z, t)$ and $\xi_i(z, t)$ are both small quantities in a sense yet to be defined. Let us consider the velocity gradient tensor at $y = 0$, $u_{i,j}(0, z, t)$. We may write

$$\begin{pmatrix} 0 & u_{1,2} & \underbrace{u_{1,3}} \\ 0 & \underbrace{u_{2,2}} & \underbrace{u_{2,3}} \\ 0 & u_{3,2} & \underbrace{u_{3,3}} \end{pmatrix}, \tag{56}$$

where we have indicated by an underbrace terms that will vanish for vanishing wall deformation $\xi \to 0$. Note that for the rigid-wall case $\xi \equiv 0$, we have $u_{2,2} = 0$ from continuity, and $u_{i,3} = 0$ because the velocity vanishes identically at the wall. We now proceed to eliminate all those terms in (53) that are quadratic in quantities that vanish when $\xi = 0$. The precise magnitude of the errors that this step incurs will be discussed below. The boundary conditions with the quadratic terms eliminated will be referred to as the "ultra-linearized" boundary conditions,

$$\begin{array}{ccccccc} u_1 & + & U_{1,2}\xi_2 & + & u_{1,2}\xi_2 & = & \dot{\xi}_1, \\ u_2 & & & & & = & \dot{\xi}_2, \\ u_3 & & & + & u_{3,2}\xi_2 & = & \dot{\xi}_3. \end{array} \tag{57}$$

Note that the above set of boundary conditions still contains two terms, $u_{1,2}\xi_2$ and $u_{3,2}\xi_2$, that are nonlinear in the dependent variables of our problem. Because of the complications associated with this nonlinearity, we want to drop these terms, too. We estimate that doing so results in at most a 30% error. It is clear that a potential error of that magnitude is certainly not negligible, although it is consistent with other simplifications (e. g. streamwise invariance) that enter our model. A detailed comparison of results from the low-dimensional model to matching direct numerical simulations is mandatory in order to assess the effect that the simplifications in our models have on the dynamics of the system.

It is important to note that direct numerical simulation and low-dimensional modeling are thus closely intertwined and mutually complement each other in our approach: A direct numerical simulation may show a need to refine the original POD-Galerkin models, and provide improved eigenfunction bases (see below), and the modified models may then give us a better idea of how to choose optimal parameters for a drag-reducing wall coating.

† Note that this magnitude of ξ coincides with the findings in (Choi *et al.* 1997).

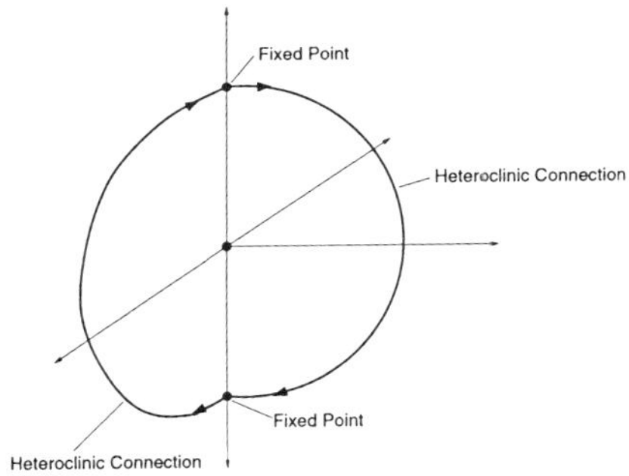

FIGURE 6. Schematic diagram of a heteroclinic cycle.

6.2.3. *Model Equations*

The boundary conditions (57) may be used to eliminate $\hat{\xi}_k$ and its derivatives from the model equations (45). Then the equations are coupled using the pressure terms in (45) and (46). The boundary conditions may also be combined to give a third equation which is essentially a relationship between $\dot{\hat{\xi}}$ and $\hat{\xi}$. The resulting model equations are

$$
\begin{aligned}
\dot{a}_k^{(1)} &= \sum_p b_{kp}^{(1)} a_k^{(p)} + \sum_{k',p,q} c_{(k',k-k')_{pq}}^{(1)} a_{k'}^{(p)} a_{k-k'}^{(q)} \\
&\quad + \sum_{r,k',p,q} d_{rk'pq}^{(1)} a_k^{(r)} Re\left(a_{k'}^{(p)} a_{k'}^{(q)*}\right) \\
\dot{a}_k^{(2)} &= \frac{\tilde{\rho}}{\phi_{2_k}^{(2)}\Big|_w} \left[\sum_p b_{kp}^{(2)} a_k^{(p)} + \sum_{k',p,q} c_{(k',k-k')_{pq}}^{(2)} a_{k'}^{(p)} a_{k-k'}^{(q)}\right] \\
&\quad - \tilde{d}\, a_k^{(2)} + \tilde{k} \left.\frac{\phi_{1_k}^{(3)}}{\phi_{2_k}^{(2)} \frac{\partial U}{\partial y}}\right|_w a_k^{(3)} \\
\dot{a}_k^{(3)} &= - \left.\frac{\phi_{2_k}^{(2)} \frac{\partial U}{\partial y}}{\phi_{1_k}^{(3)}}\right|_w a_k^{(2)},
\end{aligned}
\tag{58}
$$

where $\tilde{d} = d/\tilde{m}$, $\tilde{k} = k/\tilde{m}$, $\tilde{\rho} = \rho/(\tilde{m}\phi_{2_k}^{(2)^*}|_w)$, $\tilde{m} = m + \rho/(\phi_{2_k}^{(2)}\phi_{2_k}^{(2)^*})|_w$. Subscript w implies evaluation at the equilibrium position of the wall ($y = 0$).

Three equations are needed for each Fourier number, k, in order to satisfy the boundary conditions. We will consider a system of two Fourier wavenumbers so our system will have a dimension of 6 complex variables (12 real variables).

6.3. RESULTS

Similar POD/Galerkin Projection models have been studied by Aubry *et al.* (1988) and Armbruster, Guckenheimer & Holmes (1988) for rigid wall flows. They found that heteroclinic cycles exist in the phase space of variables. A schematic diagram of such a cycle is shown in Figure 6.

The fixed points of the cycle correspond to a situation when the amplitudes of the basis functions are not changing with time. These fixed points are joined by heteroclinic connections which form a cycle. The flow "bursts" as it moves away from a fixed point

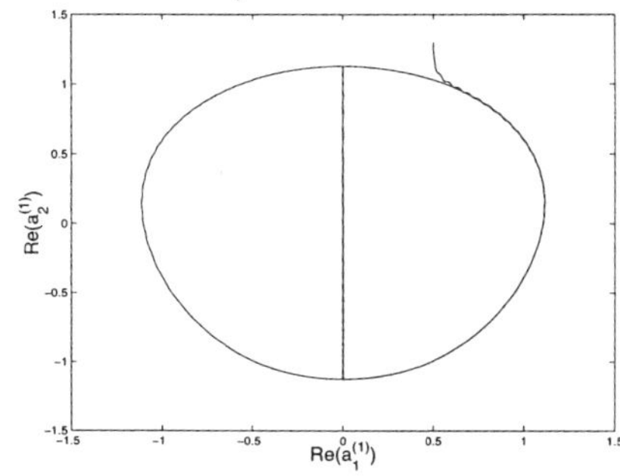

FIGURE 7. Phase portrait of the dynamics of the compliant wall model with $\tilde{k} = 5000$, $\tilde{d} = 0.1$ and $\tilde{\rho} = 0.1$.

along a heteroclinic trajectory and then reforms as it approaches another fixed point. This behavior is repeated, and since the heteroclinic cycle is attracting, the period of motion tends to infinity as time tends to infinity.

In our compliant wall models we find similar heteroclinic cycles in the phase space. In order to compare the compliant wall model with the rigid wall model we would like to compare the period of motion or the "bursting frequency" for the two. The hope is that the compliant wall model will exhibit a longer bursting period which would indicate a drag reduction. A phase portrait for the compliant wall model with parameters $\tilde{k} = 5000$, $\tilde{d} = 0.1$, $\rho = 0.1$ is given in Figure 7†.

We systematically varied the compliant wall parameters $\tilde{d}$ and $\tilde{k}$ to find optimal values for drag reduction. Since the important dynamics take place in the linear region of the fixed points, we also completed a linear analysis of the system in order to determine optimal wall parameters. The overall conclusion was that, for this model, and taking into account only the linear analysis (near the fixed points) of the corresponding dynamical system, no such optimal parameters exist. In many cases the compliant wall model performed worse than the rigid wall case. The best case that was found was when the compliant wall and rigid wall dynamics were the same.

A time series plot comparing the dynamics of the rigid wall and compliant wall models (using the same parameters as before) is given in Figure 8.

6.4. STOCHASTIC MODELING

At this point we would like to re-introduce the pressure term at the outer boundary which has previously been neglected from our equations. This term has been modeled as white noise in (Aubry *et al.* 1988) and (Stone & Holmes 1989). Although the pressure signal at the outer boundary differs from white noise, it may be approximated in this way without any ill effects.

The stochastic differential equations are given in the following form,

$$da^{(j,k)} = f(a^{(j,k)})dt + \epsilon dW,$$

where dW is a random process, f is the function of the eigenfunction coefficients derived from the Galerkin projection and ϵ is its amplitude. After introducing this stochastic term to the equations the dynamics take on a different behavior. Although the phase portraits

† We note that in this representation, the phase portraits of the rigid- and the compliant-wall model would be indistinguishable to within plotting accuracy.

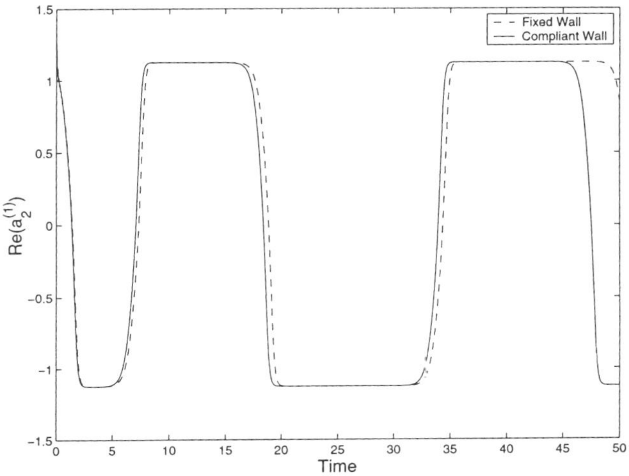

FIGURE 8. Time series of the dynamics of both rigid and compliant wall models with $\tilde{k} = 5000$, $\tilde{d} = 0.1$ and $\tilde{\rho} = 0.1$.

still show evidence of heteroclinic cycles, the trajectories no longer continue to approach this cycle as time tends to infinity (resulting in a period of motion that tends to infinity as well). Rather, the trajectories are perturbed by the stochastic terms resulting in a period of motion that is random, with a mean value. In effect, the pressure signal at the outer boundary triggers bursts in the flow and modifies the time scale of the process.

By integrating sample paths we may find an average period. Results for both the rigid and compliant wall (with $\tilde{k} = 5000$, $\tilde{d} = 0.1$, $\tilde{\rho} = 0.1$) are shown in Figure 9.

As can be seen, for a wide range of noise amplitudes, ϵ, the compliant wall model appears to do no better than the rigid wall case, as we encountered in the previous section.

6.5. NEW WORK

Our early work described above was limited by the fact that no eigenfunctions for turbulent flow in a compliant channel were available, forcing us to make do with the somewhat artificial Stokes eigenmodes described above. However, data from the direct numerical simulation of a turbulent channel flow with compliant boundaries as described above has now become available. Hence, it has been possible to carry out a POD on this data and extract eigenfunctions for the flow with a compliant wall. A new model has been developed using these basis functions. The outline of the model development is given below. As this work is still in progress results are unavailable at this time.

6.5.1. *New Model Development*

In this new model we will continue to model the wall motion as a damped mass-spring system as in (46). The equations for the flow will also be the same. However, whereas before several terms dropped out of the equations, due to some components of the eigenfunctions being equal to zero, now we must consider the general case where all components of the eigenfunctions are non-zero.

In order to satisfy the boundary conditions in (57) we must have three basis functions. Therefore we will have three dynamical equations for the amplitudes of these basis functions. In addition, we will have an equation for the wall motion in terms of ξ and

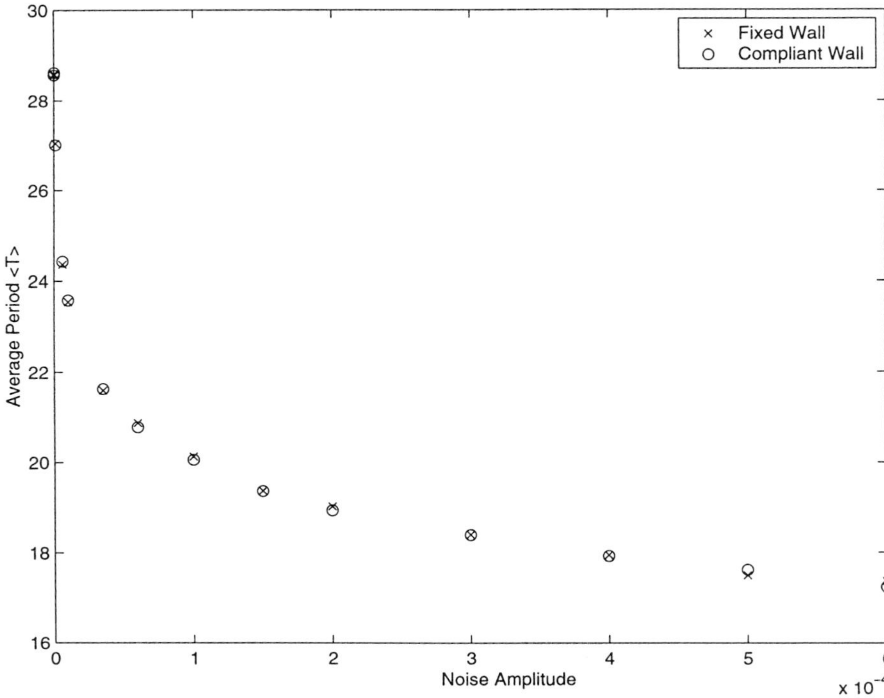

FIGURE 9. Average bursting period as a function of noise amplitude for both rigid and compliant wall models with $\tilde{k} = 5000$, $\tilde{d} = 0.1$ and $\tilde{\rho} = 0.1$.

its derivatives (as before). We will use the boundary conditions to remove these ξ terms from the equations, and couple the equation for wall motion to the equations for fluid motion through the wall-pressure term. The boundary condition which does not include a ξ term will be satisfied automatically since this condition requires the spanwise velocity to be zero, which is always the case in our simulations. This will give us a system of 3 dynamical equations for each Fourier wavenumber, k.

The resulting equations now take the following form:

$$\dot{a}^{(j,k)} = L^{(j,k)} + Q^{(j,k)} + C^{(j,k)} + \frac{1}{\rho}\varphi_2^{(j,k)*}(0)\left[M^k \sum_n \dot{a}^{(n,k)}\varphi_2^{(n,k)}(0) + D^k \sum_n a^{(n,k)}\varphi_2^{(n,k)}(0) + \frac{K^k}{-\frac{\partial U}{\partial y}\big|_{y=0}} \sum_n \dot{a}^{(n,k)}\varphi_1^{(n,k)}(0)\right], \tag{59}$$

where the terms L, Q and C represent linear, quadratic and cubic terms which are determined by the Galerkin projection process. This system of equations must be solved simultaneously (since each $\dot{a}^{(j,k)}$ appears on the left hand side and in each of the summations on the right hand side) to determine the dynamical equations which will be analyzed.

7. Conclusions

The material described above clearly represents work in progress, and has gone through its ups and downs as they are associated with any original work. We had found early indications of reduced drag both in low-dimensional models and direct simulations, only to later dismiss them as being caused by modeling flaws, or by numerical inaccuracies. However, through this process we have constantly improved both the numerical methods for our direct simulation and the basis for our low-dimensional models, and we are confident that both our approaches, taken together, will give us a much clearer picture of the interaction of turbulence with a compliant wall.

Will we be able to ultimately find compliant walls with properties such that turbulent drag can be reduced significantly? At this point, it may be wise for us to be cautious, and simply answer with a "remains to be seen". Ultimately, the question we are asking is the following: If we look at the whole range of parameters for the elastic properties of a compliant wall, are there any parameters that would result in average wall shear stresses that are lower than the ones we get for the limiting case of an infinitely rigid wall? Intuitively, to us at least, it seems unlikely that the answer to that question should be negative. We might also note that if indeed the worst-case scenario did turn out to be true, and if indeed for fundamental reasons that we do not yet understand a rigid wall turns out to be the one that produces the lowest possible amount of turbulent drag, this would constitute a very surprising and important new result on the physics of turbulence.

Until we do have definite answers to these questions, we do believe that our research in this field is worthwhile, and we remain confident that we will be able to give such answers by following the path we have taken. Thus, although our results so far have not been positive for drag reduction, it is our hope that with improved simulations and low-dimensional models which include more of the relevant physics, we will ultimately arrive at a firm understanding of the interaction of compliant surfaces with near-wall turbulence.

Acknowledgements

This work has been supported by the AFOSR under contracts F49620-96-1-0329, F49620-98-1-0397, F49620-99-1-0012, and F49620-00-1-0029.

References

REYNOLDS, O. (1883) On the Experimental Investigation of the Curcumstances which Determine Whether the Motion of Water Shall Be Direct or Sinuous. *Phil. Trans. Royal. Soc.* **174**, pp. 953–982.

LUMLEY, J. L.; BLOSSEY, P. N. (1998) Control of Turbulence. *Annu. Rev. Fluid Mech.* **30**, pp. 311–327.

ENDO, T.; KASAGI, N. (2001) Active control of wall turbulence with wall deformation. *Fluids and Thermal Engineering* **44** (2), pp. 195-203.

CHOI, H.; MOIN, P.; KIM, J. (1994) Active turbulence control for drag reduction in wall-bounded flows. *J. Fluid Mech.* **262**, pp. 75–110.

BECHERT, D. W.; BRUSE, M.; HAGE, W.; VAN DER HOEVEN, J. G. T.; HOPPE, G. (1997) Experiments on drag-reducing surfaces and their optimization with an adjustable geometry. *J. Fluid Mech.* **338**, pp. 59–87.

CHOI, H.; MOIN, P.; KIM, J. (1993) Direct numerical simulation of turbulent flow over riblets. *J. Fluid Mech.* **255**, pp. 503–539.

KRAMER, M. O. (1961) The Dolphin's Secret. *J. Am. Soc. Nav. Engr.* **73**, pp. 103–107.

BUSHNELL, D. M.; HEFNER, J. N.; ASH, R. L. (1977) Effect of Compliant Wall Motion on Turbulent Boundary Layers. *Phys. Fluids A* **20**, S31.

GAD-EL-HAK, M. 1986 Boundary Layer Interactions with Compliant Coatings: An Overview. *Appl. Mech. Rev.* **3**, pp. 511–524.

CHOI, K. S.; YANG, X.; CLAYTON, B. R.; GLOVER, E. J.; ATLAR, M.; SEMENOV, B. N.; KULIK, V. N. (1997) Turbulent Drag Reduction Using Compliant Surfaces. *Proc. R. Soc. Lond. A* **453**, pp. 2229–2240.

SEMENOV, B. N. 1991 On conditions of modelling and choice of viscoelastic coatings for drag reduction. In: *Recent Developments in Turbulence Management*, (ed. K.-S. Choi), pp. 241-262.

KIREIKO, G. V. (1991) Interaction of wall turbulence with a compliant surface. *Fluid Dynamics* **25** (4), 550.

STERNBERG, J. (1962) A theory for the viscous sublayer of a turbulent flow. *J. Fluid Mech.* **13**, pp. 241–271.

KULIK, V. M. (1986) The analysis of interaction of one-layer monolithic damping coatings with turbulent flow. *J. Eng. Phys.* **51**, N6, pp. 959–965.

HOLMES, P.; LUMLEY, J. L.; BERKOOZ, G. *Turbulence, Coherent Structures, Dynamical Systems and Symmetry.* Cambridge University Press, 1996.

ARMBRUSTER, D.; GUCKENHEIMER, J.; HOLMES, P. Heteroclinic cycles and modulated traveling waves in systems with O(2) symmetry. (1988) *Physica D* **29**, pp. 257–282.

AUBRY, N.; HOLMES, P.; LUMLEY, J. L.; STONE, E. (1988) The dynamics of coherent structures in the wall region of a turbulent boundary layer. *J. Fluid Mech.* **192**, pp. 115–173.

STONE, E,; HOLMES, P. (1989). Noise-induced intermittency in a model of a turbulent boundary layer. *Physica D* **37**, pp. 20–32.

Chapter 11

FLOW-INDUCED WAVES ON COMPLIANT SURFACES SUBJECT TO A TURBULENT BOUNDARY LAYER

K. S. YEO
Department of Mechanical Engineering
National University of Singapore
Kent Ridge Crescent, 119260
SINGAPORE

Abstract

In this paper, we are concerned with waves that form on a compliant surface when it is subjected to a turbulent flow. Two types of waves have been observed so far in the experiments: a fast travelling wave on nearly elastic surfaces, and a nearly stationary or slowly moving wave on surfaces with significant damping. In their incipient state, the fast waves have small amplitude and propagate at speeds ranging from 0.3-0.5 times the freestream speed U_∞, whereas the slow waves (also termed static divergence waves) have large amplitude and move forward at speeds of less than $0.05U_\infty$. Details of these waves were recorded in experimental studies by Gad-el-Hak *et al.* (1984), Gad-el-Hak (1986) and Hansen *et al.* (1980). This paper discusses the progress that has been made in the theoretical modelling of these waves. Emphasis is given to the recent works and the new insight that they offer about unstable interaction between turbulent flow and compliant surface. Quasipotential flow approximations have been fairly successful in predicting some of the qualitative and quantitative aspects of the observed waves. However, the quasipotential models are reliant on the availability of accurate parameter inputs from other experiments or additional modelling for them to work well. In this regard, they are not self-sufficient for general application. The effects of viscosity and turbulence are needed to produce a complete and self-consistent model for the accurate prediction of onset wave conditions. While linear stability models are generally adequate for predicting onset flow velocities for the waves, nonlinear wave effects are necessary to explain the detailed features of the observed waves.

P.W. Carpenter and T.J. Pedley (eds.),
Flow past Highly Compliant Boundaries and in Collapsible Tubes, 253-274.

1. Introduction

The motion of a fluid medium is said to be turbulent when its velocity and pressure fields fluctuate in a disorderly and chaotic manner about some mean of bulk state of the flow. The fluctuations are not completely random, however, but are composed, at a statistical level, of measurable sets or spectra of spatial and temporal scales. Turbulence is the manifestation of a process by which an unstable inertia-dominated flow seeks equilibrium with its dissipative mechanism. The turbulence mechanism involves the flow of energy from the mean flow into the larger scales motion of turbulence and thence to the energy dissipation scales, via an inertia-dominated process of progressive eddy breakdown. The dissipative mechanism, viscosity in this case, plays a dual role here: in the equilibration of the mean turbulent flow and in being the insidious cause of flow instability.

The rich and complex mechanism of a turbulent flow suggests that there may be ample oppportunities for a wide range of interactions with a compliant boundary. Indeed, there has been a keen interest in turbulent flow/compliant surface interaction since the mid 1960's, following Kramer's pioneering experiments (Kramer, 1960) to demonstrate the ability of compliant surfaces to delay laminar-turbulent transition. A review of the early work with focus on turbulent drag reduction by compliant surfaces may be found in Bushnell *et al.* (1977). The subject of turbulent flow-compliant wall interaction is still attracting a good deal of interest today because of its practical significance. Semenov and co-workers have been researching on turbulent drag reduction by compliant surface since the 1960s and the work is still ongoing (Semenov *et al.*, 1998). Working jointly with Russian researchers, Choi *et al.* (1997) recently published experimental results which showed a turbulent drag reduction of up to 7% for a fairly stiff compliant coating with $C_t/U_\infty \approx 3.5-10$ (C_t denotes the elastic shear-wave speed of the compliant material). For coatings of softer materials, $C_t/U_\infty \approx 1.1-3.1$, Lee *et al.* (1993) observed small-scale deformations on the compliant surfaces that appeared to mirror the turbulent fluctuations of the boundary layer. This interaction was found to produce a slight broadening of the buffer and viscous wall layers, which suggests a potential reduction of skin friction drag. As the coating materials become even softer, $C_t/U_\infty < 1$, the small scale surface fluctuations eventually give way to or become dominated by a series of distinct waves on the surface of the compliant coatings, Gad-el-Hak (1986). The streamwise intervals of these waves scale approximately with the thickness of the compliant layer. Clearly the mechanism of interaction between the turbulent boundary layer and the compliant wall is sensitive to the stiffness of the latter.

In this paper, we shall be focusing primarily on flow-induced waves on soft compliant surfaces subject to a turbulent boundary layer. Instead of drag reduction, the flow-induced waves may increase the skin friction drag on the surface, and are thus of practical concern. We shall begin, in the following section, by summarizing the experimental observations on the occurrence and physical characteristics of these waves. With that as the background, we then proceed to review the progress that has been made in the theoretical modelling of these waves. Emphasis is placed on the recent works and the new insight that they offer into the wave mechanics of turbulent flow/compliant wall interaction.

2. Experimental Observations of Surface Waves

The experiments of Hansen, Hunston, Ni & Reischman (1980), Gad-el-Hak, Blackwelder & Riley (1984) and Gad-el-Hak (1986) show that surface waves on compliant layers under a turbulent boundary layer may be divided into two principal types: the *slow* waves and the *fast* waves. The slow waves, whose existence was first noted by Boggs & Hahn (1962), are a series of large-amplitude spanwise-aligned waves that typically propagate with wave speeds $c < 0.05U_\infty$. They are found on compliant layers possessing a significant level of material damping. The slow waves have also been termed *static divergence* (SD) waves because of their small wave speed and large amplitude. Static divergence waves have also been observed on bottleneck porpoises and human subjects under condition of burst swimming (Aleyev, 1977). The fast waves, on the other hand, are series of small-amplitude wave trains that travel with wave speeds in the range of $0.3U_\infty < c < 0.5U_\infty$. So far, they have only been observed and documented by Gad-el-Hak (1986) on nearly elastic layers. The slow and fast waves set in when the flow velocity U_∞ exceeds certain fairly well defined threshold values that are dependent on the compliance of the layer; the latter being a function of the material shear-wave speed C_t and the thickness h of the layer. The existence of well-defined threshold or onset velocities suggests that these waves may have originated from some mechanisms of unstable interaction between the flow and the compliant surface.

2.1. SLOW WAVES – STATIC DIVERGENCE

Hansen *et al.* (1980), Gad-el-Hak *et al.* (1984) studied the occurrence of the slow or static divergence waves on embedded layers of highly damped viscoelastic materials subject to a zero-pressure-gradient turbulent boundary layer. The compliant layers were formulated from mixtures of pastisol and PVC to produce surfaces of different compliance. Gad-el-Hak *et al.* cited material relaxation time of about 1 second for their layers, and a similar quality of damping may be applicable to the materials in Hansen *et al.*'s experiments. Hansen *et al.* carried out their experiments in a recirculating free-surface water tunnel, while Gad-el-Hak *et al.* did theirs in a towing tank. The onset flow velocities of the flow or SD waves were measured in these experiments. Gad-el-Hak *et al.* also employed a high-resolution optical system to record the displacement history of the compliant surface, which yielded detailed profile information on the surface waves.

Hansen *et al.* distinguished three types of SD waves in their experiments (see Figure 1). As the flow velocity U_∞ was increased, they observed that spanwise surface ridges first began to appear near the leading edge of the compliant layer (designated Type I SD waves). The waves became more evident and extended further downstream from the leading edge when U_∞ was increased above their initial onset velocity. The Type I waves were followed by a Type II and Type III SD waves at a slightly higher U_∞; the latter two waves occurred almost simultaneously and had onset U_∞ that were on the average 17.5% above those of the Type I waves. The Type II SD waves were slightly crescent-shaped ridges that occurred at the spanwise edges of the compliant layers, while the Type III SD waves were found in the central region of the compliant surface, seemingly independent of edge conditions. The Type I waves appeared to be triggered by small surface irregularities at the leading edge of the compliant layer. By ensuring a smooth seam between the layer and the mounting plate in one of their experiments, Hansen *et al.* were able to suppress the waves. With carefully fabricated joints between the compliant layers and the plate, Gad-el-Hak *et al.* had observed only the Type III waves.

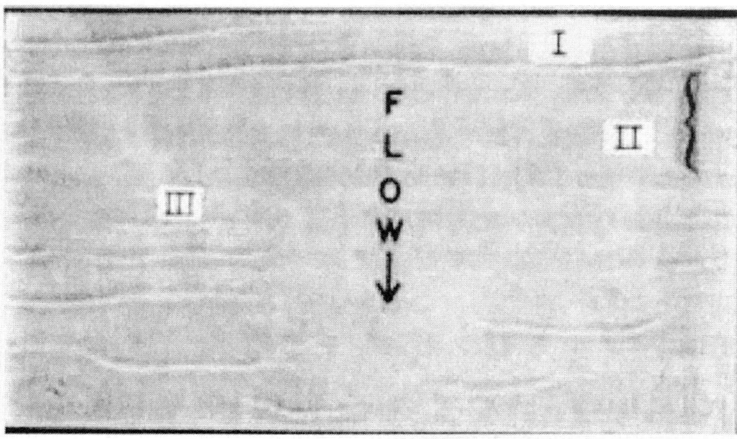

Figure 1. Static divergence waves on a viscoelastic layer. (from Hansen *et al.*, 1980.)

The static divergence waves were almost stationary at their inception. Their wave speeds $c < 0.01U_\infty$ at the onset of the waves according to Table 1 of Gad-el-Hak *et al.* and $0.03U_\infty$ for U_∞ exceeding between 8 and 20% of the onset flow velocity according to Table 3 of Hansen *et al.* Gad-el-Hak *et al.* also noted that prior to the occurrence of the SD waves, the mean velocity field and the r.m.s. fluctuations of the turbulent boundary layer on the compliant layers did not differ perceptibly from those on a rigid surface.

The SD waves have fairly well defined wavelength, and possess a highly asymmetrical non-sinusoidal waveform with relatively sharp crests and shallow troughs in between, see Figure 2. The observations of Gad-el-Hak (1986) on the birth or initial development of the SD waves are particularly significant. They noted that the appearance of the SD waves was always preceded by a highly transient small-amplitude wave train. The evolution of these incipient wave trains into the large-amplitude SD waves occurred so rapidly that they could not be captured on film. The SD waves equilibrated at amplitudes that were of the order of the thickness of the compliant layer (see Figure 2), indicating that nonlinear wave effects are probably important in establishing the final form of the observed SD waves. The above account shows quite clearly that the SD waves on highly damped viscoelastic compliant layers are most likely the outcome of a strong instability of the flow-compliant wall system.

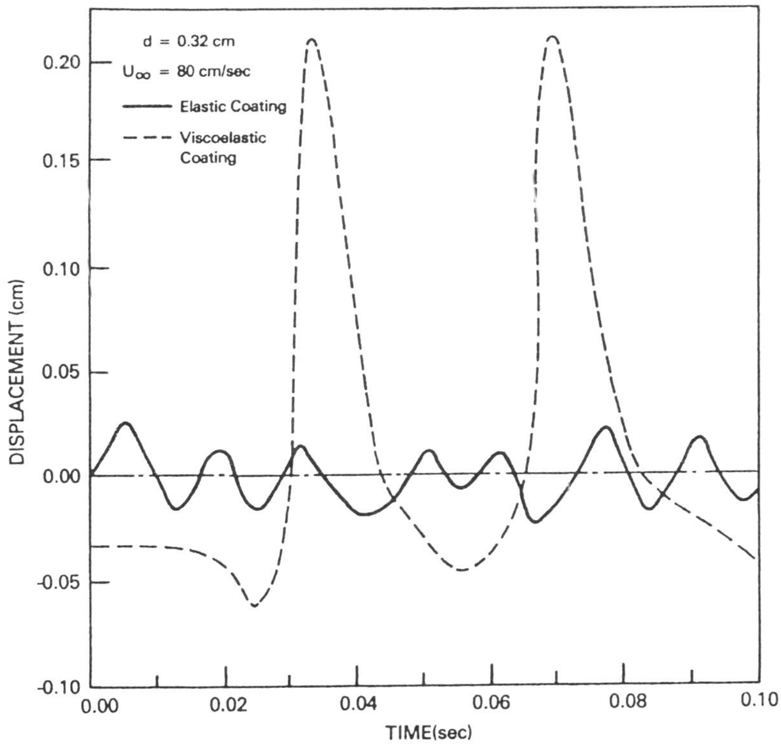

Figure 2. Typical surface displacement histories: elastic coating: ——— ; viscoelastic coating: – – – –. Coating thickness 0.32 cm. and U_∞ = 80 cm/sec. (from Gad-el-Hak, 1986 with permission of author.)

Indeed both Carpenter & Garrad (1986) and Riley, Gad-el-Hak & Metcalfe (1988) have pointed out that SD is an absolute form of instability. An absolute instability typically arises from the unstable coalescence of an upstream and a downstream wave mode. The coalescence gives rise to a non-propagating wave mode that grows exponentially in time. Nonlinear effects in the flow and the compliant wall would eventually arrest the growth of the SD modes. Gad-el-Hak *et al.* noted that the large amplitude SD wave ridges acted as a series of roughness elements on the surface of the compliant layer, and that its presence resulted in enhanced levels of turbulence and a significant thickening of the boundary layer. The occurrence of the slow or SD waves will clearly produce an increase in skin-friction drag and flow noise. This has important implications for the use of compliant layers in underwater applications. Gad-el-Hak *et al.* also carried out a series of tests under laminar flow conditions, but failed to find any SD waves even when the flow velocities were raised to twice the onset values of the SD waves under a turbulent boundary layer.

2.2. FAST WAVES

Much less is known experimentally about the fast waves, which were found on nearly elastic compliant layers produced from a gelatinous material. Gad-el-Hak (1986) remains the principal source of experimental data concerning these waves. Visualization and accurate surface displacement measurements showed that the fast waves took the form of a fairly regular wave train of small amplitude when the flow velocity exceeded certain threshold values that depended on the stiffness of the material and the thickness of the layer. These waves propagate at speeds of up to $0.5U_\infty$ and have profiles that are fairly symmetrical, if not quite sinusoidal (see Figure 2).

3. Theoretical Models

Scientists and engineers frequently create theoretical models to help them understand and explain complex physical phenomena. A successful model also allows the users to make reliable quantitative predictions. Benjamin (1960, 1963) and Landahl (1962) were among the earliest researchers to carry out theoretical investigations into the problem of flow past compliant surfaces. Applying normal-mode temporal stability analysis to flow over simple membrane or plate surfaces, Benjamin and Landahl were able to obtain and classify all the important unstable waves that may arise from the interaction of a flow with a compliant boundary. In particular, Benjamin (1960) showed theoretically that the well-known Tollmien-Schlichting instability of a laminar boundary layer is stabilized by wall compliance, but destabilized by wall dissipation. Inviscid (inertial) interaction between the flow and the surface, on the other hands, leads to two new instabilities; a travelling-wave instability (termed TWF by Carpenter & Garard, 1986) and a static instability (SD). These instabilities are associated with the wave-bearing modes and static deformation modes of the compliant surface respectively. A third class of inertial instability, a strong instability of the Kelvin-Helmholtz type, may also occur on very soft surfaces when the hydrodynamic forces outweigh the restoring force of the surface at all wave speeds. This is usually not a critical mode of instability except under the special situation when the two earlier classes of inertial instability are prohibited (such as in uniform flow past a non-dissipative surface). We shall not go into the details of these and other early works here, but instead proceed with a summarised account of the background theory following the presentation of Yeo & Dowling (1987). From there we shall move on to consider the quasipotential flow model of Duncan *et al.* (1985), which had been successful in explaining various qualitative features of turbulent flow-compliant wall interaction, and then to progressively more complex and sophisticated models that offer improved agreement with the experiments.

Besides temporal (normal-mode) instability, we shall have occasions to consider convective and absolute instabilities as an alternative description of the stability behaviour of the flow-wall system. The classification of instabilities as convective or absolute was derived from the asymptotic spatio-temporal theory of Briggs (1964) (see Bers, 1983). An instability is said to be *convective* if it grows only as it propagates away from the source of its excitation. An *absolute* instability, on the other hand, grows with time in an expanding neighbourhood of the source of its initiation. A convective instability thus has the appearance of a spatially amplifying travelling wave, whereas an absolute instability behaves like a non-propagating (stationary) temporally growing wave. An absolute instability is clearly a very strong instability. This theoretical dichotomy of convective and absolute instabilities corresponds directly with the physical attributes of the fast waves and the slow or SD waves. The spatio-temporal theory presents a more rigorous interpretation of wave behaviour because it takes into account the causal factor, which is not considered in conventional normal-mode temporal theory. Nevertheless, conventional temporal growth theory has remained an important tool for stability analysis because of its simplicity and because temporal instability is a necessary pre-condition for occurrence of both convective and absolute instabilities.

3.1. INVISCID AND QUASIPOTENTIAL FLOW MODELS

The problem of semi-bounded shear flow over a compliant layer is depicted schematically in Figure 3, where the Cartesian coordinate frame has its positive x_1- and x_3-axes pointing in the streamwise and vertical directions respectively as shown. The perturbation streamfunction $\phi(x_3)\exp[i(\alpha x_1 - ct)]$ of an inviscid shear flow with basic velocity distribution $U(x_3)$ satisfies the Rayleigh's equation

$$(U-c)(\phi''-\alpha^2\phi)-U''\phi=0\,, \tag{1}$$

where superscript prime $'$ denotes derivative with respect to x_3. The α and c denote the wavenumber and wave speed of the perturbation respectively. Let $\hat{\eta}_3|_w \exp[i(\alpha x_1 - ct)]$ represent the corresponding perturbation wave on the compliant surface, where $\hat{\eta}_3|_w$ denotes the complex amplitude of vertical displacement at the surface. The continuity of normal velocity at the surface then implies that

$$\phi_w=(c-U_w)\hat{\eta}_3|_w\,, \tag{2}$$

where the subscript w indicates evaluation at the mean position of the surface along $x_3=0$. The perturbation normal stress and pressure acting on the compliant suface are then given by

$$\hat{\sigma}_{33}|_w=-p_w=(U_w-c)\phi_w'-U_w'\phi_w\,, \tag{3}$$

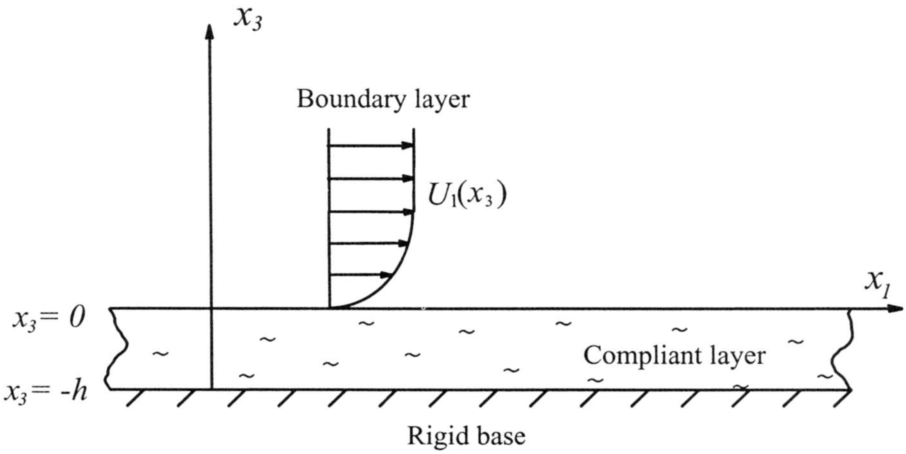

Figure 3. Boundary-layer flow over a compliant layer.

For a uniform potential flow with $U(x_3) = U_\infty$, $\phi(x_3) = A\exp(-\alpha x_3)$, where A is a complex constant. The normal flow stress is then given by

$$\hat{\sigma}_{33}\big|_w = \hat{\eta}_3\big|_w \alpha(U_\infty - c)^2 . \tag{4}$$

For a thin shear layer (small-α approximation), we have

$$\hat{\sigma}_{33}\big|_w = \hat{\eta}_3\big|_w \alpha\left[(U_\infty - c)^2 - icS\right], \tag{5}$$

where cS denotes the out-of-phase component due to the critical-layer effect (Benjamin 1963).

The dynamical response of compliant surface to normal stress $\hat{\sigma}_{33}\big|_w$ can generally be cast in the following generic form (Yeo & Dowling, 1987)

$$\hat{\sigma}_{33}\big|_w = \hat{\eta}_3\big|_w (E - \alpha^2 c^2 I - i\alpha c D) , \tag{6}$$

where the quantities E, I and D are positive quantities related to the stored energy, kinetic energy and energy dissipation of the compliant wall respectively. E, I and D are mostly functions of wavenumber α for surface-based wall models such as membrane and plate surfaces. In particular, for membrane and plate surfaces on viscoelastic foundation, I = mass per unit area of compliant surface and D = foundation damping constant. $E = T\alpha^2 + k_F$ for a tensioned membrane with tension T on a foundation with elastic constant k_F; $E = R\alpha^4 + k_F$ for a bending plate with flexural rigidity R on the same foundation. For the more complex volume-based viscoelastic layer, E, I and D are also dependent on c. It may then be more convenient to obtain the complex surface stiffness $[\hat{\sigma}_{33}/\hat{\eta}_3]_w$ (see (10) below) by directly solving the wall governing equations with the appropriate boundary conditions than to find the component quantities E, I and D. The general representation of wall dynamics (6) is useful, however, for the application of variational considerations to obtain generic stability results.

The stability eigenvalue problem for thin shear flows (and potential flow, $S=0$) over compliant surfaces is then obtained by equating the normal stresses (5) and (6) to yield

$$c^2(\alpha I+1)+c(i\upsilon-2U_\infty)+(U_\infty^2-\alpha^{-1}E)=0\,, \tag{7}$$

where $\upsilon=(D-S)$ is related to the irreversible flow of energy from the system. By analysing (7) and applying a variational argument, Yeo & Dowling derive the following necessary condition for instability for a general compliant surface:

$$U_\infty > \min_{\alpha,\eta}\left\{\sqrt{\alpha^{-1}E}\right\} \quad \text{or} \quad U_\infty > \min_{\alpha,\eta}\left\{c_S\right\}, \tag{8a,b}$$

where the minimum is taken over all α and admissible displacement fields. $\alpha^{-1}E$ is related to the potential energy stored in one wavelength of the compliant wall, while c_S denotes the speed of free-surface waves on the wall. The two quantities for a compliant wall may be deduced by variational argument from (6) (see Yeo & Dowling). The stability limits (or onset velocities) thus obtained for the SD waves (8a) and fast waves (8b) were found to be quite conservative when compared with the experimental results of Hansen *et al.*, Gad-el-Hak *et al.* and Gad-el-Hak. The conservative estimates of the onset velocities is due mainly to the fact that the normal stress fluctuations for potential and thin shear flows given by (4) and (5) significantly overestimate the true magnitude of normal stress fluctuations in a boundary layer.

This over-estimation is rectified in the quasipotential flow approximation of Duncan *et al.* (1985). Duncan *et al.* studied the occurrence of slow or SD waves on viscoelastic layers by assuming the normal stress fluctuations of the flow acting on surface to be given by that of a potential flow (4) scaled by a complex factor $K_p e^{i\theta_p}$

$$\hat{\sigma}_{33}\big|_w=\hat{\eta}_3\big|_w\,\alpha(U_\infty-c)^2K_pe^{i\theta_p}\quad . \tag{9}$$

Values of K_p and θ_p were obtained from the literature to simulate the pertubative stress actions of laminar and turbulent boundary layers on the compliant surface: K_p (<1) scales the amplitude of the hydrodynamics stress $\hat{\sigma}_{33}\big|_w$ down to the level encountered in a boundary layer, while θ_p sets its phase relative to the vertical displacement $\hat{\eta}_3\big|_w$ of the surface. The flow stress $\hat{\sigma}_{33}\big|_w$ is matched with the stress solution of the wall

$$\hat{\sigma}_{33}\big|_w=\hat{\eta}_3\big|_w\,\alpha Gg(\alpha,c,\tau_t) \tag{10}$$

to yield the stability eigenvalue relation, where G is the complex shear modulus (Voigt model) of the wall material given by

$$G = \rho_w C_t^2(1 - i\omega\tau_t) = \rho_w C_t^2 - i\omega d \quad , \tag{11}$$

where $\omega = \alpha c$ is the frequency; and ρ_w, d and τ_t are the density, the damping coefficient and relaxation time of the wall material. $g(\alpha, c, \tau_t)$ is given by equation (2b) in Reutov & Rybushkina (1998).

Duncan *et al.*'s model predicted small wave speeds c and some qualitative features for the SD waves that are in general agreement with the observation of Gad-el-Hak *et al.* Their numerical study yielded an onset flow velocity of $(U_\infty / C_t)_{onset} = 2.86$ for their turbulent boundary layer ($K_p = 0.25, \theta_p = -10^\circ$), which is somewhat lower than all the observed values. For a laminar boundary layer ($K_p = 0.067$ and $\theta_p = -30.4^\circ$), Duncan *et al.* obtained a significantly higher $(U_\infty / C_t)_{onset} = 5.92$, which is in general agreement with the observation that the SD instability of a laminar boundary layer will have much higher onset velocities. The turbulent and laminar flow values of K_p and θ_p were obtained from the experiments of Kendall (1970) and the computational simulation of Balasubramanian & Orszag (1983) respectively. For laminar boundary layers, analytical formulae for K_p and θ_p have also been rigorously derived by Carpenter & Gajjar (1990).

The stability criterion for SD wave (8a) becomes

$$U_\infty > \sqrt{E(\alpha K_p \cos\theta_p)^{-1}} = C_t\sqrt{2\rho_w (K_p \cos\theta_p)^{-1}} \tag{12}$$

when Duncan *et al.*'s quasipotential flow stress model (9) is applied (Yeo 1990). This criterion yields a value $(U_\infty / C_t)_{onset} = 2.85$ and 5.88 for the turbulent and laminar boundary layers of Duncan *et al.* These values provide lower bounds for the onset velocity of the static wave modes for the two boundary layers. The close comparison with Duncan *et al.* shows that the variational optimization leading stability conditions (6) and (12) give fairly sharp lower bound results when the flow stress is accurately modelled. Duncan (1988) also applied the quasipotential approximation of the normal flow stress (9) to turbulent boundary layer over two-layer viscoelastic coatings.

A more complex quasipotential flow approximation was introduced by Reutov & Rybushkina (1998) to predict the occurrence of waves on viscoelastic layers subject to a turbulent boundary layer. They adopted the same dynamic wall model of a viscoelastic layer (10) as Duncan *et al.* above. The fluctuating flow stress $\left.\hat{\sigma}_{33}^{RT}\right|_w$ acting on the compliant surface was first derived by employing a procedure that Reutov & Troitskaya (1995) had devised for a wind-wave problem. The basic turbulent velocity profile $U(x_3)$ was computed from the equilibrium turbulent boundary layer theory of Mellor &

Gibson (1966) using a three-layer total (molecular and turbulent) viscosity model. $\hat{\sigma}_{33}^{RT}\big|_w$ is dependent on the c, α and the local Reynolds Re. A quasipotential flow approximation to $\hat{\sigma}_{33}^{RT}\big|_w$ was then constructed in the form of

$$\hat{\sigma}_{33}\big|_w = \hat{\eta}_3\big|_w \alpha(fU_\infty - c)^2 + \delta\,\hat{\sigma}_{33}^{RT}\big|_w \quad , \tag{13}$$

comprising the contribution from a reduced potential flow fU_∞ and a residual component $\delta\hat{\sigma}_{33}^{RT}\big|_w$. The residual component contains the out-of-phase stress responsible for irreversible energy flow. The velocity reduction parameter $f(\alpha, Re) < 1$ was determined based on the least square minimization of the real part of $\delta\hat{\sigma}_{33}^{RT}\big|_w$ over the range relevant of α and Reynolds number Re. Matching (13) to (10) yields the stability eigenvalue problem. The theory predicted onset velocities for both SD and fast waves that fit the data of Gad-el-Hak *et al.* and Gad-el-Hak well. The predicted wavelengths for the SD waves were approximately half the experimental values.

The formation of the fast and slow waves on viscoelastic coatings was reconsidered by Yeo, Khoo & Zhao (1999) from the perspective of their origins as convective and absolute instabilities respectively. The quasipotential flow approximation (9) of Duncan *et al.* was adopted in this study. They found that the absolute mode of instability (slow, SD waves) is always preceded by a convective mode (fast waves) of instability, except in cases where there is no convective instability. Moreover, for values of K_p and θ_p corresponding to the turbulent boundary layers of Gad-el-Hak *et al.*, the onset flow velocity $(U_\infty / C_t)_{onset}$ of the convective mode was found to converge rapidly onto the onset velocity of the absolute mode when the damping of the compliant layer is increased. Thus for compliant layers with a significant level of damping, the absolute mode of instability (SD mode) will set in simultaneously with the convective mode (fast wave mode). When that happens, it is expected that only the dominant SD mode will be observed. The wave speed c of the convective mode also tends towards very small values as the convergence of the modes occurs. For uniform potential flow over a compliant layer with non-zero material damping, where there is no convective TWF instability, Yeo *et al.* (1999) found that the incipient and follow-on instabilities are always absolute.

3.2. VISCOUS FLOW AND TURBULENCE MODEL

Quasipotential models, such as those we have discussed above, may only predict inertial instability since the inviscid background flow does not support viscous stresses. A non-inflexional viscous boundary layer over a rigid surface may be unstable when convecting perturbation waves in the flow are reinforced by viscosity-induced waves emanating from the wall, giving rise to the well-known Tollmien-Schlichting waves. The effects of fluid viscosity on an unstable mode of inertial origin are not always negligible. Its inclusion may be needed for making accurate quantitative prediction. For models that do not account directly for viscous effects, additional inputs (empirical or theoretical) are usually needed to produce improved predictions. Thus empirical inputs on pressure scalings and phase shifts were used by Duncan *et al.*; while the more complex quasipotential pressure model of Reutov *et al.* (1998) incorporated functions of wavenumber and Reynolds number that were derived by curve-fitting computed turbulent boundary layer data over the relevant parameter ranges. The model of Reutov *et al.* (1998) was shown to match the experimental results well, but is nevertheless non self-sufficient in this regard

A more complete model for waves on compliant surfaces subject to a turbulent boundary layer would have to take into account the viscous and turbulent nature of the flow. A viscous-flow model of turbulent flow over elastic and viscoelastic layers was used by Evrensel & Kalnins (1988). Their formulation of the stability eigenvalue problem is equivalent to that of the customary Orr-Sommerfeld treatment for laminar boundary layer over compliant layers. The basic turbulent flow field was approximated by the 1/7th power law with a semi-empirical correction for the velocity gradient near the wall. The slow or SD waves were interpreted as temporally unstable waves with zero or small real wave speed c_r. There was generally good consistency between their results and those of Gad-el-Hak *et al.* and Gad-el-Hak at the qualitative level. Limited numerical details and results were provided in their paper. However, the two given values of onset velocity for SD waves on highly damped layers show fairly good agreement with the experiments.

Every turbulent mean flow is associated with its own unique distribution of turbulent or Reynolds stresses. From the viewpoint of turbulence modelling, the turbulent mean velocity profile and the Reynolds stresses of the flow are coupled consequences of the turbulent-averaged momentum equations (and continuity equation) and the closure model. Although a turbulent mean velocity profile was used in their stability model, Evrensel & Kalnins did not consider the associated turbulent stresses. In the absence of turbulent stresses, the turbulent ensembled equations are simply reduced to the Navier-Stokes equations. However, the turbulent mean velocity profile is not a solution of the latter. There is thus a theoretical inconsistency. The stability model of Evrensel & Kalnins may be regarded as a quasi-laminar approximation, so that even if it has shown good agreement with some experimental results, it may not be reliable under more varied flow conditions.

To address this apparent incongruity, Yeo, Zhao & Khoo (2001) recently approached the stability problem for turbulent boundary layer on a compliant layer by starting from the unsteady turbulent-averaged equations of continuity and momentum:

$$\frac{\partial \langle U_1 \rangle}{\partial x_1} + \frac{\partial \langle U_3 \rangle}{\partial x_3} = 0 , \tag{14}$$

$$\frac{\partial \langle U_1 \rangle}{\partial t} + \langle U_1 \rangle \frac{\partial \langle U_1 \rangle}{\partial x_1} + \langle U_3 \rangle \frac{\partial \langle U_1 \rangle}{\partial x_3} = -\frac{1}{\rho}\frac{\partial \langle P \rangle}{\partial x_1} + \nu \nabla^2 \langle U_1 \rangle + \frac{\partial}{\partial x_3}\langle -u_1' u_3' \rangle , \tag{15a}$$

$$\frac{\partial\langle U_3\rangle}{\partial t}+\langle U_1\rangle\frac{\partial\langle U_3\rangle}{\partial x_1}+\langle U_3\rangle\frac{\partial\langle U_3\rangle}{\partial x_3}=-\frac{1}{\rho}\frac{\partial\langle P\rangle}{\partial x_3}+\nu\nabla^2\langle U_3\rangle+\frac{\partial}{\partial x_1}\langle -u_1'u_3'\rangle\ , \tag{15b}$$

where $\langle\ \rangle$ denotes a suitable ensemble averaging operation. $\langle U_1, U_3\rangle$ and $\langle P\rangle$ represent the ensemble-averaged velocity and pressure fields respectively. Only the dominant component $\langle -u_1'u_3'\rangle$ of the Reynolds (turbulent) stress tensor has been retained in equations (15a-b). Turbulence closure was achieved with an eddy-viscosity model

$$\langle -u_1'u_3'\rangle=\nu_t\left(\frac{\partial\langle U_1\rangle}{\partial x_3}+\frac{\partial\langle U_3\rangle}{\partial x_1}\right),\qquad \nu_t=l^2\frac{\partial\langle U_1\rangle}{\partial x_3}\ , \tag{16a,b}$$

with the mixing length l prescribed by a well-established composite van Driest's law:

$$l(x_3)=\begin{cases} kx_3\left[1-\exp\left(-\dfrac{x_3}{A}\right)\right] & \text{for}\quad x_3\le x_3^c \\ 0.09\delta & \text{for}\quad x_3> x_3^c \end{cases}\ , \tag{17}$$

where δ is the boundary layer thickness, $k=0.41$, $A=26(\nu/u_*)$ and x_3^c denotes the point at which the two expressions for $l(x_3)$ have equal value (White, 1991). Linear perturbation analysis then yields the following *turbulent* Orr-Sommerfeld equation

$$(U_1-c)(\psi''-\alpha^2\psi)-U_1''\,\psi=\frac{1}{i\alpha R_{\delta^*}}(\psi^{(4)}-2\alpha^2\psi''+\alpha^4\psi)+\frac{1}{i\alpha}\sum_{j=0}^{4}F_j\psi^{(j)}\ , \tag{18}$$

where the last term in (18) contains the effects of the Reynolds stress.

The turbulent mean velocity profiles were computed from an integration of the van Driest's law with the appropriate wake correction. This was done to maintain a high degree of consistency between the governing equations and the basic velocity profiles. Alternatively, the mean velocity profiles may be computed by marching/integrating the turbulent boundary layer equations with the turbulence model (16-17), as had been done by Hou (1996). The above stability model for the turbulent-mean fields is valid for events whose spatial and temporal scales are considerably larger than the fine-scale motions of turbulence; the latter contribute directly to the production of the turbulent stress that was modelled. The fast and SD waves clearly satisfy this condition. Wave propagation in the compliant layer is governed by the viscoelastic Navier's equations. Interaction between the flow and the compliant layer occurs through the continuity of velocity and stress at their interface, Yeo (1988).

The above wave model for turbulent flow bears similarity to the organized wave theory of Reynolds & Hussain (1972), but with one important difference: the turbulence model is an integral part of the total momentum equations (15a-b) before the introduction of wave perturbation, whereas Reynolds & Hussain had invoked turbulence closure via an eddy-viscosity model only at the level of the organised wave.

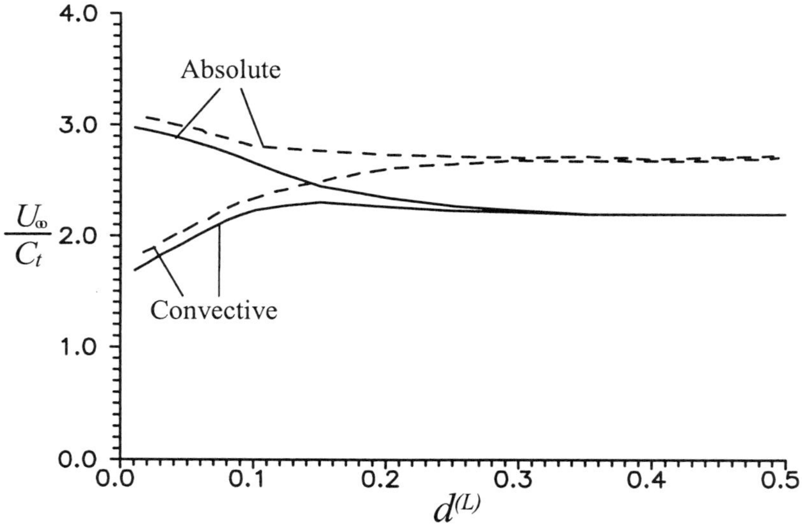

Figure 4. Onset velocities for convective and absolute instabilities on compliant layers with thickness $h^{(L)} = 1.0$. $R_{\delta^*} = 1200$, ———— ; $R_{\delta^*} = 2000$, - - - - . (From Yeo *et al.* 2001.)

Yeo *et al.* (2001) applied the turbulent Orr-Sommerfeld equation (18) to study the onset of surface waves on viscoelastic compliant layer as consequences of convective and absolute instabilities. The study confirms many of the qualitative observations that were made in the previous quasipotential flow study of Yeo *et al.* (1999) concerning the fast and SD waves. Figure 4 shows that the convective or fast-wave instability always comes before the onset of an absolute or SD instability. As damping in the wall is increased to a high level, the onset velocities of the two instabilities converge, so that for a highly damped viscoelastic layer, the convective mode may only have a fleeting existence as it is quickly converted (by mode coalescence) into the dominant absolutely unstable SD mode. There is some evidence of this transformation in the account given by Gad-el-Hak (1986) with regards to the rapid growth of the incipient SD modes from a highly transient wave train. It is interesting to note that the wave velocity of the convective mode tends to zero as the two instability modes converge at high damping levels.

The onset velocities $(U_\infty / C_t)_{onset}$ for the slow or SD waves are compared with the experimental results of Hansen *et al.* and Gad-el-Hak *et al.* in Figure 5. The theoretical predictions of the turbulent wave theory are in very good agreement with the Type I wave results of Hansen *et al.* The Type III wave results of Hansen *et al.* also agree very well with the theoretical predictions when the waves were preceded by the Type I waves. In absence of the latter, as in experiment 5 of Hansen *et al.* and in Gad-el-Hak *et al.*, the observed onset velocities are some 16 to 20% above the predicted values. This suggests that the Type I waves might have acted as a trigger for the downstream Type III waves. Agreement with Gad-el-Hak *et al.* is improved, however, when we take into consideration that a discrete set of test velocities had been used in their experiments (as suggested by Table 1 of their paper) – the dotted bars in the figure mark the range associated with the possible onset of the SD waves at a flow speed between the recorded onset flow speed and the next lower discrete test speed. Also, according to Professor Gad-el-Hak (2001, personal communication), the very first appearance of the Type III waves in Gad-el-Hak *et al.* (where Type I waves were absent) was somewhat intermittent and a small excess of flow velocity was generally needed to establish a series of sufficiently well defined waves whose properties could be measured. This may

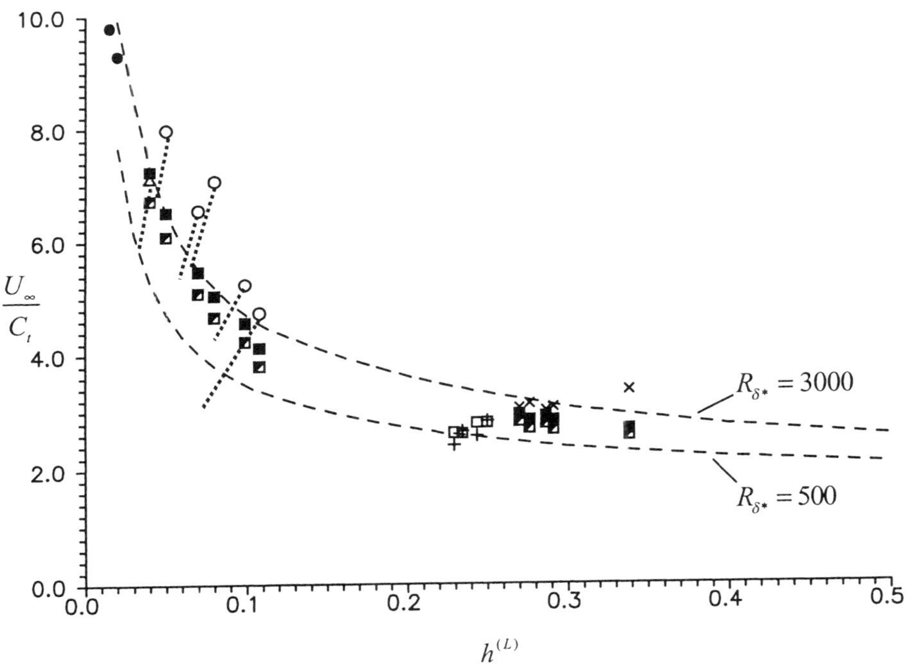

Figure 5. Onset velocities for slow or SD waves on damped compliant layers. Experimental results of Hansen *et al.* (1980): Type I, + ; Type III, × . Experimental results of Gad-el-Hak *et al.* (1984): isotropic, O ; non-isotropic, Δ . Theoretical predictions for compliant layer with $d^{(L)} = 0.5$: at start of layer ($R^s_{\delta^*}$), □ ; at mid-layer ($R^m_{\delta^*}$), ◪ ; at end-layer ($R^e_{\delta^*}$), ■ . Theoretical onset velocity curves at fixed R_{δ^*} , – – – – . Selected data of Gad-el-Hak *et al.* based on $R^m_{\delta^*} = 2000$, ● . The dotted bar (········) denotes range associated with discrete test speeds.

partially explain the higher values of the experimental $(U_\infty / C_t)_{onset}$ compared to the theoretical ones in the figure. Corresponding results for fast waves are given in Figure 6. The theoretical predictions are in good agreement with the experimental results of Gad-el-Hak (1986) when the experimental scatter is taken into account.

Yeo *et al.* (2001) also found that both the fast waves and the SD waves tend to occur first at the end upstream of a compliant layer and to extend downstream as the flow speed is increased. This is consistent with the sequential appearance of the Type I and Type III waves as observed by Hansen *et al.* The wavelengths of the SD waves in Gad-el-Hak *et al.* are typically 1.8 to 2.2 times the theoretically predicted values; while the wavelengths of the fast waves in Gad-el-Hak are up to 50% larger.

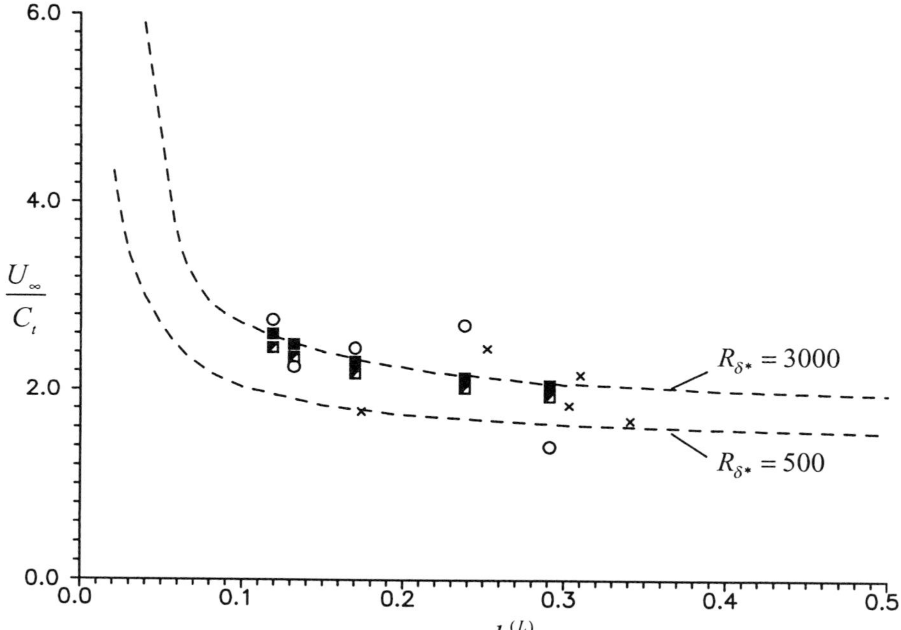

Figure 6. Onset velocities for fast waves on elastic compliant layers. Experimental results of Gad-el-Hak (1986): O . Theoretical predictions for compliant layers with $d^{(L)} = 0.001$: at mid-layer ($R^m_{\delta^*}$), ◪ ; at end-layer ($R^e_{\delta^*}$), ■ . Theoretical onset velocity curves at fixed R_{δ^*} , - - - - . Selected data of Gad-el-Hak based on $R^m_{\delta^*} = 2000$, X . (From Yeo *et al.* 2001)

Similar computations for a laminar (Blasius) boundary layer, Figure 7, did not show the convergence of the onset velocities for the convective and absolute modes even at high material damping levels. The convective mode precedes the absolute mode by a significant onset velocity margin even at high damping, which may explain why distinct SD waves have never been observed under a laminar (flat-plate) boundary layer. Gad-el-Hak *et al.* (1984) in fact noted that their viscoelastic compliant layers did not sustain any SD waves under laminar flow condition even when the flow speeds were increased to twice the onset values of the waves under a turbulent boundary layer. However, SD waves were observed under an artificially created turbulent wedge in an otherwise laminar boundary layer. Yeo *et al.* (2001) also found that the most unstable SD mode is two-dimensional, which agrees with the spanwise character of the observed SD waves.

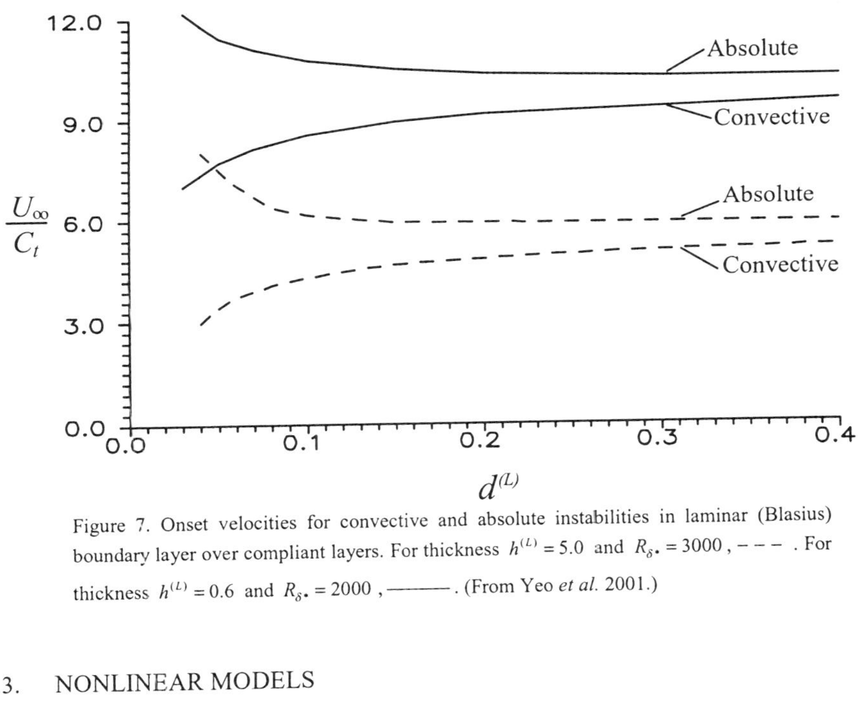

Figure 7. Onset velocities for convective and absolute instabilities in laminar (Blasius) boundary layer over compliant layers. For thickness $h^{(L)} = 5.0$ and $R_{\delta^*} = 3000$, – – – . For thickness $h^{(L)} = 0.6$ and $R_{\delta^*} = 2000$, ———. (From Yeo *et al.* 2001.)

3.3. NONLINEAR MODELS

The theories that we have described so far are based on linear stability analysis. They have been successful to varying degrees in predicting the onset of the wave instabilities and many of the observed trends. The simpler quasipotential flow models generally require additional inputs for them to function. The success of the quasipotential flow models, partial though they may be, indicates that the fast and SD waves are primarily of hydroelastic origin. This may not be surprising given the leading role of inertial effects in a turbulent boundary layer. For the accurate prediction of onset flow velocities, however, it is necessary to include the effects of fluid viscosity and even of turbulence. The inclusion of these also makes for a more complete and self-sufficient theory.

The linear theories have been successful because the instabilities are of small amplitude in their incipient state. However, none of the linear theories are able to predict the detailed physical features of the observed waves; such as the peaky crests and shallow troughs of the SD waves, as recorded by Gad-el-Hak *et al.* For a start, the computed wavelengths of the SD waves by the linear theories are generally too short, typically half the measured values. The computed wavelengths for the fast waves, though in better agreement with the experiments, are still on the low side. A strong instability such as the SD, which produces large amplitude deformation on the compliant surface, is certain to trigger nonlinear effects in both the fluid and the coatings. The nonlinear effects are necessary to arrest the explosive growth of the SD waves, and they are probably involved in shaping the eventual profile of the SD waves. There is also a hint of nonlinearity in the non-sinusoidal waveform of the observed fast waves.

Two recent studies have attempted to account for the effects of nonlinearity on the growth of the SD waves. The first of these, due to Lucey *et al.* (1997), is based on the direct numerical simulation of a perturbed uniform potential flow over a nonlinear spring-backed flexural plate of finite length L. The nonlinear plate equation incorporates the effect of wave-induced tension:

$$\sigma_{33} = \rho_w h\ddot{\eta} + k_\mu \dot{\eta} + R\eta_{,xxxx} + k_F\eta - T\eta_{,xx}, \qquad (19)$$

where ρ_w and h are density and thickness of the plate respectively. k_μ and k_F are the damping and stiffness coefficients of the spring foundation respectively. The flexural rigidity R and the wave-induced tension T are given by

$$R = \frac{Eh^3}{12(1-\nu_p^2)}, \qquad T = \frac{Eh}{L(1-\nu_p^2)}\int_0^L \left[(1+\eta_{,x}^2)^{1/2} - 1\right]dx, \qquad (20a,b)$$

where the E and ν_p are respectively the Young's modulus and Poisson's ratio of the plate material. The perturbation velocity potential $\phi(\mathbf{x},t)$ obeys the Laplace equation $\nabla^2\phi = 0$. The unsteady Bernoulli pressure drives the evolution of perturbations in the flow and the wall. This is computed by a time-domain boundary-integral procedure (an unsteady panel method). The Bernoulli pressure, which refers to the perturbation pressure computed from Bernoulli's equation, is nonlinear in ϕ; and nonlinearity also arises from the finite displacement of the flow boundary (flow-wall interface).

Lucey *et al.* found that on highly damped plates, small-amplitude unstable disturbances evolve into saturated nonlinear divergence waves that have sharp peaks and shallow troughs. This similarity in waveform to the SD waves observed by Gad-el-Hak *et al.* (1984) is quite remarkable considering that the experimental SD waves were formed on viscoelastic compliant layers, which differ dynamically from plates. The simulated waves also propagate downstream at a very slow speed and share many behavioural features with the experimental SD waves. Their analysis showed that although damping has an attenuating effect on the rate of wave growth during the linear phase, a high level of wall damping is essential for the formation of divergence waves. Moreover, nonlinear effects of fluid flow also appear to be necessary for the development of peaky waves because similar linear flow results had showed only sinusoidal waves.

The important role of nonlinear effects in producing the characteristics peak-valley profile of the SD waves was further emphasized in a very recent study by Reutov & Rybushkina (2000). The physical model comprises a potential flow over a viscoelastic coating emulated as a thin plate with wave-induced tension as the source of structural nonlinearity. A weakly nonlinear wave-interaction approach was adopted in which the wave perturbation was assumed to be composed of a set of wave harmonics

$$\eta(x_1,t) = \sum_{\alpha} a_\alpha(t) e^{i\alpha x_1} . \qquad (21)$$

The perturbation flow stress acting on the wall due to the multiple wave harmonics was obtained by extending their linear quasipotential model (Reutov & Rybushkina, 1998) to include nonlinear terms due to wave interaction at the flow-wall interface:

$$\hat{\sigma}_{33}\big|_w = \hat{\sigma}_{33}^{(qp)}\big|_w + \hat{\sigma}_{33}^{(nl)}\big|_w . \qquad (22)$$

Here, $\hat{\sigma}_{33}^{(qp)}\big|_w$ is the linear quasipotential normal stress given by (13) above; and $\hat{\sigma}_{33}^{(nl)}\big|_w$ denotes the nonlinear terms deduced from the wave interaction between the compliant surface and the uniform base flow, subject to the reduced velocity approximation of the linear theory. The weakly nonlinear analysis was carried out about the quasi-steady state of the flow and the wall ($c = 0$) based on the small wave slope approximation ($\alpha\hat{\eta}_3 << 1$). Their analysis shows that quadratic (resonance) interactions in the Landau-type amplitude equations lead to rapid growth of the SD waves, whereas the cubic decay terms associated with mean-flow deformation in the turbulent boundary layer, rather than structural nonlinearity, are responsible for the equilibration of the waves at a finite amplitude. It is the author's opinion that the latter conclusion is likely to be the result of the small wave-slope approximation, which severely limits the role of structural nonlinearity in constraining the growth of the surface waves. High damping plays an essential role in their theory in triggering explosive wave interactions. It does so by bringing about the phase synchronization of the fundamental mode and its higher harmonics.

The unstable waves arising from the nonlinear wave interaction did not immediately attain the waveform and large amplitude of the SD waves upon the loss of stability. Instead, waves with the characteristics waveform and amplitude of the SD waves (see Figure 8) were obtained only when the flow velocity was raised by some significant margin above the critical flow speed. This appears to be at some variance with the experiments, which generally showed a sudden increase in the amplitude of the SD waves upon their occurrence. The supercritical phase (also termed the soft-excitation phase by the authors) or velocity margin was reduced and disappeared altogether, however, when they increased the velocity reduction parameter f of the quasipotential approximation to a sufficiently large value of 0.71. The authors (Reutov *et al.*, 2000) attributed the discrepancies to the inadequacy of the quasipotential approximation. A temporal stability formulation had been applied in their study. It is possible a nonlinear spatial-temporal formulation, which takes into account the contributions of spatial factors in wave interaction (as in linear absolute instability), may help to resolve some of the differences.

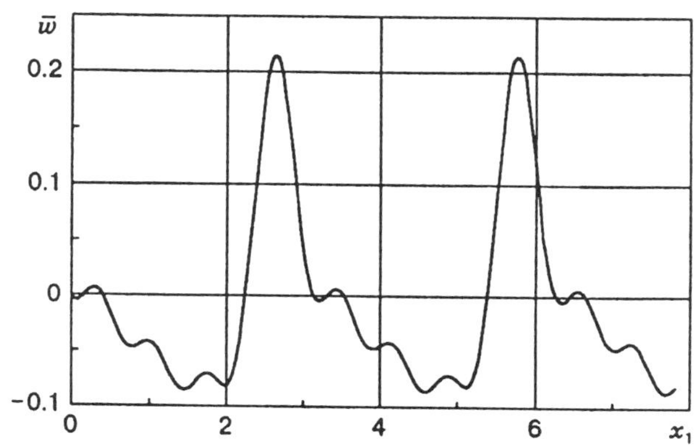

Figure 8. Computed surface displacement for quasipotential flow over a damped nonlinear compliant surface showing the wave peaks and shallow troughs. (From Reutov & Rybushkina, 2000, with permission of authors.)

4. Concluding Remarks

Recent developments in quasipotential and viscous-turbulent flow models have improved the accuracy in predicting the occurrence of fast and SD waves on compliant surfaces subject to a turbulent boundary layer. The quasipotential models generally require accurate parameter inputs that must come from other experiments or additional modelling. This renders them non self-sufficient for general application. Recent studies show that the SD waves on highly damped compliant surfaces are indeed associated with an absolute instability arising from the coalescence of modes with wave speeds that approach zero. SD waves are unlikely to occur under a laminar (Blasius) boundary layer because they require a higher onset velocity than the convective waves even at very high levels of wall damping. Linear stability models are adequate for predicting the onset conditions for the surface waves. This may be sufficient for most practical applications. However, to fully characterize the surface waves and to determine quantities such as the amplitude of the SD waves, one would have to look beyond the linear theory. Recent studies indicate that the characteristic asymmetric peak-valley profiles of the observed SD waves may be explained by invoking nonlinear wave interaction between a potential or quasipotential flow and a weakly nonlinear thin plate. High damping is also essential for the occurrence of SD waves in the nonlinear case. High damping seems to play a key role in bringing about a phase synchronization (resonance) of the fundamental wave mode with its higher harmonics that results in explosive wave growth. Further development of the nonlinear models will be required to allow direct comparisons with the experiments, however, and the effects of viscosity and turbulence may have to be included for accurate prediction of wave characteristics.

5. References

ALEYEV, Yu. G. 1977 *Nekton*. The Hague, W. Junk.

BALASUBRAMANIAN, R. & ORSZAG, S. A. 1983 Numerical studies of laminar and turbulent drag reduction. *NASA Contract Rep.* 3669.

BENJAMIN, T. B. 1960 Effects of a flexible boundary on hydrodynamic stability. *J. Fluid Mech.* **6**, 161.

BENJAMIN, T. B. 1963 The three-fold classification of unstable disturbances in flexible surfaces bounding inviscid flows. *J. Fluid Mech.* **16**, 436.

BERS, A. 1983 *Handbook of Plasma Physics*. Chapter 3. North-Holland Publish Company.

BOGGS, F. W. & HAHN, E. R. 1962 Performance of compliant skins in contact with high velocity flow in water. In *Proc. 7th Joint Army-Navy-Air Force Conf. On Elastomer Research and Development*. San Francisco, U.S.A. Office of Naval Research 2, 443.

BRIGGS, R. J. 1964 *Electron-Stream Interaction with Plasmas*. Monograph No. 29, MIT Press. Cambridge, Massachusetts.

BUSHNELL, D. M., HEFNER, J. N. & ASH, R. I. 1977 Effect of compliant wall motion on turbulent boundary layers. *Phys. Fluids* **20**, S31.

CARPENTER, P. W. & GAJJAR, J. S. B. 1990 A general theory for two- and three-dimensional wall-mode instabilities in boundary layers over isotropic and anisotropic compliant walls. *Theor. & Comput. Fluid Dyn.* **1**, 349.

CARPENTER, P. W. & GARRAD, A. D 1986 The hydrodynamic stability of flow over Kramer-type compliant surfaces. Part II. Flow-induced surface instabilities. *J. Fluid Mech.* **170**, 199.

CHOI, K. S., YANG, X., CLAYTON, B. R., GLOVER, E. J., ATLAR, M., SEMENOV, B. N. & KULIK, V. M. 1997 Turbulent drag reduction using compliant surfaces. *Proc. Roy. Soc. Lond.* A **453**, 2229.

DUNCAN, J. H. 1988 The dynamics of waves at the interface between a two-layer viscoelastic coating and a fluid flow. *J. Fluids & Struct.* **2**, 35.

DUNCAN, J. H., WAXMAN, A. M. & TULIN, M. P. 1985 The dynamics of waves at the interface between a viscoelastic coating and a fluid flow. *J. Fluid Mech.* **158**, 177.

EVRENSEL, A. & KALNINS, A. 1988 Response of a compliant slab to viscous incompressible fluid flow. *Trans. ASME, Ser. E: J. Appl. Mech.* **55**, 660.

GAD-EL-HAK, M. 1986 The response of elastic and viscoelastic surfaces to a turbulent boundary. *Trans. ASME, Ser. E: J. Appl. Mech.* **53**, 206.

GAD-EL-HAK, M., BLACKWELDER, R. F. & RILEY, J. J. 1984 On the interaction of compliant coatings with boundary-layer flow. *J. Fluid Mech.* **140**, 257.

HANSEN, R. J., HUNSTON, D. L., NI, C. C. & REISCHMAN, M. M. 1980 An experimental study of flow-generated waves on a flexible surface. *J. Sound & Vib.* **68**, 317.

HOU, J. T. 1996 The stability analysis of turbulent mean flow over a compliant wall. *MEng. Thesis*, National University of Singapore, Singapore.

KENDALL, J. M. 1970 The turbulent boundary layer over a wall with progressive waves. *J. Fluid Mech.* **14**, 259.

KRAMER, M. O. 1960 Boundary layer stabilization by distributed damping. *J. Am. Soc. Naval Engng.* **73**, 25.

LANDAHL, M. T. 1962 On the stability of laminar incompressible boundary layer over a flexible surface. *J. Fluid Mech.* **13**, 609.

LEE, T., FISHER, M. & SCHWARZ, W. H. 1993 Investigation of stable interaction of a passive compliant surface with a turbulent boundary layer. *J. Fluid Mech.* **257**, 373.

LUCEY, A. D., CAFOLLA, G. J., CARPENTER, P. W. & YANG, M. 1997 The nonlinear hydroelastic behaviour of flexible walls. *J. Fluid & Struct.* **11**, 717.

MELLOR, G. L. & GIBSON, D. M. 1966 Equilibrium turbulent boundary layers. *J. Fluid Mech.* **24**, 225

REUTOV, V.P. & RYBUSHKINA, G. V. 1998 Hydroelastic instability threshold in a turbulent boundary layer over a compliant coating. *Phys. Fluids* **10**(2), 417.

REUTOV, V.P. & RYBUSHKINA, G. V. 2000 Generation of nonlinear waves on a viscoelastic coating in a turbulent boundary layer. *J. Appl. Mech. And Tech. Phys.* **41**(6), 1003.

REUTOV, V.P. & TROITSKAYA, Yu. I. 1995 On the nonlinear effects arising in the interaction of water waves with turbulent wind. *Izv. Russ. Acad. Sci Atmos. Ocean. Phys.* **31**, 6. (English translation from Russian).

REYNOLDS, W. C. & HUSSAIN, A. K. M. F. 1972 The mechanics of an organized wave in turbulent shear flow. Part 3. Theoretical models and comparisons with experiments. *J. Fluid Mech.* **54**, 263.

RILEY, J. J., GAD-EL-HAK, M. & METCALFE, R. W. 1988 Compliant Coatings. *Ann. Rev. Fluid Mech.* **20**, 393.

SEMENOV, B. N. & SEMENOVA, A. V. 1998 Recent advances in interference analysis of compliant boundary actions on near-wall turbulence. In *Proc. Intl. Symp. Seawater Drag Reduction*. 22-23 Jul. 1998, Newport, Rhodes Island, USA. 189.

WHITE, F. M. 1991 *Viscous Fluid Flow*. McGraw-Hill.

YEO, K. S. 1988 The stability of boundary-layer flow over single- and multi- layer viscoelastic walls. *J. Fluid Mech.* **196**, 359.

YEO, K. S. 1990 The hydrodynamic stability of boundary-layer flow over a class of anisotropic compliant walls. *J. Fluid Mech.* **220**, 125.

YEO, K. S. & DOWLING, A. P. 1987 The stability of inviscid flow over passive compliant wall. *J. Fluid Mech.* **183**, 265.

YEO, K. S., KHOO, B. C. & ZHAO, H. Z. 1999 The convective and absolute instability of fluid flow over viscoelastic compliant layers. *J. Sound & Vib.* **223**(3), 379.

YEO, K. S., ZHAO, H. Z. & KHOO, B. C. 2001 Turbulent boundary layer over a compliant surface – absolute and convective. *J. Fluid Mech.* **449**, 141.

Chapter 12

EXPERIMENTAL RESEARCH ON TURBULENT FLOW OVER COMPLIANT WALLS

KWING-SO CHOI
School of Mechanical, Materials, Manufacturing Engineering and Management
University of Nottingham
University Park, Nottingham NG7 2RD, UK

Abstract

Since Kramer's successful experiments in late 1950s and early 1960s, there have been a series of study on compliant coating for drag reduction in the following three decades. As a result, it is now believed that compliant coatings, if suitably fabricated mimicking dolphins' skin, could delay the transition of the laminar boundary layer to turbulence, therefore resulting in a substantial drag reduction. As for the use of compliant coatings in the turbulent boundary layer, however, experimental investigations offered mixed results on a possibility of drag reduction. No single experimental result has been successfully repeated by others. As we went into the 1990s, there has been resurgence in compliant coating research, partly because the former Soviet Union's research data became available to the West. Since then, there have been a number of tests in USA, UK and Russia, demonstrating a strong possibility that compliant coatings could reduce the skin-friction drag of turbulent boundary layers. In this paper, some of the recent experimental work on compliant coating is reviewed, with a current understanding of the mechanisms of turbulent drag reduction.

P.W. Carpenter and T.J. Pedley (eds.),
Flow past Highly Compliant Boundaries and in Collapsible Tubes, 275-292.

1. Previous Work

The original idea of obtaining a reduction in drag force by using compliant surfaces came from Kramer (1957, 1960, 1961, 1962, 1965) whose experiments showed substantial reductions in drag of up to 60 percent using a compliant coating modelled on dolphins' skin. His study was based on the observations by Gray (1936) that dolphins can swim at an exceptionally fast speed and, therefore, the dolphins' muscles would have to generate more than seven times greater power than other types of mammals. This is known as Gray's paradox. Kramer believed, however, that this anomaly was due to the special material properties of dolphins' skin rather than their muscles. He was convinced that the compliant surfaces could delay the transition of the laminar boundary layer to turbulence, resulting in a substantial reduction in drag. In his review paper, Benjamin (1964) theoretically examined the possibility of obtaining drag reduction using compliant surfaces in the turbulent boundary layer instead of using them in the laminar boundary layer, where dolphins are believed to be able to reduce the skin-friction drag. Since then, many experimentalists have tried to find suitable compliant materials to achieve turbulent drag reductions. However, all the investigators who later tried to repeat Kramer's work failed to substantiate such a claim despite careful and comprehensive experimental programmes.

Dinkelacker (1966) performed an experiment with turbulent pipe flow to see if there were any effects of the compliant coatings on the turbulence characteristics. Although the results seem to indicate certain changes associated with a turbulent drag reduction, the overall picture of the outcome from this study was rather inconclusive due to a large number of uncertainties in the experimental setup. A series of wind-tunnel experiments carried out at the University of Oklahoma (Looney & Blick, 1966; Blick & Walters, 1968; Blick *et al.*, 1969) were reported to have shown turbulent drag reductions of up to 50 percent together with a 25 percent reduction in the Reynolds stress. But these results could not be repeated in other investigations, including those by Hefner & Weinstein (1976) and McMichael *et al.* (1980). Chung & Merrill (1984, see Chap. 8 for further details) conducted an experiment using a rotating disc with a silicone-polymer coating with which substantial drag reduction was observed. In the absence of velocity measurements, however, it is not certain whether the flow was laminar or turbulent in this experiment. Taylor (1983) and Falco & Chu (1984) carried out experiments where they claimed to have obtained drag reductions in turbulent boundary layers. The compliant coatings used in these investigations were very soft, so that deformations of the compliant surfaces could have produced pressure gradients, affecting the drag measurements.

Lee *et al.* (1993) conducted an investigation of turbulent boundary layer over a compliant surface, where a significant reduction in turbulence intensity was shown by the upward shift of the logarithmic velocity profile and the reduction in the Reynolds stress. The result of their flow visualisation also indicated that the spanwise spacing of low-speed streaks was increased over the compliant surface compared to the rigid surface. Unfortunately, no information on the amount of drag reduction is available from this experiment since the skin-friction coefficient over the compliant surface was not measured. Nevertheless, the observed changes in the near-wall structures seem to suggest that there was a turbulent drag reduction by the compliant surface. Using the optimum compliant coating proposed by

Semenov (1991), Kulik *et al.* (1991) conducted a test in a lake and found a significant turbulent drag reduction of up to 20 %. It must be noted here that the Russians conducted their test in a very similar way to Kramer by measuring the skin-friction drag of the compliant surface while towing a cylindrical body attached to a boat through open water. The experimental conditions of these experiments were, therefore, not fully controlled.

In order to verify these results in a better-controlled environment, a water-tunnel experiment was conducted by Choi *et al.* (1997) where compliant coatings similar to those used by Kulik *et al.* (1991) were tested. The results of the floating balance measurement showed that the turbulent skin-friction drag was reduced by up to 7 % for one of the compliant coatings tested. The RMS values of skin-friction fluctuations obtained from a glue-on hot film sensor showed reductions of up to 6 % over the range of test speeds from 1.5 m/s to 4 m/s, while the wall-pressure intensity measured by a pressure transducer immediately downstream of the compliant coating revealed reductions of as much as 19 % compared with values for the smooth rigid surface.

2. Turbulent Boundary Layers

Choi *et al.* (1999, 2001) carried out a further experimental study of turbulent drag reduction by compliant coating in a high-speed water flume at the University of Liverpool. The basic open channel arrangement of this facility allowed easy access to the test section that measured 1.37 m wide x 0.84 m deep x 3.66 m long, thereby minimising the setting-up time of the experiments. The flow velocity of the flume was controlled from a minimum speed of 0.03 m/s to the maximum of 6.1 m/s. The Reynolds number based on the momentum thickness was $Re_\theta = 5.4 \times 10^3$ at the flow speed of 2 m/s. A 2.1 m-long slender body of revolution of 0.175 m diameter equipped with a 0.66 m-long floating cylindrical element to measure the skin-friction drag was made available by the Institute of Thermophysics, Russia for this investigation. The outer surface of the floating-element section was covered by a layer of compliant coating to be tested, except for the upstream and downstream edges of the section. This was the same test body used in the previous investigation of compliant coatings for turbulent drag reduction (Choi *et al.* 1997). For a flow speed of 2 m/s and above, the boundary layer over the test body corresponded with fully developed turbulent flow over the entire length of the floating-element section, which was located 0.78 m downstream of the nose.

The compliant coatings tested were made of silicone rubber to a uniform thickness of 7 mm. The measurements of viscoelastic properties were made at the Institute of Thermophysics using small samples taken from each of the compliant coatings. Over the frequency range from 0.65 kHz to 69 kHz, the storage modulus of elasticity $E' = 3.23 \sim 3.63 \times 10^6$ Pa and the loss tangent $\tan\delta = 0.084 \sim 0.092$ for Coating 1, and $E' = 2.07 \sim 2.93 \times 10^6$ Pa and $\tan\delta = 0.23 \sim 0.26$ for Coating 2. From the previous investigations (Choi *et al.* 1997), Coating 1 is known to produce turbulent drag reduction of up to 7 %, while Coating 2 gives a small increase in skin-friction drag for velocities higher than 4m/s.

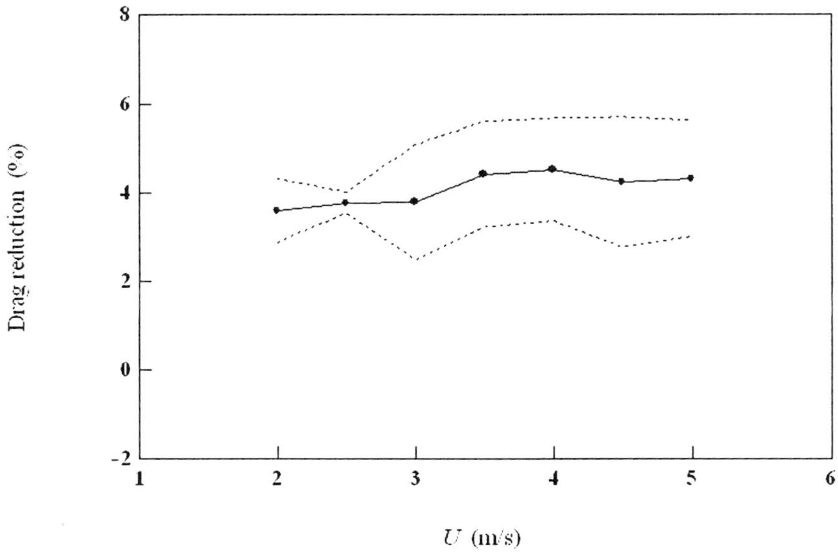

Figure 1. Turbulent drag reduction from Coating 1.

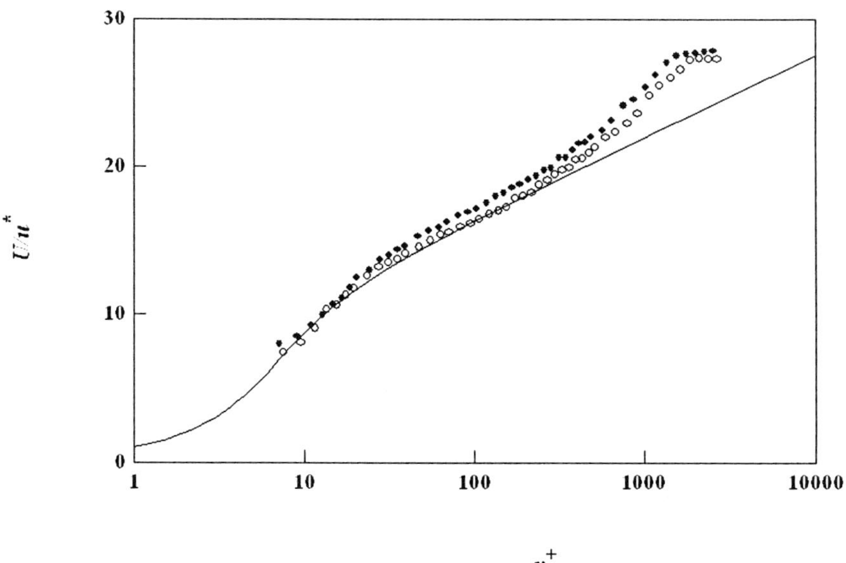

Figure 2. Logarithmic velocity profiles for the boundary layer over the test model with a free-stream velocity of 2 m/s: !, compliant coating; ∀, rigid surface; —— Spalding's curve (1961) with $\kappa = 0.41$ and $C = 5.0$.

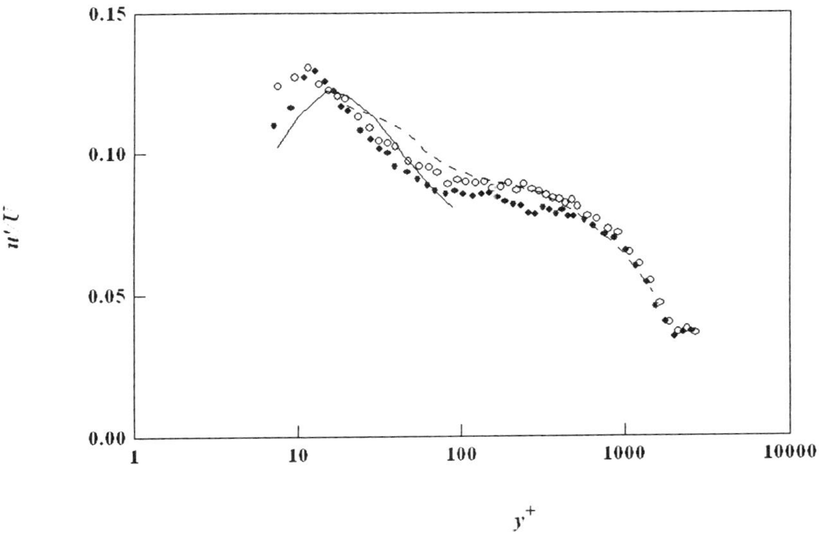

Figure 3. Turbulence intensity profiles of the streamwise velocity fluctuation in the boundary layer over the test model with a free-stream velocity of 2 m/s: !, compliant coating; ∀, rigid surface; —— pipe flow by Durst *et al.* (1995), Re = 7.44 x 10^3; - - - channel flow by Wei & Willmarth (1989), Re = 39.6 x 10^3.

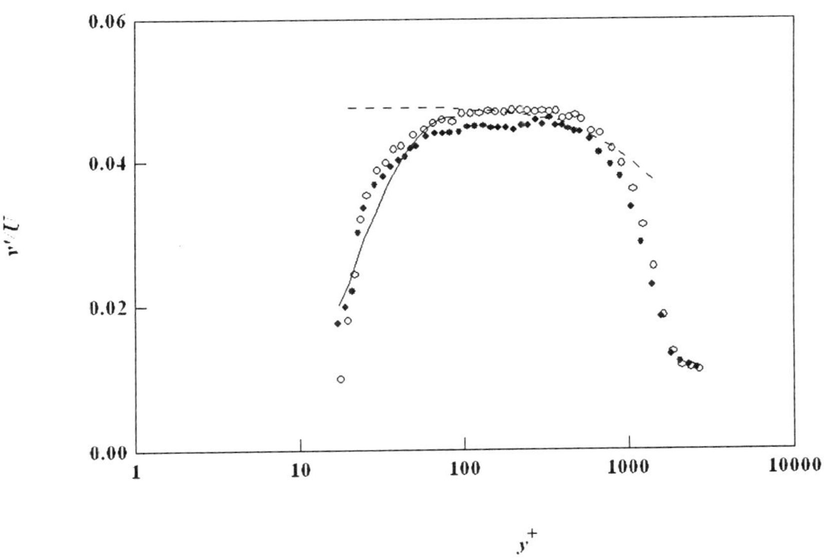

Figure 4. Turbulence intensity profiles of the *v*-component velocity fluctuation in the boundary layer over the test model with a free-stream velocity of 2 m/s: !, compliant coating; ∀, rigid surface; —— pipe flow by Durst *et al.* (1995), Re = 7.44 x 10^3; - - - channel flow by Wei & Willmarth (1989), Re = 39.6 x 10^3.

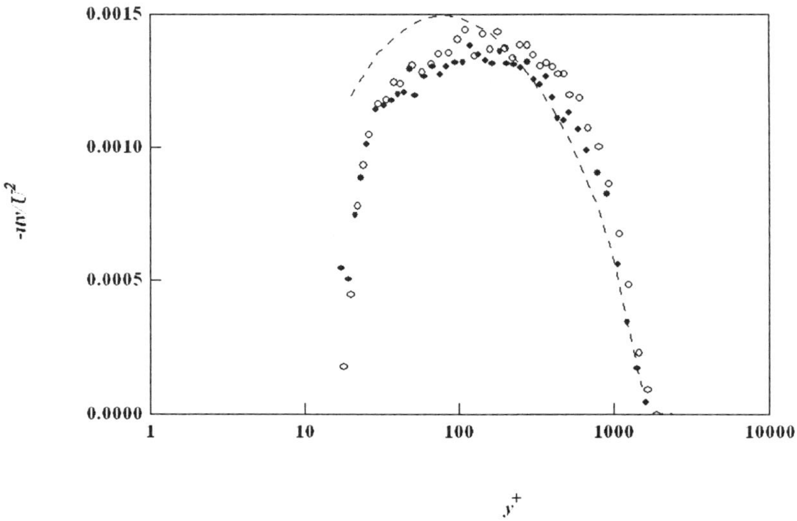

Figure 5. The Reynolds stress profiles in the boundary layer over the test model with a free-stream velocity of 2 m/s: !, compliant coating; ∀, rigid surface; - - - channel flow by Wei & Willmarth (1989), Re = 39.6 x 10^3.

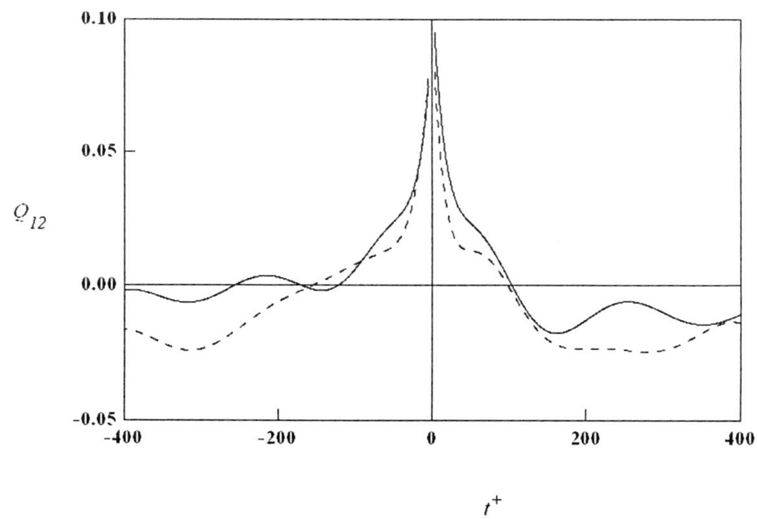

Figure 6. Conditional-correlation coefficient of the u- and v-component velocities at $y^+ = 26$ with a free-stream velocity of 2 m/s: - - - compliant coating; —— rigid surface.

The results given in Figure 1 for Coating 1 confirmed yet again the ability of compliant coatings to reduce turbulent skin-friction drag. This time, however, the amount of drag reduction from Coating 1 has been reduced somewhat from 7% to 5%, and Coating 2 produced approximately 3% of drag reductions for an entire range of test speed up to 5 m/s (not shown here). These differences, although small, may be attributed to the differences in the flow environment where the measurements were carried out, and also to the ageing of the coating materials. The ageing tests conducted by Kulik and Semenov (1996) on a similar compliant coating show, however, that the changes in the viscoelastic properties due to ageing do not significantly affect the drag-reduction characteristics for at least five years.

Figure 2 shows an upward shift of the logarithmic velocity profile of the turbulent boundary layer over the compliant coating (●) for a flow speed of 2 m/s, compared with the data over the rigid surface (O). Here, the velocity profile in the wall region over the rigid surface, including the buffer layer and the logarithmic-law region, is in close agreement with Spalding's curve (1961) with the von Kármán constant $\kappa = 0.41$ and the intercept value $C = 5.0$. Those data points near the wall, which were clearly affected by the reflection of the laser beams as well as by the test-model vibration (Choi *et al.* 1999) are excluded here. The friction velocities used to plot the logarithmic velocity profiles were obtained from the direct measurement of skin-friction drag. Here, the upward shift of the logarithmic velocity profile can be considered as an adjustment of the balance between the turbulent energy production and the viscous dissipation in the near-wall region of the boundary layer, which is reflected by the thickening of the viscous sublayer arising from an increase in the smallest size of turbulent eddies (Lumley 1973). This shift in the logarithmic velocity profile implies, therefore, that the turbulence energy production is reduced in the near-wall region of a turbulent boundary layer.

Figure 3 shows the profiles of turbulence intensity of the streamwise velocity fluctuation for a flow speed of 2 m/s. Reductions in the u-component turbulence intensity of up to 7% are evident in this figure across the entire thickness of the boundary layer over the compliant coating. There is also some evidence for an increase in the skewness and kurtosis profiles near the edge of the viscous sublayer (not shown here). The present results for flow over the rigid surface are in good agreement with the pipe-flow data by Durst *et al.* (1995), the channel flow data by Wei & Willmarth (1989) and the boundary layer data by Gupta & Kaplan (1972). It must be noted here that the pipe-flow study by Durst *et al.* (1995) was conducted at a very low Reynolds number, therefore, a care must be taken in a comparison of data outside the viscous sublayer and the buffer layer. The Reynolds number ($Re_\theta = 1.9 \times 10^3$) of the boundary layer data by Gupta & Kaplan (1972) is also considerably lower than the Reynolds number of this study ($Re_\theta = 5.4 \times 10^3$).

The results of the two-component LDA measurements over a compliant coating also indicate the reductions in the turbulence intensity of the normal-velocity fluctuation as well as in the Reynolds stress across the boundary-layer thickness. For the measurements of the v-component velocity and the Reynolds stress, reliable measurements were only possible for $y^+ > 22$ owing to the interference of the laser beam with the test body. Near-wall measurements were even more difficult with an increase in flow speed. Figure 4 shows that the RMS value of the v-component velocity fluctuation is reduced by up to 7%, similar to the amount of reduction in the RMS value of the u-component velocity fluctuation (see Figure 3). It is known that the Reynolds stress is nearly constant within the inertial sublayer of a turbulent boundary layer where the logarithmic velocity profile is established, and is given by $-\rho\overline{uv} \approx \rho u^{*2}$ (Tennekes & Lumley 1972). Therefore, the reduction of the Reynolds stress within the inertial sublayer should correspond to the reduction of the turbulent skin-friction drag for flow over the compliant coating. As shown in Figure 5, the reduction in the Reynolds stress over the compliant surface is up to 5%, agreeing well with the results of direct skin-friction measurement where a similar measure of drag reduction is observed (Figure 1). Again, the agreement of the present data for flow over the rigid surface with the existing set of data by Durst *et al.* (1995), Wei & Willmarth (1989) and Gupta & Kaplan (1972) is good, considering the difference in the Reynolds number in each experiment.

The conditional correlation coefficient Q_{12} for large t^+ ($-400 < t^+ < 400$) is given in Figure 6, where the least-square fit of a polynomial to the data over the compliant coating is shown together with the rigid-surface data. The results seem to indicate that the first minimum in the conditional cross-correlation coefficient, which corresponds to the ejection event with positive v-velocity following the sweep with positive u-velocity, is located at around $t^+=$ 150 for the rigid surface. The minimum is, however, shifted to $t^+= 250$ for the compliant surface, suggesting that the period between the sweeps and the ejections is increased by the compliant coating leading to a reduction in the burst frequency (Choi *et al.* 1999). A preliminary result of a DNS study of a turbulent boundary layer over a compliant surface, conducted at Cornell University (Parsons *et al.* 1999, see also Chap. 10), also suggests that the burst frequency is reduced by the compliant coating.

3. Turbulent Pipe Flows

A pipe-flow facility was used by Tamilarasan & Choi (2001) for an investigation of turbulent drag reduction by compliant coating (Figure 7). A centrifugal pump drew water from the discharge tank into the header tank in the upstream of the pipe facility. The header tank was designed to provide a constant head of water to drive the flow through the system. The inside of the tank was fitted with a series of three horizontal mesh screens to suppress any large-scale disturbances before the flow reached the inlet port. The higher of these two ports was used in the present study to provide flow rates up to $U_B = 1.8$ m/s.

Commercially available, drawn plastic pipes with 0.100 m internal diameter, 12.3 m long from Durapipe S&LP were used for the development section of the pipe-flow facility. The manufacturing accuracy on the inner diameter of the pipes was up to ±1 mm, while the surface finish was quoted as $R_a < 5$ μm in terms of the average surface roughness. The test section consisted of three independent sections each of 0.94 m in length. The centre section B was designed to be easily removable so that a rigid or compliant section could be mounted at this position depending on the test being run. The test sections for the rigid sections and the 4 mm compliant section were machined from grade 304 stainless steel tubes, while other sections were made from aluminium. The measurements of bulk flow speed in the pipe flow facility were made using Khrone IFM1010K electromagnetic flow meter, which provided accurate non-intrusive measurements of the flow rate. After the flow meter, the final component in the "main loop" is a 90° radius bend connected directly to the discharge tank.

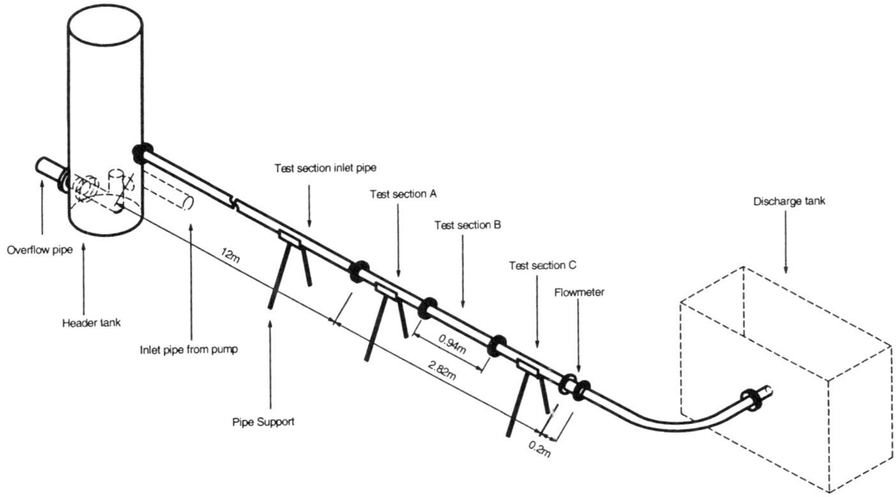

Figure 7. Pipe flow facility for drag reduction test of compliant coating.

Each test-section piece had eight pressure tapping holes, four placed circumferentially equidistant from each other at two longitudinal locations either end of the pipe. These were positioned 40 mm from the flange face in both cases. Validyne DP15 variable-reluctance pressure transducer was used for the measurement of pressure drop, which has the accuracy of measurement of ±0.25% full scale. When actual measurements were taken, the flow velocity was kept constant using a combination of the pump controller setting and the overflow system in the header tank. Using this method, the fluctuations in the velocity readings were kept within ±0.45% at the lowest Reynolds number considered and ±1.3% at the highest Reynolds number considered.

A novel testing approach was utilised to gain dynamic material properties of prepared coatings, which were made of silicone rubber, in order to ascertain their likely behaviour over a range of frequencies corresponding to the turbulence spectrum for the flow speeds of interest. Under these conditions the coating had a storage modulus range of 1.26×10^6 Pa < $E' < 1.48\times10^6$ Pa and a loss tangent range of $0.085 < \tan\delta < 0.17$. Substantial drag-reduction levels were observed from the present investigation across the range of Reynolds numbers tested, $5\times10^4 < \mathrm{Re}_D < 1.75\times10^5$ (see Figure 8). A maximum of 10% drag reduction was observed, with peak values of this order observed for three out of four thickness values of the coating. It is argued (Choi *et al.* 1997) that the compliant coating responds to the pressure pulses of turbulent boundary layers during the burst events taking place in the near-wall region. Therefore, it is expected that the natural frequency of the coating (Kulik *et al.* 1991)

$$f_n = \sqrt{\frac{E'}{\rho}} \Big/ 4H$$

is similar to the characteristic frequency of the pressure pulses when drag reduction is observed in the turbulent boundary layer over the compliant coating. Here, ρ is the density of the compliant coating material and H the coating thickness. The corresponding non-dimensional frequency is given by

$$f_n^+ = f_n \, \nu / u^{*2}$$

where ν is the kinematic viscosity of the fluid and u^* the friction velocity. When the amount of drag reduction is plotted against the non-dimensional natural frequency of the coatings (Figure 9), drag characteristic curves fell into a single line with drag-reduction peaks in a range of $0.35 < f_n^+ < 0.61$. Other studies (Choi *et al.* 1997; Choi *et al.* 1999; Kulik *et al.* 1991) indicated that the maximum drag reductions are in a range of $0.0141 < f_n^+ < 0.61$.

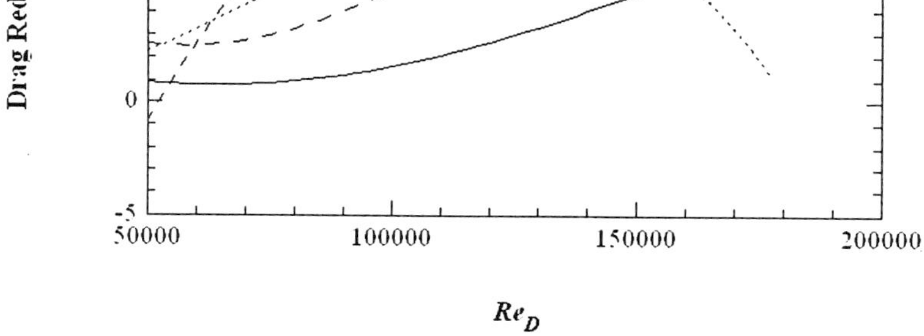

Figure 8. Drag reduction in turbulent pipe flows using compliant coatings with different thickness as a function of the Reynolds number: ———, 2 mm; — — —, 4 mm; – – – – –, 6 mm; ··············, 8 mm.

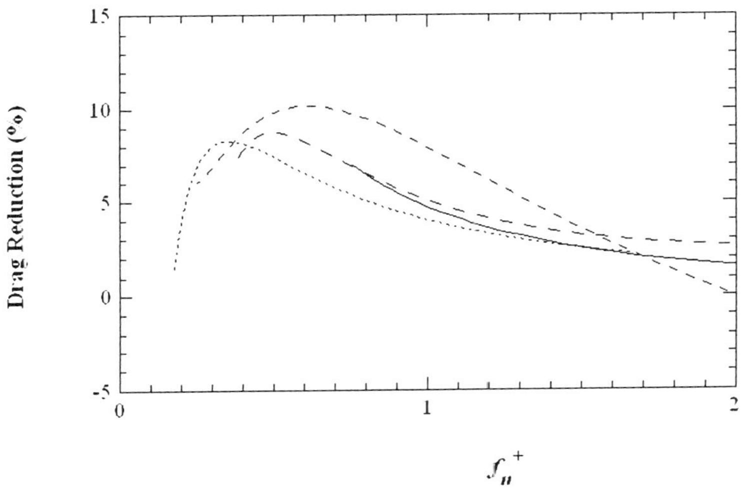

Figure 9. Drag reduction in turbulent pipe flows using compliant coatings with different thickness as a function of the natural frequency of the coatings: ———, 2 mm; — — —, 4 mm; - - - - -, 6 mm; ……………, 8 mm.

4. Conditions for Turbulent Drag Reduction

4.1. HYDRODYNAMIC SMOOTHNESS

As one of the conditions for turbulent drag reduction, Semenov (1991) suggested that the *dynamic* surface roughness of compliant coatings must be small, below the value to be considered as hydro dynamically smooth. This is to say that the magnitude of surface deformation of compliant coatings interacting with the turbulent boundary layer should be much less than the viscous sublayer thickness. For typical compliant coatings in turbulent boundary layers, it has been estimated that the non-dimensional amplitude of wall deformation is $\eta^+ < 1$ (Kulik & Morozova 2001). For typical application in turbulent boundary layers, therefore, the dynamic roughness of compliant coatings is not an issue to be considered.

4.2. NATURAL FREQUENCY

Another condition for turbulent drag reduction is that the natural frequency of compliant coatings must be chosen in such a way to give an appropriate response to the fluctuating wall pressure in the turbulent flow (Semenov 1991). For homogeneous, single-layer material with a modulus of elasticity E, density ρ and thickness H, the fundamental frequency of the coating's longitudinal vibration is given by $f_0 = \sqrt{E/\rho}/4H$ (Rossing & Fletcher 1995). Therefore, the corresponding non-dimensional period of the coating's fundamental frequency is given by $t_0^+ = f_0^{-1} u^{*2} / \nu$. The effect of the viscoelasticity on the vibrational characteristics is usually small (Semenov 1996), which can often be neglected. The values for f_0 and t_0^+ for previous investigations of compliant coatings by Kulik *et al.* (1991), Lee *et al.* (1993) and Choi *et al.* (1997) are summarised in table 1. It can be seen that the non-dimensional period of the first harmonic of the compliant coating falls within $5 < t_0^+ < 74$ in all of these experiments.

This range of non-dimensional period of first harmonic is very similar to the drag-reduction region ($50 < t_0^+ < 150$) of compliant coatings calculated by Semenov (1991) based on Sternberg's theory (1962). It was noted by Choi *et al.* (1997) that this corresponds to the period between the sweep events associated with the high-intensity, wall-pressure fluctuations and the subsequent ejection events, suggesting that the surface movement of compliant coatings may weaken the strength of upwash during the ejection events.

Table 1. Boundary-layer parameters and the material properties of the compliant coating.

	Kulik *et al.* (1991)	Lee *et al.* (1993)	Choi *et al.* (1997)
U m/s	6.0 ~ 15.0	0.15 ~ 0.51	2.0 ~ 6.0
ρ kg/m^3	2.1×10^3	1.0×10^3	2.1×10^3
E Pa	3.7×10^6	0.68×10^3	2.8×10^6
H mm	2.5 ~ 7.0	38.0	7.0
C_T m/s	24.0	0.47	20.9
f_0 Hz	1.5×10^3 ~ 4.2×10^3	5.7	1.3×10^3
t_0^+	26 ~ 74	5 ~ 63	5 ~ 44
U/C_T	0.25 ~ 0.63	0.32 ~ 1.1	0.096 ~ 0.29

4.3. RESONANCE

Resonance of compliant coatings seems to be vital for them to respond effectively to the pressure pulse in the turbulent boundary layer for turbulent skin-friction reduction. For example, the maximum static deformation of the compliant coating studied by Choi *et al.* (1997) is estimated to $\eta^+ = 0.12$, when the intensity of the wall-pressure fluctuation in the boundary layer is given by $p = 0.0055\rho U^2$ (Willmarth 1975). Therefore, it is unlikely that the compliant coating is able to interact with the near-wall structure of turbulent boundary layer unless the compliant coating is resonated during the sweep events to give greater wall amplitude. It is known that the period of the pressure pulse during the sweep events is about $t_0^+ = 20$ (Johansson *et al.* 1987; Nepomuceno & Lueptow 1997), which is within the range of non-dimensional period of the first harmonic of compliant coatings (see discussions above). This suggests a strong possibility that the pressure pulse is causing a resonance to the coating as it interacts with the surface during the sweep events.

Recent analysis by Kulik & Morozova (2001) seems to suggest, however, that a small wall deformation of compliant coatings is sufficient for turbulent drag reduction as long as the wall-normal velocity of the coating is large enough. They estimated that non-dimensional velocity of wall deformation can be up to $v^* = 0.04$ for typical compliant coatings in turbulent boundary layers.

4.4. SHEAR VELOCITY

The surface of compliant coatings must be free from hydrodynamic instabilities, although this is a necessary, but not sufficient, condition for turbulent drag reduction. A numerical study conducted by Duncan (1986) suggested that the compliant surface is always stable when the non-dimensional flow speed U/C_T is less than 1.2 irrespective of the damping characteristics of the coating. Here, U is the freestream velocity of the boundary layer, C_T is the shear velocity given by $C_T = \sqrt{G/\rho}$, G represents the shear modulus and ρ is the density of the coating. If the compliant coating is made of homogeneous, single-layer silicone rubber, the shear velocity can be given in terms of the modulus of elasticity E as $C_T \approx \sqrt{E/3\rho}$ since Poisson's ratio for rubbery material is $\mu \approx 0.5$ for small deformations (Ferry, 1980). Table 1 shows that the non-dimensional flow speed U/C_T is always less 1.1 for recent studies of compliant coating that showed turbulent drag reduction. This indicates that the compliant surfaces that have been successful in reducing drag were free from all hydrodynamic instabilities.

4.5. CONVECTION VELOCITY

Duncan (1986) also showed in his numerical study that the deformation of compliant coatings is localised and confined within the vicinity of the pressure pulse when the wall surface is free from hydrodynamics instabilities. It is also known, at least for a short streamwise distance along the wall, that the convection velocity of the wall-pressure signature is similar to that of the velocity signature in the near-wall region of the boundary layer (Willmarth & Wooldridge 1962; Bull 1967). It is, therefore, expected that the localised deformation of compliant coatings is convected downstream at the same convection speed as the near-wall structure of the turbulent boundary layer, making a continuous interaction with the near-wall turbulence events. This may be a reason why compliant coatings can produce a relatively large amount of turbulent drag reduction of up to 18% (Kulik *et al.* 1991) with seemingly a small amount of surface deformation.

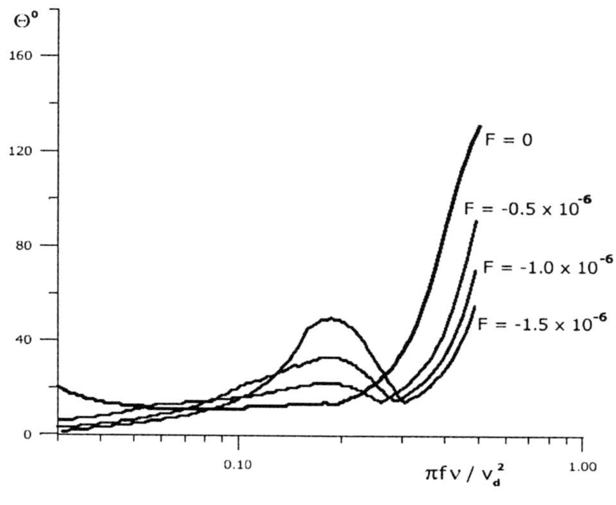

Figure 10. Phase-frequency region of drag reduction by compliant coating for turbulent boundary layers under favourable pressure gradient.

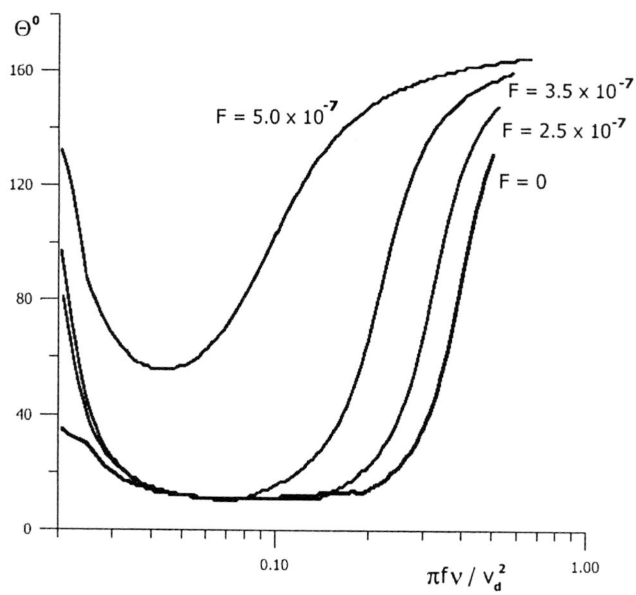

Figure 11. Phase-frequency region of drag reduction by compliant coating for turbulent boundary layers under adverse pressure gradient.

4.6. PRESSURE GRADIENT

The influence of pressure gradient on turbulent drag reduction over compliant coatings was recently studied by Semenov & Semenova (2001). The theoretical study was conducted for the turbulent boundary layer with $Re_\theta = 5000$ for the pressure gradient between $-1.5x10^{-6} < F < 5x10^{-7}$, where the pressure gradient parameter is defined as $F = (dP/dx)(\nu/\rho U_0^3)$. From this study, they showed that the phase-frequency region of turbulent drag reduction is widened only to higher frequencies for the turbulent boundary layer under favourable pressure gradient (Fig. 10). On the other hand, the region of turbulent drag reduction is shown to reduce in both higher as well as lower frequencies for the turbulent boundary layer in adverse pressure gradient (Fig. 11).

5. Concluding Remarks

Experimental study has been carried out by Choi *et al.* (1999, 2001) to develop further knowledge of the near-wall turbulent structure over a compliant surface for turbulent drag reduction. Their results confirmed the basic conclusions obtained in the previous study (Choi *et al.* 1997), clearly demonstrating the ability of compliant coatings to reduce turbulent skin-friction drag. Relative to flow over a rigid surface the turbulence intensities of the *u*-velocity fluctuation were reduced and the logarithmic velocity profile shifted upwards for the compliant coating. At the same time, evidence was obtained to show an increase in the levels of the skewness and kurtosis near the edge of the viscous sublayer. There were also reductions in the turbulence intensity of the normal velocity fluctuation as well as in the Reynolds stress across the boundary layer thickness over a compliant coating, supporting the results of direct skin-friction measurements.

Using a pipe-flow facility, Tamilarasan & Choi (2001) observed a substantial turbulent drag reduction of up to 10% by compliant coating across the range of Reynolds numbers tested, $5\times10^4 < Re_D < 1.75\times10^5$. When the amount of drag reduction by compliant coating was plotted against the non-dimensional natural frequency of the coatings, drag characteristic curves fell into a single curve with drag-reduction peaks in a range of $0.35 < f_n^+ < 0.61$. Other studies (Choi *et al.* 1997; Choi *et al.* 1999; Kulik *et al.* 1991) indicated that the maximum drag reductions are in a range of $0.0141 < f_n^+ < 0.61$.

As a condition for turbulent drag reduction, Semenov (1991) suggested that the *dynamic* surface roughness of compliant coatings must be small, below the value to be considered as hydro dynamically smooth. For typical compliant coatings in turbulent boundary layers, it has been estimated that the non-dimensional amplitude of wall deformation is $\eta^+ < 1$ (Kulik & Morozova 2001). Another condition for turbulent drag reduction is that the natural frequency of compliant coatings must be chosen in such a way to give an appropriate response to the fluctuating wall pressure in the turbulent flow (Semenov 1991).

It has been shown (Choi *et al.* 1997) that the non-dimensional period of the first harmonic of the compliant coating falls within $5 < t_0^+ < 74$ in all of these experiments. This range of non-dimensional period of first harmonic is very similar to the drag-reduction region ($50 < t_0^+ < 150$) of compliant coatings calculated by Semenov (1991) based on Sternberg's theory (1962). This corresponds to the period between the sweep events associated with the high-intensity, wall-pressure fluctuations and the subsequent ejection events, suggesting that the surface movement of compliant coatings may weaken the strength of upwash during the ejection events.

The influence of pressure gradient on turbulent drag reduction over compliant coatings was studied by Semenov & Semenova (2001), who showed that the phase-frequency region of turbulent drag reduction is widened only to higher frequencies for the turbulent boundary layer under favourable pressure gradient. On the other hand, the region of turbulent drag reduction is shown to reduce in both higher as well as lower frequencies.

This work was supported by EPSRC Research Grant (GR/K82734) with matching funds from MoD. The support of the Leverhulme Trust (RF7G/10988) is also acknowledged.

6. References

Benjamin, T.B. 1964 Fluid flow with flexible boundaries. In *Proc. 11th International Congress on Application of Mathematics* (ed. H. Görtler), pp. 109-128. Berlin: Springer-Verlag.

Blick, E.F. & Walters, R.R. 1968 Turbulent boundary layer characteristics of compliant surfaces. *J. Aircraft* **5**, 11-16.

Blick, E.F. *et al.* 1969 Compliant coating skin friction experiments. *AIAA Paper* 69-165.

Buchhave, P. 1979 The measurement of turbulence with the burst-type laser Doppler anemometer – errors and correction methods. Ph.D. Thesis, SUNY Buffalo.

Bull, M.K. 1967 Wall pressure fluctuations associated with subsonic turbulent boundary layer flow. *J. Fluid Mech.* **28**, 719-754.

Choi, K.-S., Yang, X., Clayton, B.R., Glover, E., Atlar, M., Semenov, B.N. & Kulik, V.M. 1997 Turbulent drag reduction using compliant surfaces *Proc. Roy. Soc. Lond.* A**453**, 2229-2240.

Choi, K.-S, Le Huu Nho, E., Clayton, B.R. & Escudier, M.P. 1999 Turbulence structure over a compliant surface and a possible drag-reduction mechanism. *Paper presented at the American Physical Society Meeting, Division of Fluid Dynamics, New Orleans, 21-23 Nov.,* AF-04.

Choi, K.-S, Le Huu Nho, E., Clayton, B.R. & Escudier, M.P. 2001 Turbulent boundary-layer structure over a compliant surface. *Paper presented at the IUTAM Symposium on Flow in Collapsible Tubes and past other Compliant Boundaries, University of Warwick, 26-30 March.*

Chung, K.H. & Merrill, E.W. 1984 Drag reduction by compliant surfaces measured on rotating discs. In *Proc. Compliant Coating Drag Reduction Program Review Meeting*. Washington, DC: Office of Naval Research.

Dinkelacker, A. 1966 Preliminary experiments on the influence of flexible walls on boundary layer turbulence. *J. Sound Vib.* **4(2)**, 187-214.

Duncan, J.H. 1986 The response of an incompressible, viscoelastic coating to pressure fluctuations in a turbulent boundary layer. *J. Fluid Mech.* **171**, 339-363.

Durst, F., Jovanovic, J. & Sender, J. 1995 LDA measurements in the near-wall region of a turbulent pipe flow. *J. Fluid Mech.* **295**, 305-335.

Falco, R.E & Chu, C.C. 1984 Measurements of local skin-friction reduction on elastic compliant surfaces. In *Proc. Compliant Coating Drag Reduction Program Review Meeting*. Washington, DC: Office of Naval Research.

Ferry, J.D. 1980 *Viscoelastic properties of polymers*, New York: Wiley.

Gray, J. 1936 Studies of animal locomotion. VI The propulsive powers of the dolphin. *J. Exp. Biology* **13**, 192-199.

Gupta, A.K. & Kaplan, R.E. (1972) Statistical characteristics of Reynolds stress in a turbulent boundary layer. *Phys. Fluids* **15(6)**, 981-985.

Hefner, J.N. & Weinstein, L.M. 1976 Reexamination of compliant wall experiments in air with water substrates. *J. Spacecr. Rockets* **13**, 502-503.

Johansson, A.V., Her, J.-Y. & Haritonidis, J.H. 1987 On the generation of high-amplitude wall-pressure peaks in turbulent boundary layers and spots. *J. Fluid Mech.* **175**, 119-142.

Kramer, M.O. 1957 Boundary layer stabilisation by distributed damping. *J. Aero. Sci.* **24,** 459-460.

Kramer, M.O. 1960 Boundary layer stabilisation by distributed damping. *J. Amer. Soc. Nav. Engr.* **72**, 25-33.

Kramer, M.O. 1961 The dolphin=s secret. *J. Amer. Soc. Nav. Engr.* **73**, 103-107.

Kramer, M.O. 1962 Boundary layer stabilisation by distributed damping. *J. Amer. Soc. Nav. Engr.* **74**, 341-348.

Kramer, M.O. 1965 Hydrodynamics of the dolphin. In *Advances in hydroscience* (ed. V.T. Chow), vol. 2, pp. 111-130. New York: Academic Press.

Kulik, V.M. & Morozova, S.L. 2001 Response of a compliant coating to fluctuating wall pressure. *Paper presented at the IUTAM Symposium on Flow in Collapsible Tubes and past other Compliant Boundaries, University of Warwick, 26-30 March.*

Kulik, V.M., Poguda, I.S. & Semenov, B.N. 1991 Experimental investigation of one-layer viscoelastic coatings action on turbulent friction and wall pressure pulsations. In *Recent developments in turbulence management* (ed. K.-S. Choi), pp. 263-289. Dordrecht: Kluwer.

Kulik, V.M. & Semenov, B.N. 1996 The measurement of dynamic properties of viscoelastic materials for turbulent drag reduction. In *Emerging techniques in drag reduction* (ed. K.-S. Choi, K.K. Prasad & T.V. Truong), pp. 207-217. London: MEP.

Lee, T., Fisher, M. & Schwarz, W. H. 1993 Investigation of the stable interaction of a passive compliant surface with a turbulent boundary layer. *J. Fluid Mech.* **257**, 373-401.

Looney, W.R. & Blick, E.F. 1966 Skin friction coefficients of compliant surfaces in turbulent flow. *J. Spacecr. Rockets* **3**, 1562-1564.

Lumley, J.L. 1973 Drag reduction in turbulent flow by polymer additives. *J. Polymer Sci. D: Macromol. Rev.* **7**, 263-290.

McMichael, J.M., Klebanoff, P.S & Mease, N.E. 1980 Experimental investigation of drag on a compliant surface. In *Viscous Flow Drag Reduction* (ed. G.R. Hough), pp. 410-438. Washington, DC: AIAA.

Nepomuceno, H.G. & Lueptow, R.M. 1997 Pressure and shear stress measurements at the wall in a turbulent boundary layer on a cylinder. *Phys. Fluids* **9(9)**, 2732-2739.

Parsons, L.M., Blossey, P.N., Rempfer, D. & Lumley, J.L. 1999 Low-dimensional model of the interaction of near-wall turbulence with a compliant boundary. *Paper presented at the American Physical Society Meeting, Division of Fluid Dynamics, New Orleans, 21-23 Nov.*, AF-05.

Rossing, T.D. & Fletcher, N.H. 1995 *Principles of Vibration and Sound.* New York: Springer-Verlag

Semenov, B.N. 1991 On conditions of modelling and choice of viscoelastic coatings for drag reduction. In *Recent developments in turbulence management* (ed. K.-S. Choi), pp. 241-262. Dordrecht: Kluwer.

Semenov, B.N. 1996 Analysis of four types of viscoelastic coating for turbulent drag reduction. In *Emerging techniques in drag reduction* (ed. K.-S. Choi, K.K. Prasad & T.V. Truong), pp. 187-206. London: MEP.

Semenov, B.N. & Semenova, A. 2001 Interference action on compliant boundaries on flow in collapsible tubes. *Paper presented at the IUTAM Symposium on Flow in Collapsible Tubes and past other Compliant Boundaries, University of Warwick, 26-30 March.*

Spalding, D.B. 1961 A single formula for the "law of the wall" *J. Appl. Mech.* **28**, 455-458.

Sternberg, J. 1962 A theory for the viscous sublayer of a turbulent flow. *J. Fluid Mech.* **13**, 241-271.

Tamilarasan, M. & Choi, K.-S. 2001 Drag reduction in turbulent pipe flows using compliant coating. *Paper presented at the IUTAM Symposium on Flow in Collapsible Tubes and past other Compliant Boundaries, University of Warwick, 26-30 March.*

Taylor, T.D. 1983 Experimental and calculational results for compliant surface drag reduction. In *Proc. Compliant Coating Drag Reduction Program Review Meeting*. Washington, DC: Office of Naval Research.

Tennekes, H. & Lumley, J.L. 1972 *A First Course in Turbulence*. Cambridge: MIT Press.

Wei, T. & Willmarth, W.W. 1989 Reynolds-number effects on the structure of a turbulent channel flow. *J. Fluid Mech.* **204**, 57-95.

Willmarth, W.W. 1975 Structure of turbulence in boundary layers. *Adv. Appl. Mech.* **15**, 159-254.

Willmarth, W.W. & Wooldridge, C.E. 1962 Measurements of the fluctuating pressure at the wall beneath a thick boundary layer. *J. Fluid Mech.* **14**, 187-210.

Chapter 13

DOLPHIN HYDRODYNAMICS

VIKTOR V. BABENKO
Department of Hydrobionics and Boundary-Layer Control
Institute of Hydromechanics
Ukrainian National Academy of Sciences
252057, Kiev, Ukraine

PETER W. CARPENTER
School of Engineering, University of Warwick
Coventry, CV4 7AL, U.K.

Abstract

The research on the hydrodynamics of compliant walls was originally inspired by the dolphin. In particular, *Gray's paradox* – the belief that specific power output of the propulsive muscles must greatly exceed the mammalian norm in order to achieve the observed swimming speeds if some form of laminar-flow control were not involved – has led to a great deal of research on the use of wall compliance for drag reduction. The full assessment of Gray's paradox is a highly multidisciplinary undertaking. Accordingly, in this chapter we review all the aspects of dolphin hydrodynamics that might have a bearing on an investigation of Gray's paradox. There are sections on the principles of bionics, the shapes of the body and fins, the structure and characteristics of dolphin skin, swimming speeds, kinematics and dynamics of swimming, boundary-layer characteristics, drag, drag-reducing behaviour, active flow control, energetics, technological developments, and the chapter ends with a brief re-assessment of Gray's paradox.

P.W. Carpenter and T.J. Pedley (eds.),
Flow past Highly Compliant Boundaries and in Collapsible Tubes, 293-323.

1. Introduction

Belief that some dolphin species possess an extraordinary laminar-flow capability has inspired and sustained much of the research on the hydrodynamics of compliant walls. Certainly mankind has long admired the swimming skills of this fleet creature. Our community of scientists and engineers are inclined to consider the work of Gray (1936) as the starting point for scientific research on dolphin hydrodynamics. In fact, there was much interest in the subject prior to Gray, most notably by Aristotle and the early British aeronautical engineer, Sir George Cayley (c. 1800). This early work is well reviewed by Fish & Rohr (1999).

In his analysis of dolphin energetics and hydrodynamics, Gray assumed that dolphins could maintain a sustained swimming speed of 10 m/s. In order to estimate the hydrodynamic drag he then followed the usual practice of marine engineers at that time and modelled the dolphin's body as a one-sided flat plate with the same surface area and length (c. 2 m) as the dolphin. The corresponding Reynolds number based on body length was about 20×10^6. Even in a very low-noise environment the value, Re_{xt}, of the Reynolds number for transition from laminar to turbulent flow does not exceed 2 to 3×10^6 for flow over a flat plate. Accordingly, Gray assumed that if conventional hydrodynamics were involved, the flow over the dolphin's body would be mostly turbulent and consequently it would experience a large drag force. So large, in fact, that at 10 m/s its muscles would have to deliver about seven times more power per unit mass than any other mammalian muscle. The mismatch between the required and expected muscle output has become known as *Gray's Paradox*. This led Gray and others since to argue that the dolphin must be capable of maintaining laminar flow by some extraordinary means. (Gray also carried out a similar calculation on a slightly slower and smaller porpoise with much the same outcome.)

Resolving Gray's Paradox may appear at first sight to be fairly straightforward. This apparent simplicity, reflected in the brief description given above, is deceptive. More than 65 years later a satisfactory resolution still eludes us. Progress has been made in establishing that artificial dolphin skins, in other words compliant walls, do have a laminar-flow capability (see Gad-el-Hak 2002). But a full appraisal of Gray's Paradox requires a multifaceted and multidisciplinary approach. In this chapter we review in turn each of the main topics required to re-assess Gray's Paradox. The shapes of the body and fins have a bearing on the hydrodynamics and are discussed in Section 2. This is followed by a description of the structure and characteristics of dolphin skin. We assess the evidence for the existence of structures or characteristics that could be hydrodynamically relevant. Gray assumed a swimming speed of 10 m/s. But, even today, the existing data on dolphin swimming speeds are surprisingly unreliable and are reviewed in Section 4. In Section 5 we consider the kinematics and dynamics of swimming. Plainly the efficiency of the dolphin's propulsive unit, its tail flukes, is important for determining the power required to overcome drag. But propulsive efficiency is by no means a straightforward concept for the dolphin. Section 6 considers the evidence available for the state of the boundary-layer flow around the dolphin. Is it laminar or turbulent? Does the flow remain attached or does it separate? Next, in Section 7, measurements and predictions of the dolphin's drag are considered. This is followed by a review of the various drag-reducing mechanisms and modes of behaviour (e.g., bow-wave riding) that have been proposed for the dolphin. Section 9 investigates the energetics involved from a physiological viewpoint. The key question being the power that can be delivered by the propulsive muscles. The possibility of interactions between the various biological systems of the dolphin is considered in Section 10, especially the question of whether there is any evidence for active flow control. Section 11 briefly reviews technological developments inspired by the dolphin. And, finally, the chapter concludes with a re-assessment of Gray's Paradox.

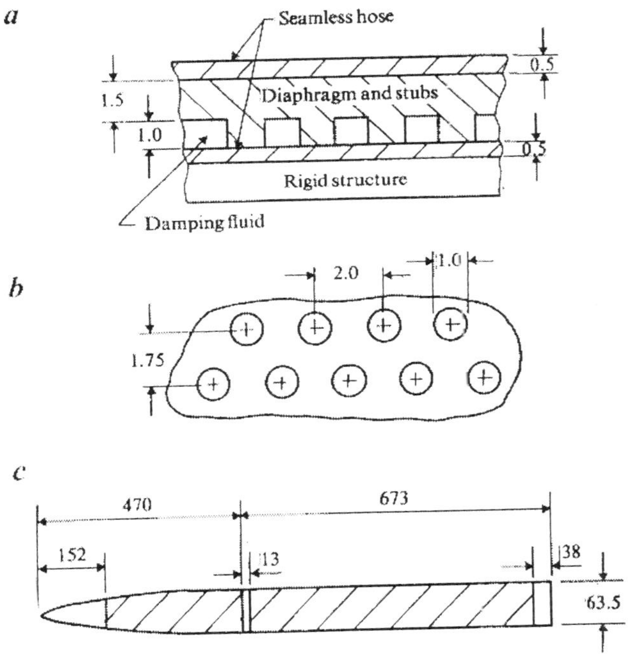

Figure 1. Kramer's compliant coating and model. All dimensions in mm. (Drawings based on those given in Kramer 1960a.) (a) cross-section; (b) cut through stubs; (c) model: shaded regions were coated. (Based on Fig. 1 of Carpenter & Garrad 1985).

Interest in dolphin hydrodynamics and in related bionic research has grown greatly since Gray's (1936) study. In many respects the most influential contribution was due to Kramer (1957, 1960). His observations of dolphins while crossing the Atlantic onboard liners in the 1950's led him to resurrect his earlier concept described in his patent of 1938. This was a drag-reducing device that imitated the structure of the hairy or feathery coats of aquatic animals and birds. It consisted of an array of fine, narrowly spaced, wires arranged parallel to the flow direction and close to the surface. Kramer hypothesized that the additional viscous effects generated in the narrow gaps between the wires would damp oscillations of the near-wall flow, thereby suppressing the generation of turbulence and postponing laminar-turbulent transition. Test work in connection with this patent was discontinued owing to the Second World War. Later Kramer (1957, 1960), inspired by his study of dolphin skin, developed his compliant coatings in an attempt to exploit the same damping concept as used in the earlier 'hairy' surfaces. The design of his compliant coatings (Fig.1) was closely related to his concept of the dolphin epidermis (Fig. 2). Drag reductions of up to 57 percent were achieved at 18 m/s in sea-water for bodies of revolution covered with his optimized compliant coatings (see Fig. 7). The success of the compliant coatings appeared to explain, partially at least, Gray's paradox. Moreover, it sparked off interest worldwide in the use of compliant walls for drag reduction and, more generally, in hydrobionics research. (It should be mentioned, though, that, as first demonstrated theoretically by Benjamin (1960), the damping concept was actually incorrect when applied to laminar-turbulent transition over compliant walls.)

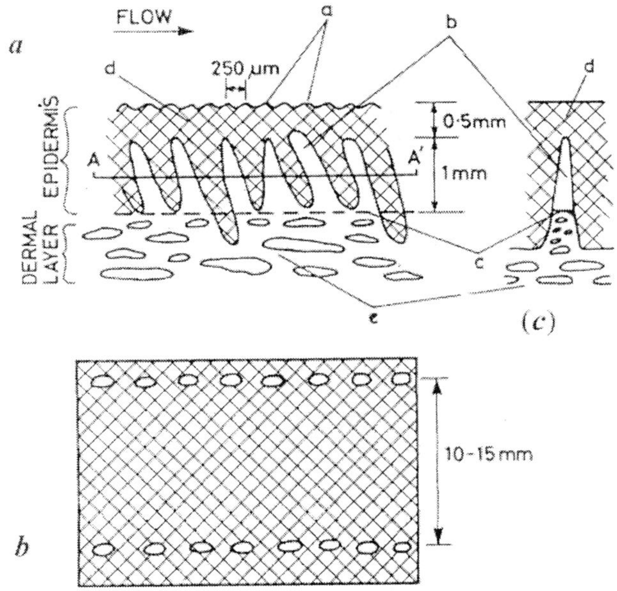

Figure 2. Structure of dolphin skin. (a) Cross-section; (b) Cut through dermal papillae at AA′; (c) Front view. Key: a, cutaneous ridges or microscales; b, dermal papillae; c, dermal ridge; d, upper epidermal layer; e, fatty tissue. [Based on Fig. 2 of Carpenter *et al.* 2000]

The study of dolphin hydrodynamics could be considered to belong to the relatively new scientific discipline of hydrobionics. Broadly, this is the study of external forms and their functions. It is a complex undertaking because it is highly multidisciplinary, including both biological and engineering sciences. Like any new scientific discipline, a suitable methodology had to be developed, particularly with regard to experimental techniques (Babenko 1981). It is important to study hydrobionics and its principles systematically if they are to be applied technologically. Various hydrobionic principles have been proposed by Babenko and his collaborators (see Babenko *et al.* 1969, 1982, Babenko 1971, 1979, 1992, Babenko & Surkina 1969, 1971, Agarkov *et al* 1973, Babenko & Nikishova 1976), by Pershin (1976, 1988), and more recently by Zerbst (1987), Reiner (1998), and Nachtigall (1998).

Following Gray (1936) there have been many reviews and monographs published on dolphin hydrodynamics and related topics, starting with Gray (1968) himself and Kramer (1965). Focke (1965) apparently provided the earliest reassessment of Gray's paradox. Hertel's (1966) monograph considered various aspects of bionics including an investigation of dolphin swimming kinematics and dynamics. He proposed various engineering devices that imitated aspects of the dolphin's hydrodynamics and mode of propulsion. Wide-ranging treatments and reviews of many aspects of bionics and animal locomotion have been published by Shuleykin (1968), Wu (1971, 2001), Lighthill (1969, 1975, 1993), Pedley (1977), Aleyev (1977) and Kozlov (1983). Reviews and monographs aimed more specifically at dolphins have been produced by Romanenko (1986), Fish & Hui (1991), and Fish & Rohr (1999). The last is an excellent, recent and comprehensive, review of all aspects of dolphin hydrodynamics.

2. Shapes of Body and Fins

To what extent are the body and fin shapes determined by the mode of propulsion? To answer this important question we have to anticipate some of the contents of Section 6. Following Lighthill (1969, 1970, 1975) and many others the anguilliform mode of propulsion can be considered fundamental. This is named after the common eel (genus *Anguilla*). It is a pure undulatory mode of propulsion in which the whole body participates, although amplitude may increase markedly towards the tail. The most important modification to this mode of propulsion is the carangiform mode. This is an adaptation for high hydromechanical efficiency at high Reynolds number. Namely, that for elongated body forms undulation of the whole body is inessential and, in fact, undesirable; it is therefore confined the posterior half, or even third, of the body.

Lighthill (1970) identifies three further stages of further evolutionary development of the carangiform mode. First, the removal of the central posterior portion of the caudal fin (e.g., herring). Provided the rear of the fin moves as a rigid whole, his theory, which shows thrust depends on the vortex-sheet strength at the trailing edge of the caudal fin, suggests no loss of thrust is incurred since the vortex-sheet strength in the gap is more or less the same as if the gap were filled in and solid.

But the smaller surface area of the fin reduces drag. The second development is typified by fish like the horse mackerel (*Carangidae*) where the sweepback of the wing-like caudal fins is reduced to increase the trailing-edge span without an increase in surface area. Finally, there is the stage of development typified by dolphins and almost all fast-swimming fish (e.g., mackerel, tunny, swordfish); namely a caudal fin or flukes with a lunate (i.e., crescent-moon) shape of high aspect ratio. This is an example of evolutionary convergence of characteristics over a wide range of different species (Lighthill 1970, 1975, van Dam 1987).

To understand the development of this extreme form of the carangiform mode and its impact on body shape, we have briefly to consider anguilloform propulsion following the slender-body analysis presented by Lighthill (1970). Suppose that a fish with a 'slender' body attempts to generate a propulsive wave that travels down its body with speed V, taking the mathematical form:

$$h = H(x)\cos\left\{\omega\left(t + \frac{\ell - x}{V}\right)\right\} \tag{1}$$

where h is the displacement of the body centreline from its straight unperturbed form $z = 0, 0 < x < \ell$; x is the streamwise co-ordinate measured along this straight unperturbed form (for a fish z is the horizontal co-ordinate perpendicular to x, for a dolphin it is aligned vertically; $x = 0$ corresponds to the nose and $x = \ell$ to the trailing edge of the tail); $H(x)$ is the local amplitude of the wave; ω its frequency; and V is its propagation speed (V must exceed the forward speed U of the fish in order to produce thrust.) The fish's muscular contractions can only determine the shape change relative to its centre of gravity. Both translation and rotation of shape are governed by two laws of motion, namely that

(a) The rate of change of the body's lateral motion equals the total side force acting on it due to the water; and

(b) The rate of change of its angular momentum about a fixed axis (say, $z = 0$, $x = \ell$) equals the moment of the side forces due to the water.

Because the fish's body has to obey these fundamental laws of mechanics, the actual wave shape produced is somewhat different from the target specified in eq. (1) and is approximately given by

$$h = H(x)\cos\left\{\omega\left(t+\frac{\ell - x}{V}\right)\right\}+\{h_1+h_3(\ell-x)\}\cos\omega t+\{h_2+h_4(\ell-x)\}\cos\omega t \tag{2}$$

where h_i (i=1,...,4) are constants and determined by applying the laws of motion (a) and (b). It turns out that both h_1 and h_2 are negative and contribute to a reduction in thrust because they give rise to a phase difference at the trailing edge of the tail between the lateral velocity of the body and that of the water. Lighthill (1970) calls this the *recoil* effect.

It is obviously desirable to minimize this recoil effect. Dolphins and high-speed fish do this in two main ways:

(i) Having a greatly reduced body cross-section in the caudal peduncle (the stalk-like structure supporting the flukes) where the wave amplitude increases rapidly; and

(ii) Further forward the body is elongated with a greatly augmented cross section.

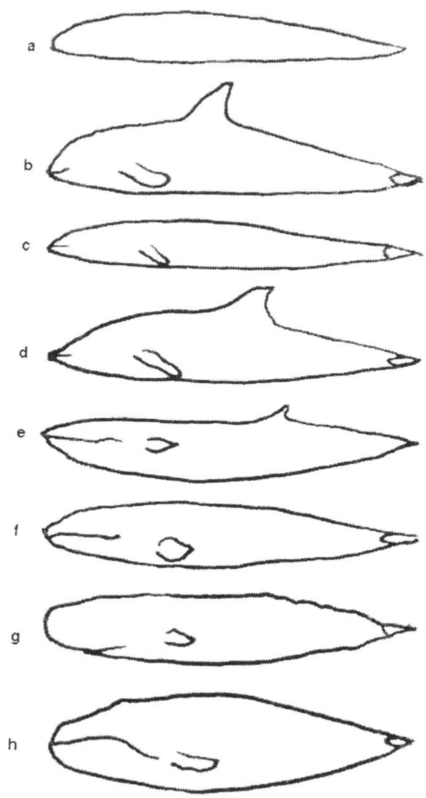

Figure 3. Vertical longitudinal profiles of various cetacean bodies compared with a laminar-flow aerofoil. a, Russian B–TSAGI series aerofoil; b, Killer whale (*Orcinus orca*); c, Northern right whale dolphin (*Lissodelphiis borealis*); d, Common dolphin (*Delphinus delphis*); e, Sei whale (*Balaenoptera borealis*); f, Gray whale (*Eschrichtus robustus*); g, Sperm whale (*Physter macrocephalus*); h, Blue whale (*Sibaldus musculus*).

These two features are well in evidence in Fig. 3. This compares the shape of a typical laminar-flow aerofoil profile (from the Russian B-TSAGI series) with the vertical longitudinal body profiles of various cetaceans. The extent to which the second feature (ii) is displayed has to be a compromise with drag requirements. If drag did not need to be considered then it would be advantageous to have the centre of gravity as far forward as possible and the body as large as possible. In fact, as suggested, in Fig. 3, the dolphin body resembles a laminar-flow aerofoil. Previous authors have also noted this similarity; for example, Hertel (1966) suggested the dolphin body shape ressembled a NACA 66-018 aerofoil and Pershin (1983) likened a killer whale to a NACA 66-026. Both aerofoils are laminar-flow designs with the maximum thickness at 60 percent chord. The length-to-thickness (ℓ/d) ratios of cetacean bodies exceed the values typical of submarines and are closer to the optimum value of around 5.

Caution must be exercised, of course, in reading too much into comparisons between the two-dimensional aerofoils and the fully three-dimensional bodies of cetaceans. Nevertheless, the pressure distributions over dolphin bodies presented by Aleyev (1977) and others, show that minimum pressure is reached close to half way along the body, much like laminar-flow aerofoils. For such aerofoils the boundary layer remains laminar over 80-90 percent of the surface, thereby ensuring that skin-friction drag and boundary-layer and wake thickness are minimized. Thus the body shapes suggest adaptation for laminar flow. That said it must be admitted that, although laminar flow is credible for dolphins, it is extremely unlikely for the larger whales.

In one sense Figure 3 is misleading because it suggests a fixed body shape. Unlike rigid bodies, cetaceans can change their body forms at will in order to execute any desired manoeuvre. The body form can be altered so as to generate lift or downforce depending on the desired manoeuvre. Manoeuvrability is greatly aided by the animal's near-zero buoyancy.

The propulsive flukes are a lateral extension of the tail and act like a pair of wings. They are composed of a cutaneous layer, subcutaneous blubber layer, a ligamentous layer, and a core of dense fibrous tissue that consitutes the bulk of the fluke (Felts 1966). As noted above, the flukes of many cetaceans are lunate in form, although that of the most widely studied species *Tursiops truncatus* has an almost straight trailing edge. Its fluke planform, however, is sharply tapering and of high aspect ratio (values range up to 8 to 10 for cetaceans). This feature plus sweep-back would tend to reduce tip vorticity and induced drag (van Dam 1987). The fluke cross-sections ressemble conventional symmetric aerofoils such as the NACA 0018 in the case of *Tursiops* (Kayan 1979). Symmetry indicates that the up and down strokes are equally important for producing thrust.

The hydrodynamic drag is roughly proportional to surface area or body length squared (more accurately $\ell^{3/2}$ for laminar flow and $\ell^{9/5}$ for turbulent flow). And the potential propulsive power should be almost proportional to the body mass or ℓ^3. Figure 4 shows that the various geometric parameters are well-correlated over a wide range of hydrobionts. This suggests some sort of optimum correlation. For cetaceans Fish & Rohr (1999) give the relation between fluke area and body length as $A_f = 0.017\ell^{1.946}$ which is roughly similar in slope to lines 1 and 2 in Fig. 4.

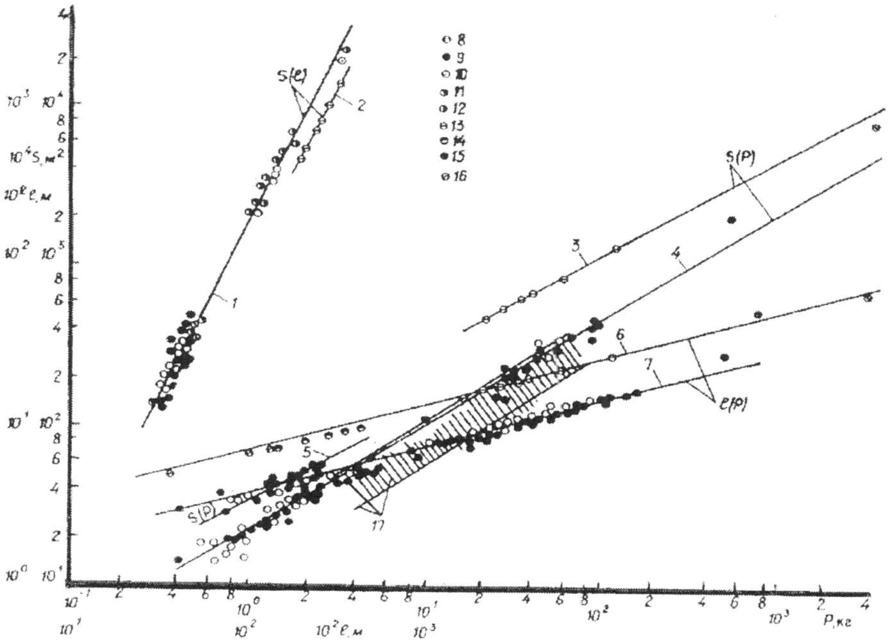

Figure 4. The relationships between the tail-fin area and body length (1,2) and body mass (3,4,5), between body length and mass (6,7) for a wide range of hydrobionts, and between wing area and body mass for birds (17). Key for data points: 8, Pelamidae; 9, various tunny; 10-15, various sharks. [Based on Fig. 5 of Babenko (1992)]

3. Structure and Characteristics of Dolphin Skin

3.1. STRUCTURE OF THE EPIDERMIS

Kramer (1957, 1960, 1965) carried out a careful study of the dolphin epidermis and his compliant coatings were designed to incorporate what he considered to be its key properties. A comparison between Figs. 1 and 2 reveals a close resemblance in terms of dimensions. But the optimal mechanical properties of dolphin skin and compliant coatings vary with flow speed (see §4.2, eqs. (4) below). And Kramer's best results in terms of drag reduction were obtained at speeds of 18 m/s that are twice the normally assumed maximum speed for dolphins. Accordingly, the ideal mechanical characterization of the Kramer compliant coatings would by no means be identical to that of dolphin skin. Moreover, according to Babenko *et al.* (1969, 1982) and Babenko & Surkina (1969) Kramer's photographs were misleading and his understanding of the structure of the dolphin epidermis faulty. In fact, it appears that there is still not a universally accepted, coherent view of the structure. Figure 2 gives a schematic composite view of the main structural features of the dolphin epidermis and upper dermal layer based mainly on the work of Babenko and his co-workers cited above, but also on that of Aleyev (1977), Stromberg (1989), and Ridgway & Carder (1993). Some authors (Babenko *et al.* 1969 and Ridgway & Carder 1993) have detected the presence of microscales or cutaneous ridges running approximately normal to the flow direction. The effect of the cutaneous ridges on the hydrodynamics is unknown. But recent unpublished numerical simulations at the University of Warwick (Ali 2002) suggest that for compliant walls similar to dolphin skin oblique components of a Tollmien-Schlichting wave-packet (analogous to those triggered in natural transition) grow significantly less rapidly when wave-like surface features analogous to the cutaneous ridges are present.

The upper epidermal layer forms a comparatively dense elastic membrane capable of transmitting all pressure fluctuations to the underlying layer. This lower layer and the dermal papillae are made of loose, more hydrated tissue, including fat cells. The size of the microscales and the angle made by the dermal papillae to the surface vary over the body (Babenko & Surkina 1969). These angles of inclination vary from 10 to 80 degrees and appear to be correlated with the gradients of the hydrodynamically generated tractions acting on the body. The dermal papillae extend upwards from and are supported in a series along the longitudinal dermal ridges. Surkina (1971b) (see also Sokolov 1955, 1973, 1987) describes how the orientation of the dermal ridges vary over the body. They appear to be aligned to the streamlines close to the surface of the dolphin's body. Sokolov (1987) described the development of the epidermis during ontogenesis. It is thought that the common characteristics shared by a wide range of species appear at an earlier stage of the embryo's development than more specialized ones. According to this principle Sokolov deduced that the dermal ridges and papillae evolved comparatively recently. The skin of dolphins millions of years ago was probably quite similar to that of present-day humans. The recent modifications are also consistent with adaptation for the reduction of skin-friction drag.

3.2. CHARACTERIZATION OF DOLPHIN SKIN

Dolphins possess a well-defined morphological adaptation that provides an intensive section-by-section blood supply to the epidermis. In addition, autonomous section-by-section innervation of the skin could be evidence of the rapid regulation of the blood supply and lymphatic systems. Accordingly, the data presented by Sokolov (1973), Babenko and Nikishova (1976), and Babenko (1992), suggest that as a result of evolution over a long period the dolphin epidermis has adapted for reducing hydrodynamic forces.

Babenko (1971) and Agarkov *et al.* (1973) developed the analogue mechanical model of the elements of the dolphin skin important for reducing the hydrodynamic forces that is depicted in Fig. 5. The corresponding approximate equation of motion takes the form:

$$(M_1 + M_2)\frac{\partial^2 \eta}{\partial t^2} + d\frac{\partial \eta}{\partial t} - (T_1 + T_2)\frac{\partial^2 \eta}{\partial x^2} + (E'_1 + E'_2 + E'_3)\eta = -p \qquad (3)$$

where M_n, T_n and E'_n are respectively the oscillating masses per unit area, tensions per unit width, and the elastic stiffnesses of the corresponding layers; η is the surface displacement in the direction normal to the surface; d is the damping factor representing the viscoelastic nature of the tissue; and p is the fluctuating hydrodynamic pressure acting on the skin.

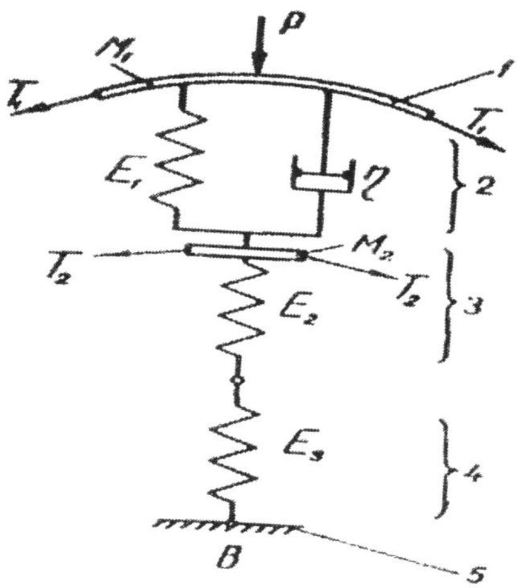

Figure 5. An analogue composite mechanical model of dolphin skin comprising: 1, outer diaphragm; 2, underlying elastomeric layer; 3, a second layer with internal tensile forces; 4, viscoelastic substrate; 5, foundation. [Based on Fig. 4 of Babenko (1971).]

The scaling of the various mechanical properties with flow speed is determined the dimensionless parameters derived by Babenko (1971, 1979) as follows:

$$k_M = \frac{U}{\mu}(M_1 + M_2), \quad k_T = \frac{T_1 + T_2}{\mu U}, \quad k_S = S\frac{\nu}{\rho U^3}, \qquad (4a, b, c)$$

$$k_\omega = \frac{\nu}{U^2}\left(\frac{S}{M_1 + M_2}\right)^{1/2}, \quad k_d = \frac{\Delta}{\pi}, \quad \frac{c_0}{U} = \frac{1}{U}\left(\frac{T_1 + T_2}{M_1 + M_2}\right)^{1/2} \qquad (4d, e, f)$$

where ρ, μ, and ν are respectively the density, dynamic and kinematic viscosity of the fluid; the stiffness, $S = E'_1 + E'_2 + E'_3$, and Δ is the logarithmic damping coefficient. The parameter k_ω is the non-dimensional cut-off frequency, and c_0 is the free-wave speed of the skin.

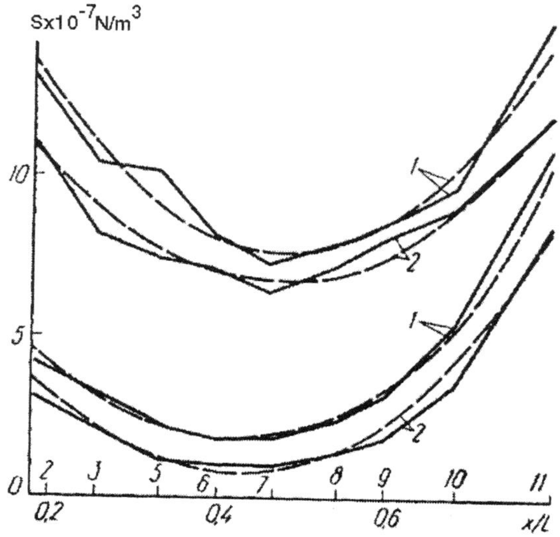

Figure 6. Distribution of skin stiffness over the bodies of (1) bottle-nosed dolphin (*Tursiops truncatus*), and (2) common dolphin (*Delphinus delphis*) at various swimming speeds. The upper set of data corresponds to cruising swimming speeds; i.e., (1) 9 m/s and (2) 7.8 m/s; and the lower set to average swimming speeds; i.e., (1) 6.9 m/s and (2) 6.0 m/s. The solid lines result from calculation and the broken lines correspond to best-fit quadratic functions of x/L. L is the total body length. [Based on Fig. 3 of Babenko (1979)]

Apparatus and techniques for measuring the skin elasticity and other mechanical characteristics of the dolphin epidermis were developed by Babenko *et al.* (1969), Babenko (1979) (see also Babenko *et al.* 1993). Measurements were made on both trained and recently captured dolphins, swimming fast and slowly. The variation of skin stiffness, S, (i.e., the ratio of the elasticity to the material thickness) over the bodies of common and slightly faster, bottle-nosed dolphins are plotted in Fig. 6. It is evident from Fig. 6 that the skin stiffness has a well-pronounced minimum over the central portions of the body. The data in Fig. 6 can be partly explained by considering the non-dimensional stiffness defined in eq. (4c). Thus if k_S always assumed its optimum value, the stiffness would vary as U^3. In fact, the stiffness data in Fig. 6 for the same species display a greater increase with swimming speed than suggested by eq. (4c), although the differences between species is more or less in accord with eq. (4c). This might suggest that the skin properties are not optimal at the lower swimming speeds. As well as controlling skin stiffness for maintaining its laminar-flow properties, the dolphin would also need increased skin stiffness at higher speeds to avoid hydroelastic divergence instability in the form of skin folding. This is known to occur at its maximum short-duration swimming speeds; for example, when about to jump out of the water (Essapian 1955, Aleyev 1977).

Babenko *et al.* also measured other properties of the epidermis, including the varying mass, the tension, and the damping coefficient. This last was determined by measuring the rebound height obtained from dropping metal balls of various masses on the surface, thereby obtaining an energy absorption coefficient. For live healthy animals the values of this quantity was found to be close to 100% at energy levels corresponding to the boundary-layer pressure fluctuations during laminar-turbulent transition and were markedly higher than those found for sick or recently deceased animals.

The free-wave speeds at various locations over the body of a live dolphin was measured by Madigosky *et al.* (1986). Over most of the body the values were 6.5 ± 0.5 m/s, with somewhat higher values in some locations. According to the optimization theory for compliant walls these values would be optimal for a swimming speed of 9 m/s (Carpenter *et al.* 2000). Fitzgerald and Fitzgerald (1995) carried out various measurements of the mechanical properties of pilot-whale blubber (see §2 of Chap. 8). Amongst other things they found that the shear-wave speed was approximately 9 m/s.

4. Swimming Speed

At least three regimes of swimming speed can be distinguished for dolphins and other aquatic animals, namely:

(a) Slow swimming that can be sustained indefinitely. The dolphins break the water surface gently, often barely showing their blowholes.

(b) Faster cruise swimming that can be maintained for 15-20 minutes. They primarily swim up to a body length or so below the surface; and there is little splashing when they come to the surface.

(c) Even faster forced swimming (sometimes called the 'running' mode) that can be maintained for no more than a few minutes. This is characterized by sequential leaps interspersed with brief subsurface swimming. Exit from and re-entry into the water is accompanied by considerable splashing.

If the dolphin were adapted for maintaining laminar flow, it seems most likely that this would primarily apply to cruise and, possibly, forced swimming, rather than slow swimming.

The data available on dolphin swimming speeds are surprisingly inconclusive and are well-reviewed by Fish & Rohr (1999). Most of the early data on dolphin swimming speeds were based on shipboard observations and were rather anecdotal. There are various ways in which a dolphin may use the wave energy associated with a ship (see §8), making ship-based observations problematic. But it is also true that dolphins reach their maximum swimming speeds at depth rather than on the surface.

There have been some studies of trained dolphins in artificial and open-water conditions (Lang & Daybell 1963, Lang 1975, Kozlov 1983, Pershin 1988), but reliable and accurate swimming speeds for free-ranging dolphins are rare. Aircraft-based observations of a school of startled free-ranging *Delphinus* reported by Fish & Rohr (1999) resulted in swimming speeds with modal and maximum values of 4 m/s and 6.5 m/s respectively. The lack of reliable data on high sustained swimming speeds has led some authorities, such as Fish & Rohr (1999), to conclude that dolphins probably do not swim at high speeds for more than a few seconds. It is fair to say, however, that a concensus remains for cruising speeds in the range 9 to 11 m/s for species such as *Tursiops truncatus*. Low drag arising from a laminar-flow capability would lead to increased cruising speeds which would confer definite advantage for hunting prey and evading predators. Even at lower swimming speeds low drag would help conserve energy. Moreover, Williams *et al.* (2000) recently reported that dolphins and other aquatic mammals adopt an energy-conserving strategy when they dive below depths of 80 metres. This involves prolonged periods (up to 80% of total dive duration) of gliding. Plainly laminar flow leading to greatly reduced drag would be also advantageous here.

5. Kinematics and Dynamics of Swimming

The mode of swimming used by dolphins and other cetaceans is an extreme adaptation of the carangiform mode whereby the caudal flukes adopt a lunate form. This was discussed briefly in §2. Theoretical analyses of this mode of propulsion have been developed by Lighthill (1970), Chopra (1974), Chopra (1976), Chopra & Kambe (1977), Kayan (1974, 1979), Kozlov (1983), Cheng & Murillo (1984), and Karpouzian *et al.* (1990). These analyses are largely based on modified unsteady wing theory. They are able to predict the thrust produced by the flukes and the work required to produce it. Chopra & Kambe (1977), followed by Fish & Rohr (1999), estimate the drag of certain dolphins by using observations of kinematic parameters to estimate the thrust produced by the flukes and assuming that the average thrust is equal to the average drag. Similar calculations due to Kayan (1974, 1979) are plotted in Fig. 7 below. Such calculations result in unexpectedly high drag values, 2 or 3 times those measured on gliding or still dolphins. However, as will be explained below, it seems to us that this approach may be unreliable.

Attractive as it seems for the purposes of analysis, it is misleading to envisage the propulsion unit of the dolphin as separate from the rest of the body. For this reason it is by no means clear how best to define propulsive efficiency for the dolphin and other aquatic animals. This difficulty is evident in Lighthill (1970) who begins in his equation (1) by defining the hydromechanical efficiency as

$$\eta = \frac{U\bar{P}}{\bar{E}} \tag{5}$$

where "$\bar{P}$ is the mean thrust required to overcome what viscous drag the animal would sustain for forward velocity U if it *remained rigid and symmetrical*; and $\bar{E}$ is the mean rate at which flexural movements do work against the surrounding water". This definition seems eminently sensible, but what if the drag of the body were reduced during propulsion, owing, say, as suggested originally by Gray

(1936), to flow being accelerated over the rear part of the body? If this were to happen, η could conceivably exceed 1! However, the great merit of this definition of efficiency is that it is based on a fixed reference, namely the drag of the rigid symmetrical body.

Later, in his equation (32), Lighthill redefines the efficiency, so that, in effect, $\bar{P}$ becomes the actual mean thrust produced and not the original hypothetical value associated with a rigid symmetrical body. This definition has the merit of always remaining less than unity. But it brings other problems. It is conceivable that a mode of propulsion exists that has high efficiency in terms of converting a large proportion of the work done into thrust, but that also results in a high level of drag. From the dolphin's point of view what is likely to be important is either minimizing energy expenditure for travelling large distances and diving to great depths, or travelling as fast as possible over relatively short distances for a fixed maximum power output. In both cases a mode of propulsion that generated a relatively small or no rise in drag coefficient above some hypothetical minimum reference value would be preferable to one with a theoretically higher propulsive efficiency (according to the second of the definitions given above) that results in a higher drag coefficient.

The main attraction of the latter definition of efficiency given above is that, unlike the former, it can be predicted by theory. However, it is misleading to determine the drag by setting it equal to the theoretical mean thrust, as done by Chopra & Kambe (1977) and Fish & Rohr (1999). This is because even in so-called steady swimming the flow field is unsteady (see Wu 2001) and the body of the dolphin is oscillating about its centre of mass owing to Lighthill's recoil effect described in §2. Thus much of the thrust has to be expended on accelerating the body and added mass of the water, even though probably only a small fraction of the body's acceleration is aligned in the direction of mean motion. Alternatively, the recoil effect can be viewed as a reduction in thrust. These estimates of body drag also ignore the contribution of the fluke's induced and viscous drags and overestimate the thrust produced by leading-edge suction. The estimates of power required for propulsion would also be much too low. Again this is because substantial work is expended (or power lost) on the inevitable body oscillations due to the recoil effect. Thus the drag estimates given by Fish and Rohr (1999) are too high and their estimates of power required too low. Ultimately, it is hard to believe that the dolphin could do no better than evolve a mode of propulsion that generates a drag many times greater than its drag when towed, even exceeding the drag generated during actively braking (see Fig. 7). The propulsive unit of the dolphin should not be analysed as an isolated system. Satisfactory results will only be obtained when dynamic analysis of the whole body during propulsion is carried out. Difficult as this would be to do, it may still not be sufficient, because systems in living organisms usually only function properly when interconnected.

6. Boundary Layer Characteristics

For most dolphin species the maximum thickness and minimum pressure are located about half-way along their bodies (Aleyev 1977). This should ensure that the boundary layer remains laminar until at least this location. Furthermore, as first pointed out by Gray (1936), the propulsive unit located at the rear of the body will tend to accelerate flow over the rear half of the body. This will also promote laminar flow and prevent flow separation. Accordingly, for a 2 m long animal transition is likely to occur at about 1 m from the nose. Thus, even for a rigid dolphin, the appropriate prediction of transitional Reynolds number with a steady swimming speed of 10 m/s would be 10×10^6, which is very much greater than Gray assumed.

Direct evidence is available of the state of the boundary layer over the dolphin body in the form of hot-film and pressure-sensor measurements of the fluctuating streamwise velocity component and pressure for various swimming conditions (Kozlov *et al.* 1974, Kozlov & Shakalo 1980, Pyatetskii *et al.* 1982, Romanenko 1986). Comparisons between the distribution of this quantity along a dolphin's body with that along a comparable rigid one suggests that laminar-turbulent transition occurs at a higher speed for the dolphin. Furthermore, the maximum amplitudes of the fluctuating velocities and pressures are lower for the dolphin as compared with a rigid body. What is most significant is that the boundary layer over the dolphin body is more regular with smaller velocity gradients at the wall. For example, pressure-sensor measurements in the boundary layer of an actively swimming dolphin indicated that, although the flow was somewhat irregular, it did not become completely turbulent. Thus these measurements suggest that the skin-friction drag for an actively swimming dolphin is significantly less than for a comparable rigid body.

Another indication of the state of the boundary layer can be found by flow visualization due to bioluminescence. Some dinoflagellates, a type of unicellular plankton, produce the bioluminescent response. Rohr *et al.* (1998) used bioluminescence to visualize the boundary-layer flow around a dolphin (see Herring 1998). The results could not conclusively discriminate between laminar and turbulent flow, but clearly showed that the boundary layer remained attached up to the tail-end of the caudal peduncle. A much earlier observation by Stevens (reported by Schmidt-Nielsen 1972) contrasted the two clean, thin, diverging lines of phosphorescence marking the wakes of dolphins with the boiling mass of turbulence characterising the wake left by seals swimming at high speed.

7. Drag

It is exceedingly difficult to determine the drag of a swimming dolphin directly. Reliable estimates can be made for gliding dolphins (inertial swimming) (Lang & Daybell 1963, Kayan 1974) or towed carcasses (Webb 1975, Aleyev 1977, Blake 1983, Williams 1987). Kayan (1974,1979) (see also Curren *et al.* 1994) carried out direct measurements of the kinematics of live swimming dolphins. All the required parameters for determining the dynamics were estimated from ciné films and direct measurements of the body and fins. The variation of the dolphin's drag with Reynolds number during active swimming was thereby estimated corresponding to a range of accelerations and inertial motion with the propulsion unit disabled. (See Fig. 7). In all cases considered the dolphin's drag coefficient, estimated using a quasi-stationary approach with the flukes modelled as flapping wings, was twice the value obtained when the thrust was assumed to be produced by flexuro-oscillatory motions of the whole body. (See discussion in §5.) It thereby appears that the greater the acceleration and the greater the effort of the body as a whole, the less the drag. Under these conditions all the organism's systems are activated to produce the required effort, and the dolphin's drag could well become less than the corresponding rigid plate with a turbulent boundary layer.

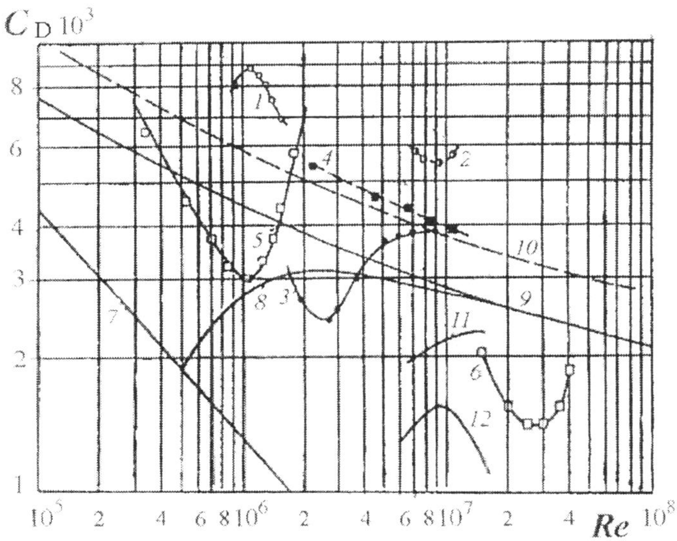

Figure 7. Drag coefficients for dolphins and models as a function of Reynolds number. 1, dolphin drag calculated using the theory of Kayan (1974, 1979) based on a flapping propellor; 2, dolphin drag during active braking; 3, drag of a 2.2 m dolphin during inertial swimming (Kayan 1974); 4, computational calculation of the drag of the corresponding rigid body; 5, drag of a towed dead dolphin (*Phocoena phocoena*) of dimensions 1 m × 0.21 m × 0.25 m; 6, drag of an equivalent model dolphin in the form of a body of revolution based on a NASA 66 profile; 7,8,9, drag of rigid flat plate with laminar, transitional, and turbulent flow respectively; 11, 12 experiments of Kramer (1957, 1960). [Based on Fig. 6 of Babenko et al. (1982)]

Fig. 7 presents comparisons between drag coefficients for dolphins and models under various circumstances. Kramer (1957, 1960), Kozlov (1983), Kozlov *et al.* (1985) and Babenko *et al.* (1993) developed artificial analogue dolphin skins. Various examples mounted on plates and bodies of revolution were investigated (see curves 6, 11 and 12 in Fig. 7). With exception of Kramer's data, the C_D curves, corresponding to dolphins or models covered with artificial dolphin skin, all exhibit a characteristic minimum at particular values of Reynolds number. This is quite unlike the curves for equivalent rigid bodies in Fig. 7, although such minima have been observed for bodies of revolution with extended favourable streamwise pressure gradients (e.g., Hansen and Hoyt 1984). Thus the minima may, as suggested by Fish and Rohr (1989), be due to the reduced form drag found when laminar-turbulent transition leads to a turbulent boundary layer over the rear of the body where there is an adverse streamwise pressure gradient. This explanation, however, would not account for the widely differing values of Reynolds number corresponding to minimum C_D. An alternative explanation is that above a critical Reynolds number flow-structure interaction causes C_D to rise. For example, it is well known (Essapian 1955 and Aleyev 1977) that dolphins exhibit skin-folding, which is a form of hydroelastic instability, at high swimming speeds. Whatever the explanation it seems that dolphins, real and artificial, have an optimum Reynolds number (swimming speed) with respect to drag.

Given the difficulties of other approaches, Gray's (1936) estimate of the dolphin body's drag based on flat-plate formulae has its merit. However, the more realistic assumption that laminar-turbulent transition occurs half way along the body should be made (see §6). This raises the transitional Reynolds number to about 10^7. In order to make specific estimates it is necessary to choose a representative dolphin, we follow Focke (1965) in selecting an individual *Tursiops truncatus* of length 2.75 m and overall body mass of 90 kg. Using the well-known Prandtl-Schlichting formula (Schlichting 1979) for turbulent boundary layers over smooth flat plates with a correction for the initial laminar flow, we obtain a skin-friction drag coefficient of $C_D \simeq 0.0015$ at a swimming speed of 11 m/s ($Re_L \simeq 30 \times 10^6$). This compares with $C_D \simeq 0.00025$ for laminar flow and $C_D \simeq 0.0025$ when the transitional Reynolds number corresponds to the maximum value, 2×10^6, found for a flat plate. None of the three values given above vary greatly with swimming speed. To translate the value of C_D into a value of drag at a specified swimming speed we need an estimate of the surface area of the dolphin body. Following Focke (1965) we approximate the body by a prolate spheroid of thickness ratio 1 : 6, giving an area of about 3.5 m^2 in the case under consideration. Thus at 9 m/s and 11 m/s for $C_D = 0.0015$, the drag would be 122 N and 182 N respectively (i.e., 0.14 and 0.2 times body weight respectively). But this estimate of drag has omitted the contributions due to the various appendages such as the fins and flukes and the additional drag caused by vortex generation where the fins meet the body and other sources of parasitic drag. Evolution should have ensured that parasitic drag has been minimized. Yasui (1980) (cited in Fish and Rohr 1999) estimate that for *Phocoena phocoena* 36% of the drag is due to appendages, whereas Lang & Pryor (1966) estimate that about 30% of the drag of *Stenella attenuata* is due to appendages. Evidently the exact proportion is species specific. *Tursiops* has a proportionally larger dorsal fin compared with *Phocoena* so we will assume about 40% of the drag is due to appendages raising the total equivalent surface area to $A = 4.9\ m^2$ and giving total drag values of 203 N and 303 N (0.23 and 0.34 body weight) corresponding to 9 m/s and 11 m/s respectively .

When a body approaches or breaks the surface its drag rises, partly owing to local streamline curvature near the surface, and partly owing to wave drag. For a body similar to a dolphin the maximum drag occurs at a depth of half a body diameter. Some authors (Fish & Rohr 1989, Au & Weihs 1980a) include a correction to the drag to allow for near-surface effects by writing γC_D where $1 < \gamma < 5$, citing Hoerner (1965) as the source for the variation of γ with depth. However, as pointed out by Gordon (1980), since γ rapidly approaches 1 with increase in depth, a dolphin wishing to minimize drag is not likely to swim very close to the surface.

8. Drag-Reduction Techniques

8.1. BOW RIDING

This is a commonly observed mode of behaviour exhibited by dolphins whereby they exploit the wave or pressure field of a vessel to obtain hydrodynamic advantage. As many observations of dolphins swimming are ship-based, it is possible that bow riding could account for any anomalously high swimming speeds. It has been widely discussed and debated in the literature, starting with Woodcock (1948) and Woodcock & McBride (1951) who suggested that when a dolphin immersed itself inclined head down in a wave, making an angle θ to the horizontal, the driving force is $(W - B)\sin\theta$ (where W and B are the weight and buoyancy force respectively). This was considered insufficient to overcome drag unless there is laminar flow. Subsequently, Hayes (1953, 1959) pointed out that, when a body is immersed in a wave and parallel to the inclined wave surface, the buoyancy force acted perpendicular to the local isobars, i.e., perpendicular to the wave surface and, therefore, the true driving force was $W\sin\theta$. Thus if we take the estimated drag values of $0.23W$ and $0.34W$ given in §7 for 9 and 11 m/s respectively, these require wave angles of 13 and 20 degrees, implying fairly substantial waves. As pointed out in §7, though, the drag rises very substantially as the body approaches the surface, so the above estimates of required angle are probably much too low.

Other authors (Scholander 1959a,b, Fejer & Backus 1960) have disagreed with Hayes and proposed alternative explanations. The former cited experimental evidence that appeared to contradict Hayes. There is little doubt that Hayes is theoretically correct and his theoretical model was supported by the experiments of Perry *et al.* (1961). His model certainly provides an explanation for the dolphins observed wave riding on the wake of a small slow vessel by Williams *et al.* (1992). However, his model is not consistent with observations (Woodcock 1948, Fejer & Backus 1960, Yuen 1961) of the behaviour of bow-wave-riding dolphins (nor, for that matter, is the alternative proposed by Scholander). The dolphins normally position themselves about 0.3 to 1.5 m ahead of the ship's stem, displaced laterally from it by 0.2 to 1.5 m, at depths varying from 0.3 to 1.8 m. Fejer & Backus proposed that thrust was generated owing to the pressure field generated by the ship. Focke (1965) estimated that a maximum thrust of about 1400 N could be generated in this way at 10 m/s which is greatly in excess of the drag, even for completely turbulent flow.

8.2. ALTERNATE LEAPING AND SUBMERSED SWIMMING

This behaviour is often known as 'porpoising'. It is characteristic of high-speed swimming when alarmed. The extra exertion required for high-speed swimming demands a greater oxygen intake and increased frequency of breathing. Thus the animal is forced to spend more time at or near the surface where the drag is much greater. It therefore makes a virtue out of necessity and adopts a mode of locomotion whereby submerged swimming alternates with parabolic leaps. This minimizes the time spent in the high-drag, near-surface region. At the simplest level, if air resistance is neglected, this is merely exchange of kinematic and potential energy. In this way the maximum height of the leap can be estimated as

$$h = \frac{V^2 \cos^2 \theta}{2g} \tag{6}$$

If it is assumed that the minimum feasible leap height is 0.5 m and that the exit angle, θ, from the water is 30 degrees, the critical or lowest feasible swimming speed for 'porpoising' according to eq.(6) is 3.6 m/s. This is a little lower than the estimates of Au & Weihs (1980a) based on a fairly simple, but more elaborate, analysis.

The estimate based on eq. (6) is too low because it neglects the energy lost when the animal exits from the water (partly owing to the water entrained by the body and lost as potential energy, but also owing to the extra drag generated due to surface tension and other effects) and the additional energy required to re-accelerate the added mass of water when the dolphin re-enters the water. Au & Weihs (1980a,b) attempt to include this lost energy in their analysis which is criticized by Gordon (1980). Some of his points are sound, but it seems incorrect to argue that the water shed by the dolphin as spray at exit from the water does not represent a loss of energy. A more accurate estimate of the critical speed at which 'porpoising' becomes advantageous would require a computer simulation that takes account of the trajectory followed by the animal, the rate of breathing required as a function of speed, the rise in drag in the near-surface region, and the additional energy losses due to exit and re-entry.

8.3. SECRETIONS

It is well-known that the presence of long-chain polymers in sufficient concentrations in the boundary layer can dramatically reduce turbulent skin-friction drag. According to Uskova *et al.* (1983) secretions from the eyes of dolphins are highly viscous complexes of protein and polysaccharides. And they demonstrated that such secretions could reduce skin friction. But Sokolov *et al.* (1969) demonstrated that eye secretions had no significant effect on drag, probably because only a small fraction of the total body surface area was affected. They also tested the effects on drag of weak concentrations (up to 1%) of the epidermal cells that are continually shed at a high rate by dolphins. Again no discernible difference was observed compared with pure sea-water.

9. Active Flow Control

It appears from §3.2 that the mechanical characteristics become hydrodynamically optimal only at high speeds. This implies that the power output available is adequate at lower swimming speeds. But at cruising speeds the power available may be insufficient without adaptations for reducing drag. In particular, the forms of the body and fins, and the specific structure of the epidermis may play an essential role. At high speed such adaptations would only be fully effective when all the animal's systems are interconnected.

The schematic in Fig. 8 depicts the arrangements for a typical body section of the various sytems and their interconnections. It is evident that each vertebra, blood and lymphatic vessel, and bundle of nerve fibres are directed radially outwards. This arrangement gives the shortest route for providing blood and allows for automatic regulation section by section to govern the blood flow in the outer layers of the epidermis. Two such layers, which are saturated in longitudinal layers of blood vessels, are evident in Fig. 8. The layer of dermal muscle can regulate the mechanical properties of the dermal layers lying above it. Furthermore the motive musculature causes the stress of the whole skin to change cyclically in accordance with the propulsive action of the flukes.

From an analysis of the interactions between the systems depicted in Fig. 8. the following modes of interactive control have been identified:

- Change in form of the body and fins;
- Change in geometric parameters of epidermis;
- Regulation of the skin's mechanical properties by means of tension induced by muscles;
- Change of skin's mechanical properties by means of the circulatory and innervation systems;
- Regulation of skin and body temperature
- Vibrations on the skin surface.

Agarkov *et al.* (1973) show how regulation of the body form occurs. Pershin (1988) and Sokolov (1987) also show how the forms of the dorsal and horizontal fins are varied. Such shape modification specifically occurs on account of the regulation of blood supply in the complex of blood vessels (Sokolov 1973, 1987).

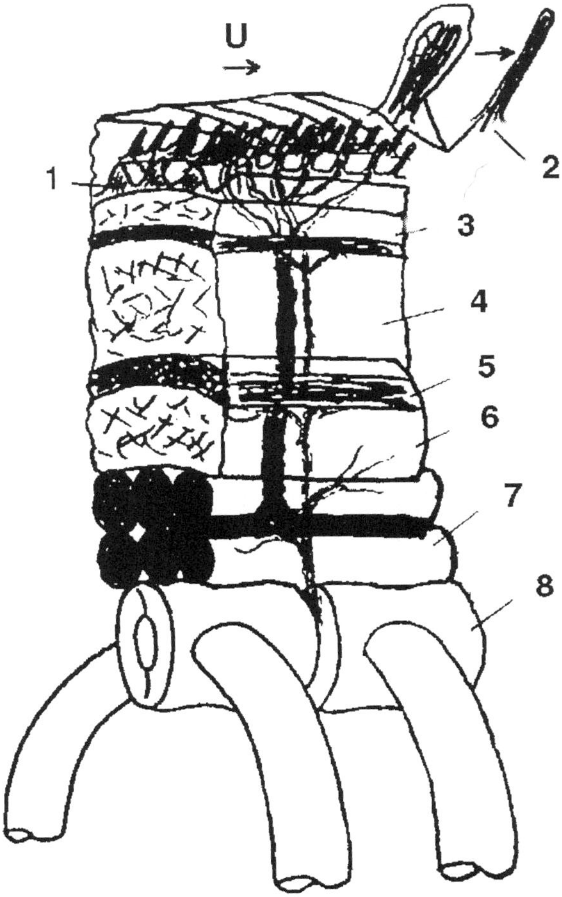

Figure 8 A schematic of the dolphin's body cross-section showing the arrangement of the circulatory, lymphatic, and innervation systems. Key: 1, Dermal ridge; 2, dermal papillae; 3, sub-papillary dermal layer; 4, dermal layer; 5, cutaneous musculature; 6, fat layer; 7, musculature; 8, skeleton.

Babenko and Surkina (1969) show how the compliance of the epidermis could be controlled by changing its tension by use of the skin muscles. There is also a functional connection between the dolphin's skeleton, particularly its mobile ribs, and the propulsive and cutaneous musculature, and indeed the whole skin, that includes the blood circulations, and lymphatic and innervation systems (Pershin 1988). When the propulsive musculature is in tension and the rear part of the body executes oscillatory motion for propulsion, the volume, cross-section and form of the body can vary on account of the high mobility and specific structure of the skeleton. Furthermore, the specific structure of the muscles and membraneous tendons in the skin allow the thickness and form of the epidermis to be varied (Agarkov *et al.* 1973). Thus during fast swimming a longitudinal fold is formed along each side of the body. Also a propulsive wave is generated along the body surface. In this way there could be an interaction between the skin and the vortical structures in the boundary layer. When the oscillatory frequency of the tail flukes changes with a change in swimming regime, such parameters as skin tension and body diameter may be adjusted in order to reduce drag. Thus, as well as its direct purpose of generating the propulsive force, the propulsive musculature may also influence drag and manoeuvrability.

Since cetaceans are warm-blooded, the elastic properties of the skin may be controlled through the circulatory and innervation systems. A schematic for the blood circulation in dolphin skin is presented by Parry (1949) and Babenko (1983). There are two layers with longitudinal arrangements of the dermal papillae (see Figs. 2 and 8). In combination with the cutaneous musculature, this allows a thermal barrier to be created in order to control the temperature of the body and epidermis. A complex of blood vessels below the papillae effectively govern the heat exchange. In the epidermis the capillaries have direct cross-connections allowing effective heat exchange between arterial and venous blood. Owing to a specific system of innervation, this exchange can be actively controlled. The cutaneous musculature can be actuated to close or completely open these blood vessels. In this way the heat flow can either be directed into the body or towards the outer skin layers in order to modify its mechanical charaetcistics.

Babenko (1983) carried out calculations of the heat flow rate per unit area for the skins of common and bottle-nosed dolphins. In this way the temperature of the outer skin surface could be determined experimentally for live dolphins. These measurements showed that for slow swimming in calm conditions the skin temperature differed only slightly (0.5 - 0.7 K) from the surroundings. But on the fins the temperature difference was much greater, up to several degrees, although not large enough for the thermal boundary layer to significantly affect hydrodynamic stability. At high swimming speeds the blood circulation increases leading to a rise in the partial pressure in the capillaries of the dermal papillae, causing them to assume a bottle-like form (Babenko & Surkina 1969). This leads to a rise in the skin elasticity; but, owing to the specific form of the dermal papillae, the damping properties of the skin will also be favourably modified (see §3.2). In this way the skin can assume an optimal state for boundary-layer stabilization (see §3.2). The vessels of the lymphatic system have a system of valves to maintain unidirectional lymph flow. It is likely that the lymphatic systems also plays a role in the active control of the epidermis's mechanical properties.

Direct measurements of the microvibrations of dolphin skin have been carried out by several investigators (e.g., Haider & Lindsley 1964, Kidun 1979, Pershin 1988, Ridgway & Carder 1993, Yaremchuk & Babenko 1998). These small tremor-like vibrations are active defomations of the skin under the control of the cutaneous musculature. It has been suggested that these may be used as a form of boundary-layer control.

10. Energetics

Here we briefly consider the question of the power output from the propulsive muscles. Muscle performance depends on the type of muscle fibre. Fast glycolytic fibres are adapted for short-burst activity with high power output and very high intrinsic contraction speeds; whereas slow oxidative fibres contract more slowly and are suitable for slow, sustained activity. Both types of fibres are found in dolphin muscles. Fast glycolytic and slow oxidative fibres are fuelled primarily by anaerobic and aerobic metabolism respectively. Depending on the metabolic pathway, the former have a maximum metabolic power output that is 2 to 17 times greater than the latter.

The sustained power output of muscles does not depend solely on the muscle mass but is also governed by the total body mass, because other body systems are required to support muscle power. Parry (1949) estimates that a sustained specific output of 40 W/kg may be possible under optimum conditions of oxygen supply and mechanical efficiency. There seems to be no reason to revise this estimate in the light of later research. However, with anaerobic operation a much greater output is possible for short durations, e.g. Weis-Fogh & Alexander (1977) give a figure of 110 W/kg.

The cost of transport is another relevant measure; it is defined as the metabolic rate divided by the body weight times swimming speed. The minimum cost of transport for dolphins (*Tursiops*) is remarkably low in comparison with other aquatic mammals (Williams *et al.* 1992). It falls between theoretical values based on the assumptions of laminar and turbulent flow. This is consistent with the earlier estimates of Schmidt-Nielsen (1972).

11. Technological Developments

Several attempts, starting with Kramer (1957, 1960, 1965) (see §1), have been made to emulate dolphin skin for technological application. Other experimental research on artificial analogue dolphin epidermis are presented in Kozlov (1983), Kozlov *et al.* (1985), and Babenko *et al.* (1993). Various types of such compliant-wall structures were investigated when mounted on plates and bodies of revolution.

The Reynolds-number range corresponding to the maximum drag reduction varied depending on the structural form of these compliant walls. The maximum drag reduction varied between 20 and 40 percent, and under certain conditions was comparable with that obtained with the injection of polymer solution into the boundary layer (Babenko 1997). The artificial dolphin skins of Kramer are described in §1 and Fig. 1. Babenko and his co-workers and Grosskreutz (1971) also developed designs with inclined supporting structures designed to permit the wall to oscillate in a direction inclined to the vertical (i.e. perpendicular to the wall). These types of anisotropic compliant walls were modeled theoretically and analysed by Carpenter and Morris (1990) who showed that their laminar-flow properties were superior to those of Kramer's coatings. Using the principles of hydrobionics Babenko and his co-workers have also developed and patented small underwater vehicles (Babenko *et al.* 2000) and a series of devices that are analogues to the dolphin's propulsive flukes. In these various ways the fundamental results of hydrobionic research have been applied to engineering design and technology.

12. Conclusions and Re-assessment of Gray's Paradox

Do revised estimates indicate a need for laminar-flow control for the dolphin? It appears that of the various parameters required for estimating dolphin performance the most reliable and uncontroversial data available are for the power output per unit mass of muscle (see §10), namely 40 W/kg for sustained output and a maximum of 110 W/kg. For the specimen of *Tursiops truncatus* under consideration the propulsive musculature has a mass of about 36 kg. From the discussion in §6, it should be clear that the high theoretical values of propulsive efficiency used by Chopra & Kambe (1977) and Fish & Rohr (1999) are over-optimistic. Focke (1965) used an estimated value of 75%. This may well still be too high, but we will use it here. Thus the propulsive power, P_p, available for sustained and forced swimming is $0.75 \times 36 \times 40 = 1.08$ kW and 2.97 kW respectively. The corresponding swimming speeds that this power can deliver is given by

$$U = \{(2P_p/(\rho A C_D)\}^{1/3} \tag{7}$$

The coefficient of drag varies with Reynolds number. However, for approximate estimates the values estimated in §7 and Fig. 7 can be used resulting in the swimming-speed estimates given below in Table 1. The swimming speeds corresponding to the high value of C_D for the gliding dolphin seems rather low. However, if it is assumed that the dolphin is swimming at its optimum swimming speed, the lower value of $C_D \simeq 0.0023$ from Fig. 7 should be used. The resulting estimated values of swimming speed correspond fairly closely to the revised (rigid-body) estimates with a more realistic transitional Reynolds number and $C_D = 0.0015$. And these estimates seem to correspond reasonably well to many of the observed swimming speeds reported in the literature. However, laminar flow would seem to be necessary in order to reach the sustained swimming speeds of 9 to 11 m/s reported by some authors. To what extent alternating leaping and submerged swimming (§8.2) would be a sufficient drag-reducing strategy alone without laminar-flow control has not yet been determined.

The evidence reviewed in the present chapter is rather mixed and often inconclusive. The need for laminar flow to achieve the expected swimming speeds now seems to be much less clearcut than some previous authors, including Gray, claimed. We will close with three examples of indirect evidence in favour of the dolphin's adaptation for laminar flow.

(i) As reported in §3.2 the free-wave speeds measured for the dolphin's epidermis are c. 6.5 m/s corresponding to a compliant wall optimized for transition delay at a flow speed of 9 m/s. The mechanical properties of dolphin skin are to some extent subject to active control (§9). Moreover, there is definite hydrodynamic disadvantage in having too flexible a skin because it allows skin folds to form (Essapian 1955) which are a type of hydroelastic instability (see §3.2). However, skin flexibility may have other functions unconnected with hydrodynamics.

(ii) When they form at high forced swimming speeds the skin folds have wave lengths typical of compliant walls optimized for transition delay at 9 m/s (Carpenter *et al.* 2000).

(iii) The dolphin epidermis contains structures in the form of the dermal ridges and papillae that could have been adapted for turbulent flow. Were this to have happened, one would expect riblet-like structures to have evolved like those found on the scales of fast-swimming sharks (Bechert *et al.* 1985). Instead, the dermal ridges, although aligned correctly, have separation distances that are orders of magnitude too great for turbulent drag reduction. Instead, the cutaneous ridges or microscales have evolved which plainly have a quite different hydrodynamic effect.

TABLE 1. Estimates of swimming speed in m/s corresponding to muscle power output

Case	*Drag coefficient*	*Swimming speed for sustained power output*	*Swimming speed for maximum power output*
$Re_t = 2 \times 10^6$	0.0025	5.6	7.9
$Re_t = 13.75 \times 10^6$	0.0015	6.6	9.3
Laminar	0.00025	12.08	16.9
Gliding Dolphin (Fig. 7)*	0.004	5.4	7.5
	0.0023	6.5	9.0

* *the actual estimated surface area of 3.5 m^2 was used in this case.*

References

AGARKOV, G.B., HOMENKO, B.G. and HADZHINSKIĬ, V. G. (1970) *Morphologies of dolphins*, (in Russian), Kiev: Naukova dumka.

AGARKOV, G.B., BABENKO, V.V. and FERENTS, Z.I. (1973) On innervation of the skin and skin musculature of a dolphin in connection with a hypothesis on the stabilization of the boundary-layer flow, (in Russian), *Problemy Bioniki*, Nauka, Moscow, 478-483.

ALEYEV, Y.G. (1977) *Nekton*, Junk, The Hague, Netherlands.

ALI, R. (2002) Private communication, University of Warwick.

AU, D. and WEIHS, D. (1980a) At high speeds dolphins save energy by leaping, *Nature* **284**,548-550.

AU, D. and WEIHS, D. (1980b) Leaping dolphins, *Nature* **287**, 759.

BABENKO, V.V. (1971) Basic performance of flexible coatings and similarity criteria, (in Russian), *Bionika* **5**, 73-75.

BABENKO, V.V. (1979) Research on the skin elasticity of a live dolphin, (in Russian), *Bionika* **13**, 43-52.

BABENKO, V.V. (1981) Towards a technique for experimental research in hydrobionics, (in Russian), *Bionika* **15**, 88-98.

BABENKO, V.V. (1983) Some features of the thermal control of the epidermis of aquatic animals, (in Russian), *Bionika* **17**, 35-39.

BABENKO, V.V. (1992) On the interaction of hydrobionts with flow, (in Russian), *Bionika* **25**, 3-11.

BABENKO, V.V. (1997) Polymers submission optimization with the help of sword-shaped tips, *Proc. 10th European drag reduction working meeting, Berlin*, pp. 19-21.

BABENKO, V.V. , GNITETSKII, N.A., and KOZLOV, L.F. (1969) Preliminary outcomes of research on the elastic properties of the skin of live dolphins, (in Russian), *Bionika* **3**, 12-19.

BABENKO, V.V., KARNASKIĬ, M.V, and KOROBOV, V.I. (1993) *Boundary layers over elastic surfaces*, (in Russian), Kiev: Naukova Dymka.

BABENKO, V.V., KOZLOV, L.F., PERSHIN, S.V.SOKOLOV, V.E. and TOMILIN, A.G. (1982) Self-regulation of the damping in the skin of cetacea during active swimming, (in Russian), *Bionika* **16**, 3-10.

BABENKO, V.V., MOROS, V.V. and MARTYNENKO, I.I. (2000) Experience of creation of underwater remotely operated vehicles in Ukraine and perspective of their application in geological oceanography. *Mineral resources of Ukraine* **1**, 21-25.

BABENKO, V.V. and NIKISHOVA, O.D. (1976) Some hydrodynamic principles of the skin-cover structure of some marine animals, (in Russian), *Bionika* **10**, 27-33.

BABENKO, V.V. and SURKINA, R.M. (1969) Some hydrodynamic features of dolphin swimming, (in Russian), *Bionika* **3**, 19-26.

BABENKO, V.V. and SURKINA, R.M. (1971) Definition of oscillating mass parameter for skin covering of some marine animals, (in Russian), *Bionika* **5**, 94-98.

BAINBRIDGE, R. (1958) The speed of swimming of fish as related to size and to the frequency and amplitude of tail beat, *J. Exp. Biol.* **5**, 17-33.

BECHERT, D.W., HOPPE, G. and REIF, W.-E. (1985) On the drag reduction of shark skin, *AIAA Paper 85-0546*.

BENJAMIN, T.B. (1960) Effects of a flexible boundary on hydrodynamic stability, *J. Fluid Mech.* **9**, 513-532.

BLAKE, R.W. (1983) *Fish locomotion*, Cambridge University Press.

CARPENTER, P.W. and GARRAD, A.D. (1985) The hydrodynamic stability of flow over Kramer-type compliant surfaces. Pt. 1. Tollmien-Schlichting instabilities, *J. Fluid Mech.* **155**, 465-510.

CARPENTER, P.W., DAVIES, C. and LUCEY, A.D. (2000) Hydrodynamics and compliant walls: Does the dolphin have a secret?, *Current Science* **79**, 758-765.

CARPENTER, P.W. and MORRIS, P.J. (1990) The effects of anisotropic wall compliance on boundary-layer stability and transition, *J. Fluid Mech.* **218**, 171-223.

CHENG, H.K. and MURILLO, L.E. (1984) Lunate-tail swimming propulsion as a problem of curved lifting line in unsteady flow. Part 1. Asymptotic theory, *J. Fluid Mech.* **143**, 327-350.

CHOPRA, M.G. (1974) Hydrodynamics of lunate-tail swimming propulsion, *J. Fluid Mech.* **64**, 375-391.

CHOPRA, M.G. (1976) Large amplitude lunate-tail theory of fish locomotion, *J. Fluid Mech.* **74**, 161-182.

CHOPRA, M.G. and KAMBE, T. (1977) Hydrodynamics of lunate-tail swimming propulsion. Part 2, *J. Fluid Mech.* **79**, 49-69.

CURREN, K., BOSE, N. and LIEN, J. (1994) Swimming kinematics of a harbor porpoise (*Phocoena phocoena*) and an Atlantic white-sided dolphin (*Lagenorhynchus acutus*), *Marine Mammal Science* **10**, 485-492.

ESSAPIAN, F.S. (1955) Speed-induced skin folds in the bottle-nosed porpoise, *Tursiops truncatus*, *Breviora Mus. Comp. Zool.* **43**, 1-4.

FEJER, A.A. and BACKUS, R.H. (1960) Porpoises and the bow-riding of ships under way, *Nature* **188**, 700-703.

FELTS, W.J.L. (1966) Some functional and structural characteristics of cetacean fippers and flukes, in K.S. Norris (ed.), *Whales, Dolphins and Porpoises*, Univ. of Calif. Press, Berkeley, pp. 255-276.

FISH, F.E. and HUI, C.A. (1991) Dolphin swimming – A review, *Mammalian Review*, **21**, 181-195.

FISH, F.E. and ROHR, J.J. (1999) Review of dolphin hydrodynamics and swimming performance, Tech. Rep. 1801, SPAWAR Systems Center, San Diego.

FITZGERALD, E.R. and FITZGERALD, J.W. (1995) Blubber and compliant coatings for drag reduction in water. I. Viscoelastic properties of blubber and compliant coating materials, *Material Science and Engineering* C**2**, 209-214.

FOCKE, H. (1965) Über die Ursachen der hohen Schwimmgeschwindigkeiten der Delphine, *Zeitschrift für Flugwissenschaften* **13**, 54-61.

GAD-EL-HAK, M. (2002) Chap. 9 in present volume.

GORDON, C.N. (1980) Leaping dolphins, *Nature* **287**, 759.

GRAY, J. (1936) Studies in animal locomotion VI. The propulsive powers of the dolphin, *J. Exp. Biol.* **13**, 192-199.

GRAY, J. (1968) *Animal locomotion,* Weidenfeld and Nicholson, London.

GROSSKREUTZ, R. (1971) Wechselwirkungen zwischen turbulenten Grenzschichten und weihen Wänden, *MPI für Strömungsforschung und der AVA, Göttingen, Mitt. No. 53.*

HAIDER, M. and LINDSLEY, D.B. (1964) Microvibrations in man and dolphin, *Science* **146**, 1181-83.

HANSEN, R.J. and HOYT, J.G. (1984) Laminar-to-turbulent transition on a body of revolution with an extended favorable pressure gradient forebody, *ASME J. of Fluids Eng.* **106**, 202-210.

HAYES, W.D. (1953) Wave-riding of dolphins, *Nature* **112**, 1060.

HAYES, W.D. (1959) Wave-riding of dolphins, *Science*, **130**, 1657-58.

HERRING, P.J. (1998) Dolphins glow with the flow, *Nature* **393**, 731-732.

HERTEL, H. (1966) *Structure, Form and Movement,* Reinhold, New York.

HOERNER, S.F. (1965) *Fluid-dynamic drag,* published by author, Brick Town, New Jersey.

KARPOUZIAN, G., SPEDDING, G. and CHENG, H.K (1990) Lunate-tail swimming propulsion. Part 2. Performance analysis, *J. Fluid Mech.* **210**, 329-351.

KAYAN, V. P. (1974) On the drag coefficient of the dolphin, (in Russian), *Bionika* **8**, 31-35.

KAYAN, V. P. (1979) On the hydrodynamic characteristics of the dolphin's caudal fin, (in Russian), *Bionika* **13**, 9-15.

KIDUN, S.M. (1979) An investigation into propagation speed of oscillations in the dolphin skin, (in Russian), *Bionika* **13**, 52-58.

KOZLOV, L.F. (1983) *Theoretical Bio-Hydrodynamics*, (in Russian), Naukova Dumka, Kiev.

KOZLOV, L.F., SHAKALO, V.M., BUR'YANOVA, L.D. and VOROB'EV, N.N. (1974) On the influence of non-stationarity on the flow regime in the boundary layer of a Black Sea bottle-nosed dolphin, (in Russian), *Bionika* **8**, 13-16.

KOZLOV, L.F. and SHAKALO, V.M. (1980) On the flow regime in the quasi-steady boundary layer over some cetaceans, (in Russian), *Bionika* **14**, 74-81.

KOZLOV, L.F., TSYGANYUK, A.I, BABENKO, V.V., HIKISHOVA, O.D. and VOROPAYEV, G.A. (1985) *Experimental investigations of boundary layers*, (in Russian), Naukova Dumka, Kiev.

KRAMER, M.O. (1957) Boundary layer stabilization by distributed damping, *J. Aero. Sci.* **24**, p. 459.

KRAMER, M.O. (1960) Boundary layer stabilization by distributed damping, *J. Amer. Soc. Nav. Eng.* **72**, 25-33.

KRAMER, M.O. (1965) Hydrodynamics of the dolphin, *Advances in Hydroscience* **2**, 111-130.

LANG, T.G. (1975) Speed, power, and drag measurements of dolphins and porpoises, in *Swimming and Flying in Nature*, T.Y Wu *et al.*, eds., Plenum Press, New York, pp. 553-571.

LANG, T.G and DAYBELL, D.A. (1963) Porpoise performance tests in a seawater tank, Nav. Ord. Test Station, Tech. Rep. 3063, China Lake, California.

LANG, T.G and PRYOR, K. (1966) Hydrodynamic performance of porpoises (*Stenella attenuata*), *Science* **152**, 531-533.

LIGHTHILL, J. (1969) Hydrodynamics of aquatic animal propulsion – A survey, *Ann. Rev. Fluid Mech.* **1**, 413-446.

LIGHTHILL, J. (1970) Aquatic animal propulsion of high hydromechanical efficiency, *J. Fluid Mech.* **44**, 265-301.

LIGHTHILL, J. (1975) *Mathematical Biofluiddynamics*, SIAM, Philadelphia.

LIGHTHILL, J. (1993) Biofluiddynamics: A survey, *Contemporary Math.* **141**, 1-32.

MADIGOSKY, W.M., LEE, G.F., HAUN. J., BORKAT F. and KATOKA, R. (1986) Acoustic surface wave measurements of live bottlenose dolphins, *J. Acoust. Soc. Amer.* **79**, 153-159.

NACHTIGALL, W. (1998) *Bionik, Grundlagen und Beispiele für Ingenieure und Naturwissenschafter*, Springer-Verlag, Berlin.

PARRY, D.A. (1949) The structure of whale blubber and a discussion of its thermal properties, *Quart. J. Microsc. Sci.* **90**, 13-25.

PEDLEY, T.J. (1977) (Ed.) *Scale effects in animal locomotion*, Academic Press.

PERRY, B., ACOSTA, A.J. and KICENIUK, T. (1961) Simulated wave-riding dolphins, *Nature* **192**, 148-150.

PERSHIN, S.V. (1976) Auto-control of skin damping and hydrodynamic drag reduction during active swimming of cetaceans, (in Russian), *Bionika* **10**, 33-40.

PERSHIN, S.V. (1983) Hydrobionic features for optimizing the exterior forms of the propulsion system in cetaceans, (in Russian), *Bionika* **17**, 13-24.

PERSHIN, S.V. (1988) *Fundamentals of hydrobionics*, (in Russian), Sudostroenie, Leningrad.

PYATETSKII, V.E., SHAKALO, V.M., TSYGANYUK, A.I. and SIZOV, I.I. (1982) An investigation of the flow regime of water animals, (in Russian), *Bionika* **16**, 31-36.

REINER, R. (1998) Selbstorganisation: Anwendungen eines biologischen Prinzips, *Biona-report* **8**, 13-25.

RIDGWAY, S.H. and CARDER, D.A. (1993) Features of dolphin skin with potential hydrodynamic importance, *IEEE Eng. Med. Biol.* **12**, 83-88.

ROHR, J., LATZ, M.I., FALLON, S., NAUEN, J.C. and HENDRICKS, E.W. (1998) Experimental approaches towards interpreting dolphin-stimulated bioluminescence, *J. Exp. Biol.* **201**, 1447-1460.

ROMANENKO, E.V. (1986) *Theories of swimming of fishes and dolphins.* (in Russian), Nauka, Moscow.

SCHLICHTING, H. (1979) *Boundary layer theory*, 7th. ed., McGraw-Hill, New York.

SCMIDT-NIELSEN, K. (1972) Locomotion: Energy cost of swimming, flying, and running. *Science* **177**, 222-228.

SCHOLANDER, P.F. (1959a) Wave-riding dolphins, How do they do it?, *Science* **129**, 1085-87.

SCHOLANDER, P.F. (1959b) Wave-riding dolphins, *Science* **130**, 1658

SHULEYKIN, V.V. (1968) *The Physics of the Sea*, (in Russian), Nauka, Moscow.

SOKOLOV, V.E. (1955) Structure of the epidermis of some cetaceans, (in Russian), *Byulleten' Moskovskogo Obshchestva Ispytatwleiĭ Prirody. Otd. Biologii*, **60**, 45-60.

SOKOLOV, V.E. (1973) *The mammalian epidermis*, (in Russian), Nauka, Moscow.

SOKOLOV, V.E. (1987) *Morphology and Ecology of Marine Mammals*, (in Russian), Nauka, Moscow.

SOKOLOV, V.E., BULINA, I. and RODIONOV, V. (1969) Interaction of dolphin epidermis with flow boundary layer, *Nature* **222**, 267-268.

STROMBERG, M.W. (1989) Dermal-epidermal relationships in the skin of bottlenose dolphins (*Tursipos truncatus*), *Anal, Histol. Embryol.* **18**, 1-13.

SURKINA, R.M (1971a) On the structure and function of dolphin skin muscles, (in Russian), *Bionika*, **5**, 81-87.

SURKINA, R.M (1971b) The distribution of the dermal ridges over the body of a bottle-nosed dolphin, (in Russian), *Bionika*, **5**, 88-94.

USKOVA, Ye.T., SHMYREV, V.S., RAYEVSKIY, L.N., BOGDANOVA, L.N., MOMOT, L.N., BELYAYEV, V.V. and USKOV, I.A. (1983) The nature and hydrodynamic activity of dolphin eye secretions, (in Russian), *Bionika*, **17**, 72-75.

VAN DAM, C.P. (1987) Efficiency characteristics of crescent-shaped wings and caudal fins, *Nature* **325**, 435-437.

WEBB, P.W. (1975) Hydrodynamics and energetics of fish propulsion, *Bull. Fish. Res. Bd. Can.* **190**, 37-41.

WEIS-FOGH, T. and ALEXANDER, R.McN. (1977) The sustained power output from striated muscle, in *Scale Effects in Animal Locomotion*, T.J. Pedley (ed.), Academic Press, pp. 511-525.

WILLIAMS, T.M. (1987) Approaches for the study of exercise physiology and hydrodynamics in marine mammals, in *Approaches to Marine Mammal Energetics*, A.C. Huntley, A.C. *et al.*, eds., Spec. Pub. Soc. Marine Mammals No. 1, pp. 127-145.

WILLIAMS, T.M., DAVIS, R.W., FULMAN, L.A., FRANCIS, J., LE BOEUF, B.J., HORNING, M., CALAMBOKIDIS, J. and CROLL, D.A. (2000) Sink or swim: Strategies for cost-efficient diving by marine mammals, *Science* **288**, 133-136.

WILLIAMS, T.M., FRIEDL, W.A., FONG, M.L., YAMADA, R.M., SEDIVY, P. and HAUN, J.E. (1992) Travel at low energetic cost by swimming and wave-riding bottlenose dolphins, *Nature* **355**, 821-823

WOODCOCK, A.H. (1948) The swimming of dolphins, *Nature*, **161**, 602.

WOODCOCK, A.H. and McBRIDE, A.F. (1951) Wave-riding dolphins, *J. Expt. Biol.* **28**, 215-217.

WU,T.Y. (1971) Hydromechanics of swimming of fishes and cetaceans, *Adv. Appl. Mech.* **11**, 1-63.

WU,T.Y. (2001) On the theoretical modeling of aquatic and aerial animal locomotion, *Adv. Appl. Mech.* **38**, 291-353.

YAREMCHUK, A.A. and BABENKO, V.V. (1998) On the question of comparing the bioenergetics of man and dolphin, (in Russian), *Bionika*, **27-28**, 152-154.

YUEN, H.S.H. (1961) Bow wave riding of dolphins, *Science* **134**, 1011-12.

ZERBST, E.W. (1987) *Bionik, bioligische Funktionsprinzipien und ihre technischen Anwendungen*, Teubner-Studienbucher, Berlin.

Mechanics

FLUID MECHANICS AND ITS APPLICATIONS

Series Editor: R. Moreau

Aims and Scope of the Series

The purpose of this series is to focus on subjects in which fluid mechanics plays a fundamental role. As well as the more traditional applications of aeronautics, hydraulics, heat and mass transfer etc., books will be published dealing with topics which are currently in a state of rapid development, such as turbulence, suspensions and multiphase fluids, super and hypersonic flows and numerical modelling techniques. It is a widely held view that it is the interdisciplinary subjects that will receive intense scientific attention, bringing them to the forefront of technological advancement. Fluids have the ability to transport matter and its properties as well as transmit force, therefore fluid mechanics is a subject that is particularly open to cross fertilisation with other sciences and disciplines of engineering. The subject of fluid mechanics will be highly relevant in domains such as chemical, metallurgical, biological and ecological engineering. This series is particularly open to such new multidisciplinary domains.

1. M. Lesieur: *Turbulence in Fluids*. 2nd rev. ed., 1990 ISBN 0-7923-0645-7
2. O. Métais and M. Lesieur (eds.): *Turbulence and Coherent Structures*. 1991 ISBN 0-7923-0646-5
3. R. Moreau: *Magnetohydrodynamics*. 1990 ISBN 0-7923-0937-5
4. E. Coustols (ed.): *Turbulence Control by Passive Means*. 1990 ISBN 0-7923-1020-9
5. A.A. Borissov (ed.): *Dynamic Structure of Detonation in Gaseous and Dispersed Media*. 1991 ISBN 0-7923-1340-2
6. K.-S. Choi (ed.): *Recent Developments in Turbulence Management*. 1991 ISBN 0-7923-1477-8
7. E.P. Evans and B. Coulbeck (eds.): *Pipeline Systems*. 1992 ISBN 0-7923-1668-1
8. B. Nau (ed.): *Fluid Sealing*. 1992 ISBN 0-7923-1669-X
9. T.K.S. Murthy (ed.): *Computational Methods in Hypersonic Aerodynamics*. 1992 ISBN 0-7923-1673-8
10. R. King (ed.): *Fluid Mechanics of Mixing*. Modelling, Operations and Experimental Techniques. 1992 ISBN 0-7923-1720-3
11. Z. Han and X. Yin: *Shock Dynamics*. 1993 ISBN 0-7923-1746-7
12. L. Svarovsky and M.T. Thew (eds.): *Hydroclones*. Analysis and Applications. 1992 ISBN 0-7923-1876-5
13. A. Lichtarowicz (ed.): *Jet Cutting Technology*. 1992 ISBN 0-7923-1979-6
14. F.T.M. Nieuwstadt (ed.): *Flow Visualization and Image Analysis*. 1993 ISBN 0-7923-1994-X
15. A.J. Saul (ed.): *Floods and Flood Management*. 1992 ISBN 0-7923-2078-6
16. D.E. Ashpis, T.B. Gatski and R. Hirsh (eds.): *Instabilities and Turbulence in Engineering Flows*. 1993 ISBN 0-7923-2161-8
17. R.S. Azad: *The Atmospheric Boundary Layer for Engineers*. 1993 ISBN 0-7923-2187-1
18. F.T.M. Nieuwstadt (ed.): *Advances in Turbulence IV*. 1993 ISBN 0-7923-2282-7
19. K.K. Prasad (ed.): *Further Developments in Turbulence Management*. 1993 ISBN 0-7923-2291-6
20. Y.A. Tatarchenko: *Shaped Crystal Growth*. 1993 ISBN 0-7923-2419-6
21. J.P. Bonnet and M.N. Glauser (eds.): *Eddy Structure Identification in Free Turbulent Shear Flows*. 1993 ISBN 0-7923-2449-8
22. R.S. Srivastava: *Interaction of Shock Waves*. 1994 ISBN 0-7923-2920-1
23. J.R. Blake, J.M. Boulton-Stone and N.H. Thomas (eds.): *Bubble Dynamics and Interface Phenomena*. 1994 ISBN 0-7923-3008-0

Mechanics

FLUID MECHANICS AND ITS APPLICATIONS

Series Editor: R. Moreau

24. R. Benzi (ed.): *Advances in Turbulence V.* 1995 ISBN 0-7923-3032-3
25. B.I. Rabinovich, V.G. Lebedev and A.I. Mytarev: *Vortex Processes and Solid Body Dynamics.* The Dynamic Problems of Spacecrafts and Magnetic Levitation Systems. 1994 ISBN 0-7923-3092-7
26. P.R. Voke, L. Kleiser and J.-P. Chollet (eds.): *Direct and Large-Eddy Simulation I.* Selected papers from the First ERCOFTAC Workshop on Direct and Large-Eddy Simulation. 1994 ISBN 0-7923-3106-0
27. J.A. Sparenberg: *Hydrodynamic Propulsion and its Optimization.* Analytic Theory. 1995 ISBN 0-7923-3201-6
28. J.F. Dijksman and G.D.C. Kuiken (eds.): *IUTAM Symposium on Numerical Simulation of Non-Isothermal Flow of Viscoelastic Liquids.* Proceedings of an IUTAM Symposium held in Kerkrade, The Netherlands. 1995 ISBN 0-7923-3262-8
29. B.M. Boubnov and G.S. Golitsyn: *Convection in Rotating Fluids.* 1995 ISBN 0-7923-3371-3
30. S.I. Green (ed.): *Fluid Vortices.* 1995 ISBN 0-7923-3376-4
31. S. Morioka and L. van Wijngaarden (eds.): *IUTAM Symposium on Waves in Liquid/Gas and Liquid/Vapour Two-Phase Systems.* 1995 ISBN 0-7923-3424-8
32. A. Gyr and H.-W. Bewersdorff: *Drag Reduction of Turbulent Flows by Additives.* 1995 ISBN 0-7923-3485-X
33. Y.P. Golovachov: *Numerical Simulation of Viscous Shock Layer Flows.* 1995 ISBN 0-7923-3626-7
34. J. Grue, B. Gjevik and J.E. Weber (eds.): *Waves and Nonlinear Processes in Hydrodynamics.* 1996 ISBN 0-7923-4031-0
35. P.W. Duck and P. Hall (eds.): *IUTAM Symposium on Nonlinear Instability and Transition in Three-Dimensional Boundary Layers.* 1996 ISBN 0-7923-4079-5
36. S. Gavrilakis, L. Machiels and P.A. Monkewitz (eds.): *Advances in Turbulence VI.* Proceedings of the 6th European Turbulence Conference. 1996 ISBN 0-7923-4132-5
37. K. Gersten (ed.): *IUTAM Symposium on Asymptotic Methods for Turbulent Shear Flows at High Reynolds Numbers.* Proceedings of the IUTAM Symposium held in Bochum, Germany. 1996 ISBN 0-7923-4138-4
38. J. Verhás: *Thermodynamics and Rheology.* 1997 ISBN 0-7923-4251-8
39. M. Champion and B. Deshaies (eds.): *IUTAM Symposium on Combustion in Supersonic Flows.* Proceedings of the IUTAM Symposium held in Poitiers, France. 1997 ISBN 0-7923-4313-1
40. M. Lesieur: *Turbulence in Fluids.* Third Revised and Enlarged Edition. 1997 ISBN 0-7923-4415-4; Pb: 0-7923-4416-2
41. L. Fulachier, J.L. Lumley and F. Anselmet (eds.): *IUTAM Symposium on Variable Density Low-Speed Turbulent Flows.* Proceedings of the IUTAM Symposium held in Marseille, France. 1997 ISBN 0-7923-4602-5
42. B.K. Shivamoggi: *Nonlinear Dynamics and Chaotic Phenomena.* An Introduction. 1997 ISBN 0-7923-4772-2
43. H. Ramkissoon, *IUTAM Symposium on Lubricated Transport of Viscous Materials.* Proceedings of the IUTAM Symposium held in Tobago, West Indies. 1998 ISBN 0-7923-4897-4
44. E. Krause and K. Gersten, *IUTAM Symposium on Dynamics of Slender Vortices.* Proceedings of the IUTAM Symposium held in Aachen, Germany. 1998 ISBN 0-7923-5041-3
45. A. Biesheuvel and G.J.F. van Heyst (eds.): *In Fascination of Fluid Dynamics.* A Symposium in honour of Leen van Wijngaarden. 1998 ISBN 0-7923-5078-2

Mechanics

FLUID MECHANICS AND ITS APPLICATIONS

Series Editor: R. Moreau

46. U. Frisch (ed.): *Advances in Turbulence VII.* Proceedings of the Seventh European Turbulence Conference, held in Saint-Jean Cap Ferrat, 30 June–3 July 1998. 1998 ISBN 0-7923-5115-0
47. E.F. Toro and J.F. Clarke: *Numerical Methods for Wave Propagation.* Selected Contributions from the Workshop held in Manchester, UK. 1998 ISBN 0-7923-5125-8
48. A. Yoshizawa: *Hydrodynamic and Magnetohydrodynamic Turbulent Flows.* Modelling and Statistical Theory. 1998 ISBN 0-7923-5225-4
49. T.L. Geers (ed.): *IUTAM Symposium on Computational Methods for Unbounded Domains.* 1998 ISBN 0-7923-5266-1
50. Z. Zapryanov and S. Tabakova: *Dynamics of Bubbles, Drops and Rigid Particles.* 1999 ISBN 0-7923-5347-1
51. A. Alemany, Ph. Marty and J.P. Thibault (eds.): *Transfer Phenomena in Magnetohydrodynamic and Electroconducting Flows.* 1999 ISBN 0-7923-5532-6
52. J.N. Sørensen, E.J. Hopfinger and N. Aubry (eds.): *IUTAM Symposium on Simulation and Identification of Organized Structures in Flows.* 1999 ISBN 0-7923-5603-9
53. G.E.A. Meier and P.R. Viswanath (eds.): *IUTAM Symposium on Mechanics of Passive and Active Flow Control.* 1999 ISBN 0-7923-5928-3
54. D. Knight and L. Sakell (eds.): *Recent Advances in DNS and LES.* 1999 ISBN 0-7923-6004-4
55. P. Orlandi: *Fluid Flow Phenomena.* A Numerical Toolkit. 2000 ISBN 0-7923-6095-8
56. M. Stanislas, J. Kompenhans and J. Westerveel (eds.): *Particle Image Velocimetry.* Progress towards Industrial Application. 2000 ISBN 0-7923-6160-1
57. H.-C. Chang (ed.): *IUTAM Symposium on Nonlinear Waves in Multi-Phase Flow.* 2000 ISBN 0-7923-6454-6
58. R.M. Kerr and Y. Kimura (eds.): *IUTAM Symposium on Developments in Geophysical Turbulence* held at the National Center for Atmospheric Research, (Boulder, CO, June 16–19, 1998) 2000 ISBN 0-7923-6673-5
59. T. Kambe, T. Nakano and T. Miyauchi (eds.): *IUTAM Symposium on Geometry and Statistics of Turbulence.* Proceedings of the IUTAM Symposium held at the Shonan International Village Center, Hayama (Kanagawa-ken, Japan November 2–5, 1999). 2001 ISBN 0-7923-6711-1
60. V.V. Aristov: *Direct Methods for Solving the Boltzmann Equation and Study of Nonequilibrium Flows.* 2001 ISBN 0-7923-6831-2
61. P.F. Hodnett (ed.): *IUTAM Symposium on Advances in Mathematical Modelling of Atmosphere and Ocean Dynamics.* Proceedings of the IUTAM Symposium held in Limerick, Ireland, 2–7 July 2000. 2001 ISBN 0-7923-7075-9
62. A.C. King and Y.D. Shikhmurzaev (eds.): *IUTAM Symposium on Free Surface Flows.* Proceedings of the IUTAM Symposium held in Birmingham, United Kingdom, 10–14 July 2000. 2001 ISBN 0-7923-7085-6
63. A. Tsinober: *An Informal Introduction to Turbulence.* 2001 ISBN 1-4020-0110-X; Pb: 1-4020-0166-5
64. R.Kh. Zeytounian: *Asymptotic Modelling of Fluid Flow Phenomena.* 2002 ISBN 1-4020-0432-X
65. R. Friedrich and W. Rodi (eds.): *Advances in LES of Complex Flows.* Prodeedings of the EUROMECH Colloquium 412, held in Munich, Germany, 4-6 October 2000. 2002 ISBN 1-4020-0486-9
66. D. Drikakis and B.J. Geurts (eds.): *Turbulent Flow Computation.* 2002 ISBN 1-4020-0523-7
67. B.O. Enflo and C.M. Hedberg: *Theory of Nonlinear Acoustics in Fluids.* 2002 ISBN 1-4020-0572-5

Mechanics

FLUID MECHANICS AND ITS APPLICATIONS

Series Editor: R. Moreau

68. I.D. Abrahams, P.A. Martin and M.J. Simon (eds.): *IUTAM Symposium on Diffraction and Scattering in Fluid Mechanics and Elasticity*. Proceedings of the IUTAM Symposium held in Manchester, (UK, 16-20 July 2000). 2002 ISBN 1-4020-0590-3
69. P. Chassaing, R.A. Antonia, F. Anselmet, L. Joly and S. Sarkar: *Variable Density Fluid Turbulence*. 2002 ISBN 1-4020-0671-3
70. A. Pollard and S. Candel (eds.): *IUTAM Symposium on Turbulent Mixing and Combustion*. Proceedings of the IUTAM Symposium held in Kingston, Ontario, Canada, June 3-6, 2001. 2002 ISBN 1-4020-0747-7
71. K. Bajer and H.K. Moffatt (eds.): *Tubes, Sheets and Singularities in Fluid Dynamics*. 2002 ISBN 1-4020-0980-1
72. P.W. Carpenter and T.J. Pedley (eds.): *Flow Past Highly Compliant Boundaries and in Collapsible Tubes*. IUTAM Symposium held at the Univerity of Warwick, Coventry, United Kingdom, 26-30 March 2001. 2003 ISBN 1-4020-1161-X

Kluwer Academic Publishers – Dordrecht / Boston / London